FOURIER TRANSFORM INFRARED CHARACTERIZATION OF POLYMERS

POLYMER SCIENCE AND TECHNOLOGY

Recent volumes in the series:

A Continuation Order Plan is available for this series. A continuation order will bring delivery of each new volume immediately upon publication. Volumes are billed only upon actual shipment. For further information please contact the publisher.

FOURIER TRANSFORM INFRARED CHARACTERIZATION OF POLYMERS

Edited by
Hatsuo Ishida
Case Western Reserve University
Cleveland, Ohio

PLENUM PRESS • NEW YORK AND LONDON

Library of Congress Cataloging in Publication Data

Symposium on Fourier Transform Infrared Characterization of Polymers (1984: Philadelphia, Pa.)
Fourier transform infrared characterization of polymers.

(Polymer science and technology; v. 36)
"Proceedings of a Symposium on Fourier Transform Infrared Characterization of Polymers, held August 26–31, 1984, in Philadelphia, Pennsylvania"—T.p. verso.
Held under the auspices of the Division of Polymer Chemistry, American Chemical Society.
Bibliography: p.
Includes index.
1. Infrared spectroscopy—Congresses. 2. Fourier transform spectroscopy—Congresses. 3. Polymers and polymerization—Analysis. I. Ishida, Hatsuo. II. American Chemical Society. Division of Polymer Chemistry. III. Title. IV. Series.
QD96.I5F67 1984 547.7′046 87-11183
ISBN 0-306-42582-3

Proceedings of a symposium on Fourier Transform Infrared Characterization of Polymers, held August 26–31, 1984, in Philadelphia, Pennsylvania

Printed in the United States of America

PREFACE

This book contains the proceedings of the Symposium on FT-IR Characterization of Polymers, which was held under the auspices of the Division of Polymer Chemistry, American Chemical Society (ACS) during the annual ACS meeting in Philadelphia, August, 1984. The content of each paper has been substantially extended from the papers presented during the conference.

Due to the accidental, irrecoverable loss of the entire contents of the book by the computer system used for editorial purposes, the publication of this book has been delayed more than one year over the initial scheduled date. It has been a continuous, frustrating experience for the editor as well as for the authors. An extended Murphy's law, "anything can go wrong goes multiply wrong" has been demonstrated in editor's office. It necessitated, otherwise unnecessary, repeated proof reading during which time the editor had valuable experience in familiarizing himself with each paper much more than usual. The papers in this book are state-of-the-art even after such a delay. It is the authors pride and integrity toward the quality of each paper that makes the value of this book long lasting, while responsibility of the loss of any timeliness rests at the editor's hand. For the purpose of official records, submission and acceptance dates must be stated. All papers had been submitted by September, 1984, and had been accepted for publication by November, 1984, after the critical review processes.

Since the editor had seen the first FT-IR spectrum of a polymeric material, which was recorded by a modern computerized FT-IR spectrometer, little more than a decade ago, the application of FT-IR to the polymer science field has developed at an unprecidented rate in the history of IR spectroscopy. The first FT-IR related paper of the editor was initially rejected by a reviewer because the reviewer remarked, "I do not believe in FT-IR". Nowadays, some may be willing to go as far as saying, "The IR spectrum recorded by a dispersive instrument may not be good enough". Some even propose that the use of the word "IR spectrum" should automatically indicate FT-IR spectrum. Advent of the table-top FT-IR spectrometers along with research grade spectrometers allow FT-IR to be used in all areas of industrial and academic IR studies. Of course, under any rapid growth, there is always a painful, persistent effort of the pioneers. We should thank researchers in pre-computerized FT-IR era for their valuable devotion. It is a fortuitous coincidence that the editor works at the university where Professor Michelson, the inventor of the Michelson interferometer which is the heart of FT-IR spectrometer, performed the infamous measurement of the speed of light using the interferometer exactly 100 years ago. It is also where the first paper on FT-IR characterization of polymers was written.

IR spectroscopy is one of the most valuable methods for polymer characterization. Unique sampling requirements arise from the polymeric nature of the samples. FT-IR has been extensively applied to polymers yet there has been no monograms dedicated to polymer characterization by FT-IR. The readers should enjoy a wide spectrum of articles in this book from the latest development of instrumentation to theoretical works utilizing the uniqueness of FT-IR.

The papers presented in the symposium have been rearranged in this book based on the content. Chapter I was later added to provide some background in optical theory. The contents of the remaining chapters are: Chapter II; Polarization-modulation Technique, Chapter III; New Instrumentation, Chapter IV; Application to Molecular Dynamics and Kinetics, Chapter V; Spectral Analysis and Manipulation Techniques, Chapter VI; Surface and Interface Studies, and Chapter VII; Application of Optical Theories.

It is the editor's previledge to acknowledge those who helped in editing this book. Each author's patience in spite of unacceptable delays was the major driving force for the progress. Ms. L. Piccinino and Ms. E. Raynor-Enco of Plenum Publishing Co. are both sponsoring editors who have been very patient and understanding with the delayed work. Ms. A. Lewandowski and Ms. P. Engelhorn helped to produce the book in its final form. The proof reading was done in part by R.T.Graf, J.D.Miller, H.Chatzi, R.Johnson, Y.Ishino, C.Scott, K.Nakata, K.Hoh, Y.Suzuki, J.Jang, and C.Khoo. Finally, wholehearted support from the editor's family members all made this book come to a completion.

H. Ishida
Editor

CONTENTS

INTRODUCTION TO OPTICS AND INFRARED SPECTROSCOPIC TECHNIQUES

R.T.Graf, J.L.Koenig and H.Ishida

Department of Macromolecular Science
Case Western Reserve University
Cleveland, Ohio 44106

I. Polymer Infrared Spectroscopy

Infrared spectroscopy is one of the oldest techniques for the molecular level characterization of materials, and it has of course been extensively used to study polymer systems. Excellent review articles exist for the application of both dispersive [1] and Fourier transform instrumentation [2] to polymers. The use of IR to study polymer surfaces and interfaces has also been reviewed [3]. As the number and complexity of IR techniques for examining non-routine samples has increased, there has been a growing tendency to examine samples ′in situ′. This is especially true where polymer systems are involved. Infrared spectra of such systems as filled polymers, glass reinforced plastics, fibers, and surface treated particulates, have been recorded in the past using relatively old techniques such as transmission and ATR. However, the spectral quality was low. Now it is possible to obtain high quality spectra of these systems by using such techniques as diffuse reflectance, photoacoustic, and IR microscopy.

The emphasis on measuring samples ′as is′ can lead to a dilemma in spectral interpretation. One may obtain a spectrum of an intractable sample by a suitable technique, but has one measured a spectrum of only the molecular structure of the sample, or a combination spectrum of the sample′s molecular structure and macroscopic state? Furthermore does the technique itself contribute to the measured spectrum? For many samples, these questions may only be important for quantitative work. But, for other samples even qualitative IR spectroscopy is not feasible without an understanding of the underlying physics of the experiments. In an infrared experiment, one usually measures the transmission, reflection, emission, or scattering of IR radiation, and then calculates the absorption from the measured quantity.

Usually more than one of these phenomena is present for a given experiment, and one must either experimentally minimize the others or account for them by calculations.

II. Theory of Optics: Background

The classical description of the propagation of electromagnetic radiation in free space or matter can be derived from Maxwell's equations (*). This derivation involves solution of second-order differential equations for both the electric and magnetic fields. A variety of solutions to this differential equation are possible including spherical and plane wave. For problems involving the reflection and transmission of light at planar interfaces, the plane wave solution given below is a convenient format.

$$E = E_o \exp(i\omega(n/c\ r \cdot s - t)) \quad (1)$$

n - refractive index   c - speed of light
r - position vector   s - propagation vector
E - electric vector ω - $2\pi\nu$ t - time

In this equation the electric field E is related to the initial electric field Eo by a complex exponential term which contains the frequency and refractive index (**). In a transparent medium the refractive index n is real and the electric and magnetic fields travel without attenuation at a speed of c/n. In an absorbing medium the refractive index is a complex quantity and the wave is attenuated as it propagates. The complex refractive index can be written as:

$$\hat{n} = n + ik \quad (2)$$

where the real part n is referred to as the refractive index and the imaginary part k is referred to as the absorption index. Both quantities together are often called the optical constants or the complex refractive index. For an absorbing medium equation (1) can be rewritten as

---

(*) See references 4-6 for a complete description of classical electromagnetics and optics.

(**) Note that the the time dependence of the oscillating electric field is -iwt. It is also possible to have an iwt time dependence.

(**) The refractive index is the square of the dielectric constant.

$$E = E_o \exp(i\omega(n/c\ r \cdot s - t)) \exp(-\omega/c\ k\ r \cdot s) \quad (3)$$

where the second exponential term expresses the rate of attenuation as a function of the distance travelled (r.s). The intensity of the radiation is just the square of the electric field amplitude. The base 10 logarithm of the intensity gives Beer's Law.

$$I = E.E^* = I_o \exp(4\ \pi\ k\ \nu\ d) \quad (4)$$

$$A = -\log_{10}(I/I_o) = 4\ \pi\ k\ \nu\ d/2.303 = ad \quad (5)$$

d - thickness   a - specific absorptivity

The above equations apply to the propagation of electromagnetic radiation in a single homogeneous, isotropic medium. For any real experiment, multiple phases, and hence phase boundaries will be present. Figure 1 is a schematic diagram of the simple two phase, single interface situation. The angle of reflection equals the angle of incidence, and the angle of refraction is related to the incident angle by Snell's law:

$$(6) \quad n_1 \sin(\theta_1) = n_2 \sin(\theta_2)$$

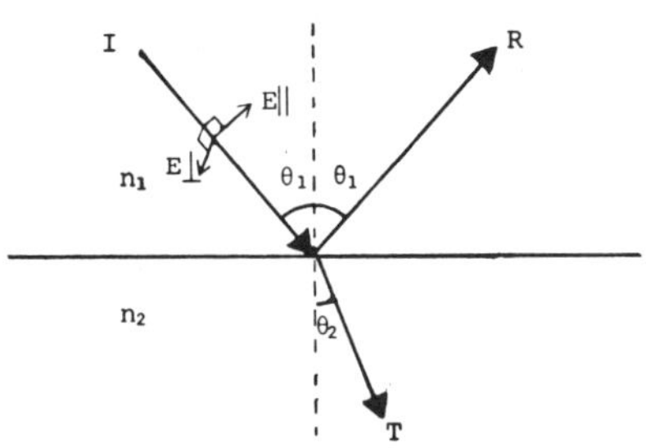

Figure 1. Schematic diagram of reflection and transmission at interface between two homogeneous and isotropic phases.

The fraction of the incident intensity (I) which is reflected (R) and transmitted (T) at a single interface can be calculated from the Fresnel relations. These equations are written in terms of the refractive indices of the two media, and the angle of incidence and refraction. There are four Fresnel relations for a single interface: one for each polarization state for both the reflected and transmitted waves. The polarization of the electric field vector can be either parallel (p) or perpendicular (s) to the plane of incidence (see figure 1). The plane of incidence is defined as the plane containing the propagation vector and the surface normal vector. For figure 1 the plane of incidence is the plane of the page. For normal incidence, the plane of incidence is undefined and the Fresnel relations simplify to the following:

$$r = \frac{n_2-n_1}{n_2+n_1} \qquad R = \frac{(n_2-n_1)^2}{(n_2+n_1)^2} \tag{7}$$

r - Fresnel reflection coefficient
R - reflected intensity

$$t = \frac{2n_1}{n_2+n_1} \qquad T = (t)^2(n_2/n_1) = \frac{4n_1n_2}{(n_2+n_1)^2} \tag{8}$$

t - Fresnel transmission coefficient
T - transmitted intensity

The Fresnel relations and Snell's law are also valid for absorbing media. In this case, since the refractive index is complex valued, the Fresnel coefficients also become complex.

The intensity of the reflected light as a function of the angle of incidence is plotted in figure 2 for two different interfaces. In this figure the first medium is a vacuum, and the second medium is either transparent (2A) or absorbing (2B). The refractive index of the dielectric is representative of ZnSe in the IR, while the optical constants of the absorbing medium are representative for a metal oxide in the IR. An incident angle of 0° corresponds to normal incidence, while an angle of 90° corresponds to grazing incidence. Note that the perpendicular component, Rs, is always greater than the parallel component, Rp, except at incident angles of 0° and 90° where they are equal. By definition both R and T are unit-less fractions of the incident intensity I. When the second medium is transparent, Rp exhibits a minimum of zero at the Brewster angle. The Brewster angle is given by:

$$\tan(\theta) = n_2/n_1 \tag{9}$$

When the second medium is absorbing, Rp exhibits a positive valued minimum at the pseudo-Brewster angle. The behavior of Rp and Rs in figure 2 is representative for all cases where the refractive index of the first medium is less than that of the second medium.

This situation is known as external reflection.

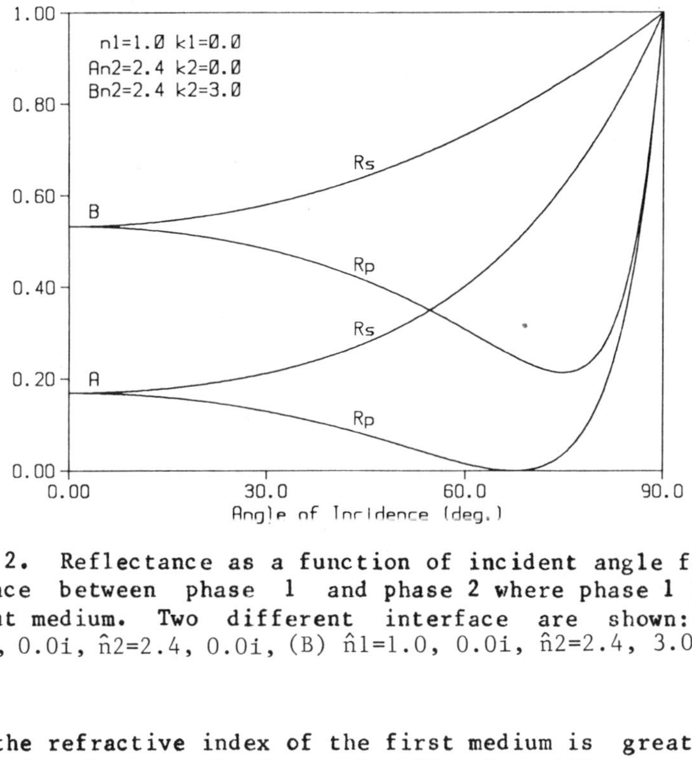

Figure 2. Reflectance as a function of incident angle for the interface between phase 1 and phase 2 where phase 1 is the incident medium. Two different interface are shown: (A) $\hat{n}1$=1.0, 0.0i, $\hat{n}2$=2.4, 0.0i, (B) $\hat{n}1$=1.0, 0.0i, $\hat{n}2$=2.4, 3.0i.

If the refractive index of the first medium is greater than the second, then total internal reflection will occur when the angle of incidence is greater than the critical angle. Below the critical angle the behavior of Rp and Rs is similar to that for external reflectance. Figure 3 shows Rp and Rs for cases where the incident medium is transparent and optically denser than the substrate medium. If the second medium is transparent, then both Rp and Rs will equal 1.0 above the critical angle, and all of the incident light is reflected. The critical angle is given by:

$$\sin(\theta) = n_2/n_1 \qquad (10)$$

For the situation in figure 2A, Rp equals to zero at the Brewster angle (20.6^o), and equals to one at the critical angle (22.0^o).

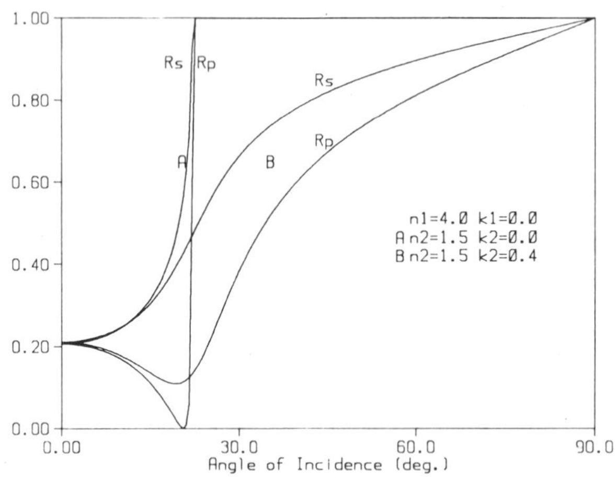

Figure 3. Internal reflectance as a function of incident angle for the interface between phase 1 and phase 2. Two different interfaces are shown: (A) $\hat{n}1=4.0,\ 0.0i$, $\hat{n}2=1.5,\ 0.0i$, (B) $\hat{n}1=4.0,\ 0.0i$, $\hat{n}2=1.5,\ 0.4i$.

If the second medium is absorbing then total internal reflection will not occur, but instead some of the incident radiation will be absorbed by the second medium. This case of attenuated total reflection is also shown in figure 3. Note, that the Rp and Rs curves have changed significantly for a relatively moderate k value of 0.4. This characteristic allows internal reflection spectroscopy to be applicable to even weak absorption bands. Typically, weak absorption bands of organic molecules in the IR will have k values in the range of 0.01-0.1, while moderate to strong bands will be in the range of 0.1-1.0. Metals are very strongly absorbing in the IR and will have k values of 10-100 across the entire mid-infrared region.

While the square of the Fresnel coefficients gives the intensity of the reflected and transmitted light at a given angle, the coefficients themselves give the amplitude and phase of the light. The following equation relates the amplitude and phase to the reflection coefficients.

$$r_z = |r_z| \exp(id_z) \qquad (11)$$

z - p,s
d_z - phase change upon reflection
$|r_z|$ - amplitude of reflected light

The differential phase (delta) and differential amplitude (psi) for reflection are given by:

$$r_p/r_s = |r_p|/|r_s| \exp(i(d_p-d_s)) \qquad (12)$$

$$= \tan(\psi) \exp(\Delta)$$

$$0^o < \Delta < 360^o \qquad 0^o < \psi < 90^o$$

Similar relations hold for the phase and amplitude of the transmitted light. Psi and delta are always real and are defined on the given intervals, while r_p and r_s may be complex.

The behavior of psi and delta as a function of the incident angle for external reflectance is given in figure 4 where two different cases are shown. In one case both phases are transparent and delta is 180° below the Brewster angle and 0° above it. Psi is 45° at incident angles of 0° and 90° and decreases to 0° at the Brewster angle. If the second medium is absorbing as shown in figure 4, then delta gradually decreases from 180° at normal incidence to 0° at grazing incidence. Psi again is 45° at 0° and 90°, but does not reach 0° at the pseudo-Brewster angle.

The behavior of psi and delta for internal reflectance is shown in figure 5. If the first medium is more optically dense than the second, and both phases are transparent, then delta will be 180° below the Brewster angle, 0° between the Brewster and critical angles, and will vary with incident angle above the critical angle. Psi is 45° at normal incidence and above the critical angle. Psi decreases from 45° to 0° when the incident angle changes from 0° to the Brewster angle. Between the Brewster and critical angles psi increases rapidly from 0° to 45°. For this example the Brewster angle is 22.6^o and the critical angle is 24.6^o.

The phenomena of phase change for reflected or transmitted light can be used to produce elliptical or circular polarization from linear polarization. Figure 6 shows that any of these polarization states can be achieved by summing two mutually perpendicular, linearly polarized waves which have an appropriate phase difference. From this diagram one can see that linear and circular polarization are special cases of elliptical polarization where the phase differences are 0° and 90° respectively.

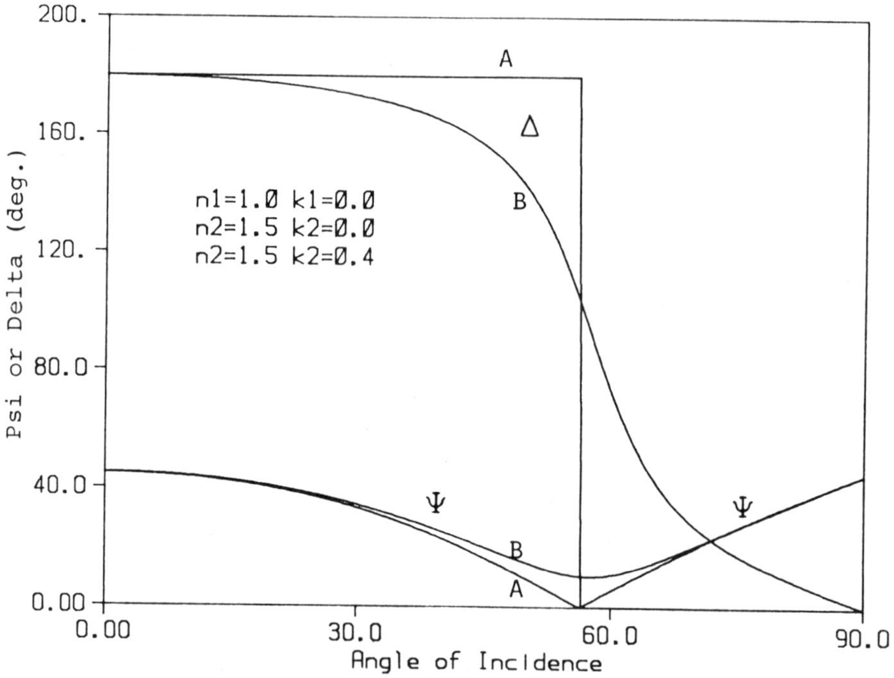

Figure 4. Psi and delta as a function of incident angle for external reflection at the interface between phase 1 and phase 2. Two different interfaces are shown: (A) $\hat{n}1=1.0,\ 0.0i$, $\hat{n}2=1.5,\ 0.0i$, (B) $\hat{n}1=1.0,\ 0.0i$, $\hat{n}2=1.5,\ 0.4i$.

When the phase difference is between 0° and 180° the tip of the electric vector will precess clockwise around the axis of the beam. When E_p and E_s are 90° out of phase, then (provided that the amplitudes are the same) the tip of the vector will describe a clockwise circle. When the phase difference is 180° then linearly polarized light is again obtained. However the axis of polarization will now be perpendicular to the case where the phase difference was 0°. When the phase difference is between 180° and 360°, the tip of the electric vector will precess counterclockwise around the beam.

Devices which can produce elliptically or circularly polarized light from linearly polarized light are referred to as linear retarders. Such devices include an internal reflection element, a metallic substrate coated with a thin dielectric film, and a birefringent dielectric element. From figure 5 it is apparent that by varying the incident angle (above the critical angle) internal reflection could give any desired phase retardation. Transmission through a birefringent element produces

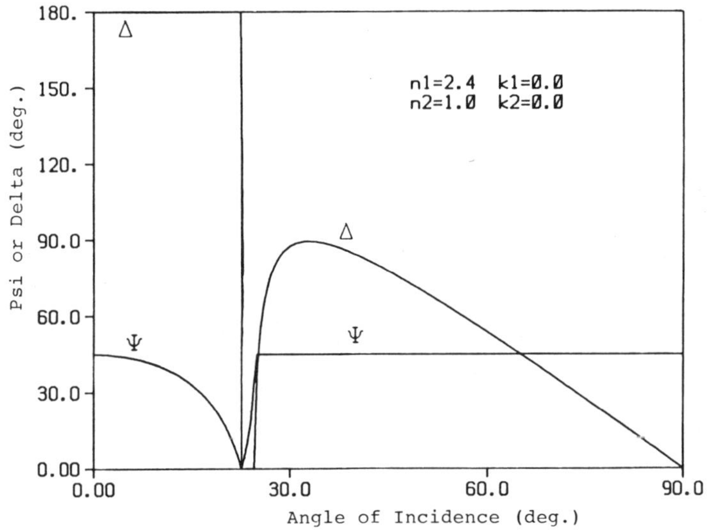

Figure 5. Psi and delta as a function of incident angle for internal reflection at the interface between phase 1 and phase 2. One interface is shown: $\hat{n}1=2.4,\ 0.0i$, $\hat{n}2=1.0,\ 0.0i$.

a relative phase retardation between the slow and fast transmission axes of the element. Therefore, linearly polarized light at normal incidence on one side of the element will emerge elliptically polarized on the other side. A dielectric coated metallic substrate induces a phase shift upon reflection which can be controlled by varying the angle of incidence or the thickness of the dielectric. Multiple reflections between the ambient-dielectric interface and the dielectric-metal interface make this

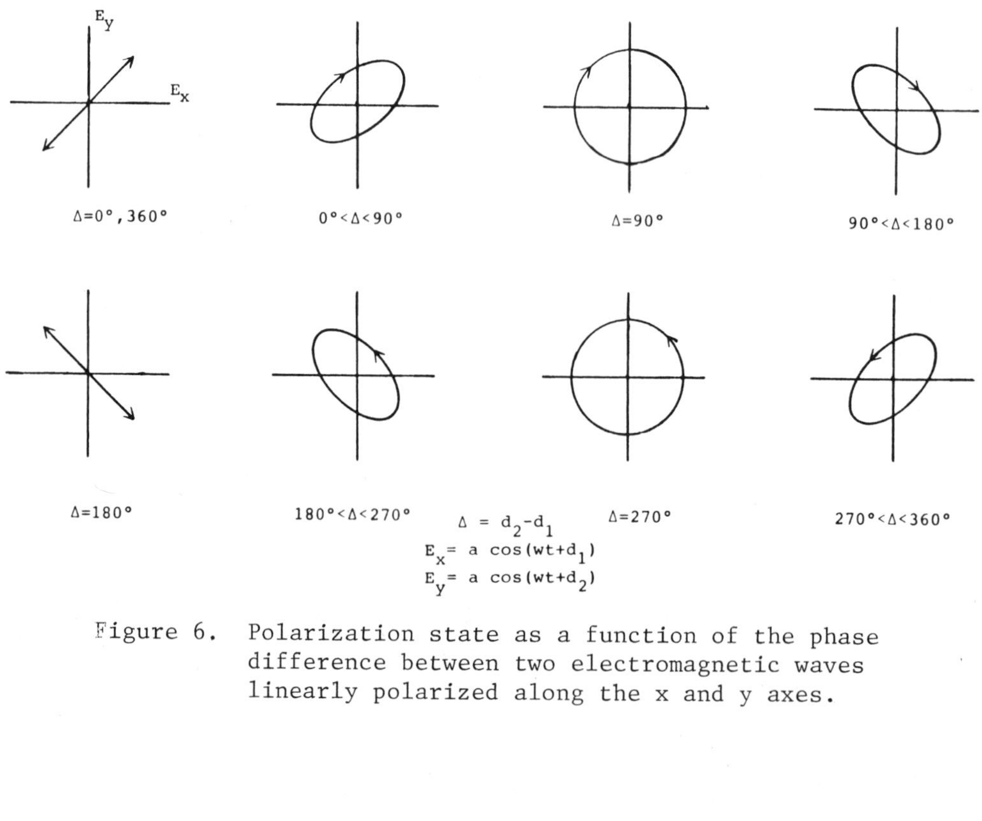

Figure 6. Polarization state as a function of the phase difference between two electromagnetic waves linearly polarized along the x and y axes.

device more suitable than bare metal only.

Multiple reflections between the front and back interfaces of a film will coherently interfere with one another, and the resulting transmittance and reflectance spectra will exhibit an interference fringe pattern. This phenomena is useful for measuring the thickness and refractive index of the film. The conditions for either constructive or destructive interference for both reflected and transmitted light is given by equation 13.

$$\nu = \frac{m}{2\ n\ d\ \cos\theta} \tag{13}$$

n - refractive index of film d - thickness
θ - angle of refraction ν - wavenumber
m - fringe order 1,2,3,... or 1/2,3/2,5/2,...

Equation 13 gives the positions of the fringe maxima for transmission and minima for reflection when integral values of m are used, and the minima for transmission and maxima for reflection when half-integral values of m are used. This protocol applies if the incident and substrate media both have higher or lower n's than the film. If the film is intermediate in refractive index, then the max/min protocol is reversed. Note that other than this the n values of the surrounding media do not enter into equation 13.

Figure 7 shows the results of some model calculations for a transparent film sandwiched between ambient media. The normal incidence reflectance and transmittance are shown for the mid-infrared region. In this ideal example the transmittance and reflectance sum to 1.0, since the film by definition is transparent. Rearrangement of equation 13 gives the thickness in terms of the refractive index and the wavenumber difference between adjacent maxima or minima (equation 14). The amplitude of the fringes in figure 7 is given in equation 15.

$$d = \frac{1}{2\ n\ d\nu\ \cos\theta} \tag{14}$$

$$dT = dR = F/(1+F) \tag{15}$$

$$F = 4R/(1-R)^2$$

$$R = (n-1)^2/(n+1)^2$$

If the amplitude and spacing of the fringe pattern can be

accurately measured, then n and d of the film can be calculated from equations 14 and 15.

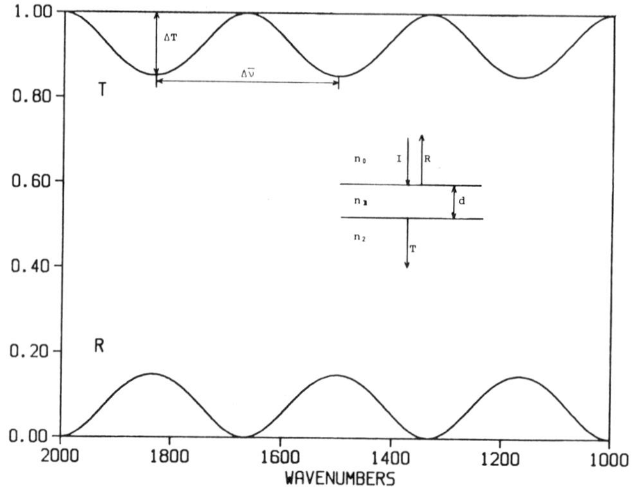

Figure 7. Interference fringe pattern from normal incidence transmittance and reflectance of thin transparent film: $\hat{n}0=\hat{n}2=1.0$, 0.0i, $\hat{n}1=1.5$, 0.0i, d=10 μm.

III. Infrared Spectroscopic Techniques

A short discussion of the various infrared techniques will be given here as an introduction. However the authors do not wish to rigorously review the theory and applications or all the recent developments.

Transmission is the oldest and best known method for obtaining infrared spectra, and is certainly the technique of choice. Liquids, solids, and gases can all be conveniently handled by transmission methods. For polymers the preferred method is to prepare free-standing thin films either by solvent casting or melt-pressing. However, either of these preparation methods can alter the crystallinity of a semicrystalline polymer, and either induce or destroy molecular orientation. For crosslinked systems, which are brittle and can be ground to a fine powder, a KBR pellet [7] or Nujol mull can be prepared. However, many polymers are too tough to grind well even at liquid nitrogen temperatures. Transmission techniques with a linear polarizer are also well suited to studying uniaxial orientation in polymer films. Changing the angle of incidence and the polarizer orientation can provide information on the polymer chain axis direction in a drawn thin film [8].

External reflectance spectroscopy was originally applied to study thin films on metal surfaces at near grazing angles of incidence [9,10]. For this application the term reflection-absorption is often used. Only at near grazing incidence is the presence of a thin (<10 nm) organic layer on a metallic substrate detectable by reflectance methods [10]. Furthermore only the radiation polarized parallel to the plane of incidence is sensitive to the surface layers [11]. These phenomena have been explained in terms of classical electromagnetic theory by Greenler [10-12], McIntyre [13] and others [14-18]. Basically, the incident and reflected radiation coherently interfere with each other and form a standing wave at the interface. For parallel polarization at high angles of incidence this standing wave produces a moderate electric field at the interface. This field can interact with any organic layer which is present. For perpendicular polarization or low incident angles the standing wave has a node at the interface, and hence there is no electric field to interact with the organic layer (see figure 8). This condition begins to break down when the film thickness reaches around 200 nanometers. A linear approximation theory has been developed for thin films (thickness<<wavelength), such that there is a linear relationship between the measured peak heights and the film thickness [13].

Sample systems other than thin films on metals have been studied by external reflectance methods, though the bands are generally weak. Self-supporting thin films, if of proper uniformity and clarity, will produce a strong interference pattern when measured by external reflectance. The patterns from different angles can be used to measure the thickness of the film [19]. Inorganic materials such as glasses which have strong absorption bands give good external reflectance spectra.

Internal reflectance or attenuated total reflectance (ATR) spectroscopy is probably the second most commonly used infrared technique after transmission. It originally was independently developed by Fahrenfort [20-21] and Harrick [22]. One of the first applications was for the study of the surface and bulk properties of semiconductors. For this application, the internal reflection element (IRE) was the material being studied itself. However, for the study of most solids and liquids, an IRE made of a suitable dielectric is brought into contact or close proximity with the material to be studied.

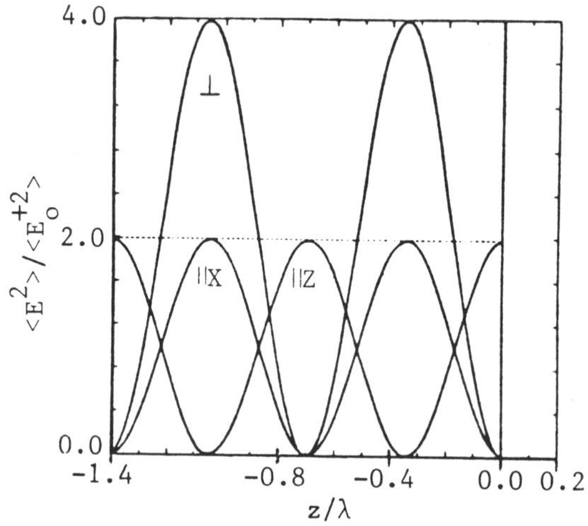

Figure 8. Mean-square electric field strengths in the standing wave at a highly reflecting metal surface. The optical parameters are: n̂0=1.0, 0.0i, n̂3=3.0, 0.0i, incident angle=45°, z is distance from surface, and lamda is the wavelength of light. Ref.: fig.10 from ref.13.

The optical contact between the sample and the dielectric element is the critical factor in obtaining a good spectrum. For soft or pliable polymers or solutions, ATR is an extremely versatile technique. Unlike transmission, the spectrum obtained is independent of sample thickness for all but the thinnest films. Thus one need not worry about having samples which are too thick. For liquids this thickness independence is even more of an advantage, since liquid transmission cells are often too thick for strongly absorbing liquids, and hence the liquid must be diluted. An ATR spectrum from a soft or liquid sample is qualitatively very similar to a transmission spectrum. The main difference will occur in the relative intensities of the bands at different wavenumbers. For hard or brittle polymers, however, it can be much more difficult to achieve good optical contact. Furthermore, it is also much more difficult to reproduce the optical contact for comparative quantitative measurements. The optical contact problem is probably the main reason why ATR is not even more widely used.

The theory and practice of internal reflection or attenuated total reflectance spectroscopy has been covered at length by Harrick [23], Hansen [24], and Crawford [25]. For internal reflectance just as was the case for external reflection, the incident and reflected light beams coherently interfere at the interface and form a standing wave. This standing wave does not extend into the rarer medium, instead a non-propagating evanescent wave is present in the rarer medium (see figure 9). The physical explanation for this evanescent wave is that the normal component of the electric field must be continuous across the interface, but at the same time there can be no net transfer of energy into the rarer medium for internal reflection. The electric field of the evanescent wave decays exponentially with distance from the interface. If the rarer medium is absorbing, then the evanescent wave will interact with the rarer medium and energy will be lost upon total reflection, hence attenuated total reflection. The depth of penetration is defined as the distance required for the electric field strength of the evanescent wave to reach 1/e of its initial value at the interface.

$$1/d_p = 2\pi\nu(\sin^2\theta - n^2)^{1/2} \qquad (16)$$

d_p - depth of penetration $\qquad n$ - $n2/n1$

Equation 16 shows that the depth of penetration depends on the relative ratio of the refractive indices of the two media, the angle of incidence, and the wavelength. The depth of penetration is the same for both polarization directions. Analogous to the cases of transmission and external reflection, simple approximate linear relations have been developed which relate the absorption coefficients of the sample to the measured reflectivity. However, for internal reflection where bulk samples have no clearly defined thickness, an effective sample thickness must be defined. This effective thickness depends on the depth of penetration, the electric field strength at the interface, and the index matching between the sample and the internal reflection element. The effective thickness is not the same for perpendicular and parallel polarizations because the surface electric field strength is greater for parallel polarization. The following equations apply for bulk samples and weak absorption bands.

$$R = 1 - a\,d_e \qquad (17)$$

$$d_e = \frac{n\,E^2\,d_p}{2\cos\theta} \qquad (18)$$

d_e - effective sample thickness

E - electric field strength at interface
a - absorptivity

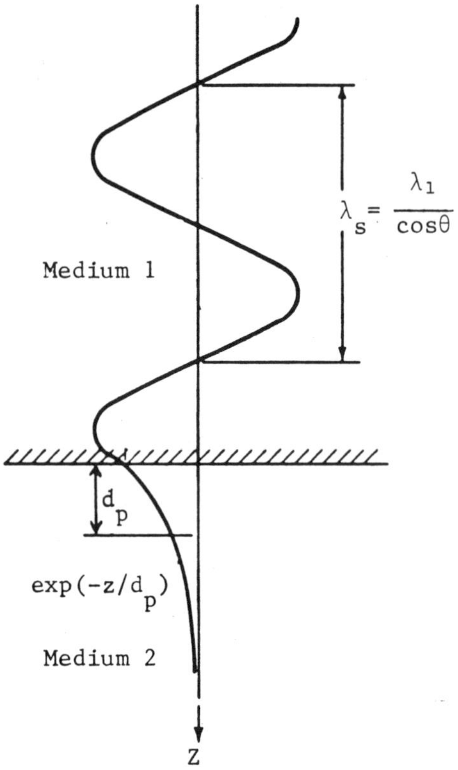

Figure 9. Standing-wave amplitudes established near totally reflecting interface: There is a sinusoidal dependence of the electric field amplitude on the distance from the surface in the denser medium 1 and an exponentially decreasing amplitude in the rarer medium 2. Ref.: p.28 from ref. 23.

Diffuse reflectance spectroscopy has recently become a powerful technique for the analysis of powders and coarse solids in the infrared. Diffuse reflectance has long been used to analyze powder samples in visible spectroscopy, but until the advent of FT-IR it was not possible to obtain good infrared diffuse spectra. Because of the high energy through-put and signal-to-noise ratio of FTIR, it is now possible to obtain infrared diffuse reflectance spectra of microgram quantities of sample. The sensitivity of diffuse reflectance infrared Fourier transform spectroscopy (DRIFTS) [26], and its quantitative accuracy for powdered samples [27] have been documented by Fuller and Griffiths. An significant advantage of DRIFTS over other techniques is the non-destructive capability of the sample preparation. For polymer systems which can be difficult to grind it may be sufficient to simply roughen the surface to be studied. For many coarse systems spectra can be obtained of the neat sample without grinding or other alteration. A wide variety of samples such as powders [27], adsorbates on silica TLC plates [26], graphite fiber - epoxy matrix composites [28], and surface treated glass fibers [29] have been studied using DRIFTS.

Theories for diffuse reflectance have been broadly classified into either continuum or statistical models [30-31] (see figure 10). Continuum theories involve the use of phenomenological constants, while statistical theories utilize fundamental quantities such as absorptivity, refractive index and particle size [32]. One of the most widely used models is the Kubelka-Munk theory [33]. It is a continuum model that gives a linear relationship between sample concentration and a function of the measured reflectance. Equation 19 gives the Kubelka-Munk relationship.

$$(1-R_\infty)^2/2R_\infty = ck'/s \qquad (19)$$

R_∞ - diffuse reflectance
k - absorption coefficient
c - molar concentration
s - scattering coefficient

The term R_∞ is the absolute reflectance of a sufficient quantity of sample such that additional sample does not alter the reflectance. The scattering coefficient s depends upon the particle size and distribution and can be difficult to calculate in practice. If the concentration range of the sample is kept small and particle size is kept constant, then s can usually assumed to be constant. The effect of variations in particle size and concentration range on diffuse reflectance spectra is shown in figure 11. The absorption coefficient k is not the same as the absorption index k mentioned earlier although the two have been related [34]. The concentration range over which this theory gives accurate results is limited [35].

Most theories for diffuse reflectance attempt to model a system of particulate absorbing particles dispersed in a particulate non-absorbing medium [30-33]. For particulate samples, most of the incident radiation is diffusely reflected, while for non-particulate samples a large portion of the radiation may be specularly reflected. For such sample systems the reflected radiation has both diffuse and specular components.

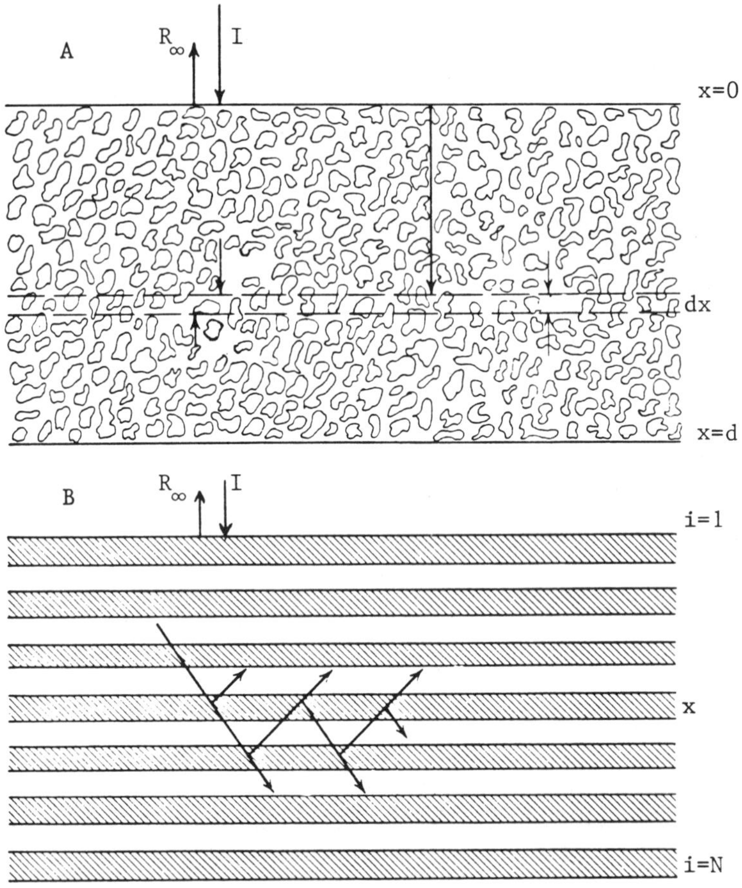

Figure 10. Schematic diagrams of diffuse reflectance experiments: (A) continuum model, (B) statistical model. Ref.:p.57 and 79 from ref.18.

Photoacoustic spectroscopy (PAS) has become a useful technique for the measurement of spectra in the mid-IR region in recent years due to improvements in instrumentation. The idea itself is quite old as it was originally proposed by Alexander Graham Bell [36]. Like diffuse reflectance, photoacoustic has the advantage of being able to examine samples in a non-destructive manner. In fact Krishnan has compared the spectra of the same sample by diffuse reflectance and photoacoustic and found them to be quite similar for powders [37]. Solids, liquids or gases can be examined with photoacoustic, though the sensitivity is low and long scanning times are usually required. The reproducibility of PAS is such that for well controlled samples it is possible to

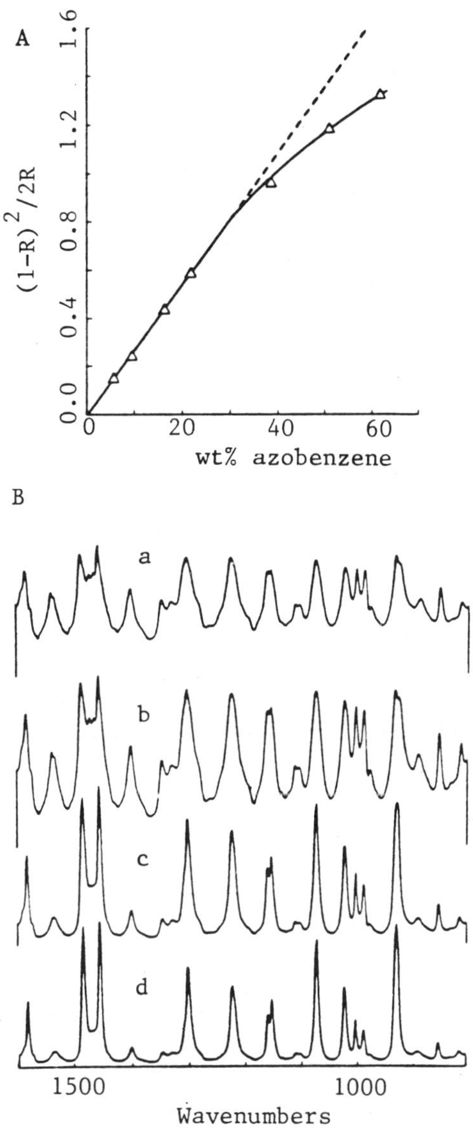

Figure 11 (A) Kubelka-Munk plot versus weight percent azobenzene in KCL for the 1960 cm^{-1} band of azobenzene. (B) Effect of particle size on the diffuse reflectance spectrum of neat azobenzene versus a KCL reference, plotted in Kubelka-Munk units. (a) d > 90 microns, (b) 75 < d < 90 microns, (c) 10 < d < 75 microns, (d) d < 10 microns. Ref.: p. 1908 from ref. 26.

perform reliable spectral subtractions [37]. Similar to ATR spectroscopy, the photoacoustic signal has a variable penetration depth which allows it in principle to be used for depth profiling experiments.

The principle of photoacoustic spectroscopy is that modulated IR radiation striking the surface of a sample will cause the surface to alternately heat and cool. This cyclic heating and cooling of the sample surface is conducted to a coupling gas in the photoacoustic cell. A standing sound wave develops which is detected by a microphone (see figure 12). If a particular frequency is not absorbed then the sample surface will not heat up and no sound wave will develop. Hence, in photoacoustic spectroscopy sound waves are used to detect infrared absorption frequencies. For a dispersive spectrometer the incident IR radiation is modulated by a beam chopper, while in Fourier transform instruments the interferometer itself acts as a modulator.

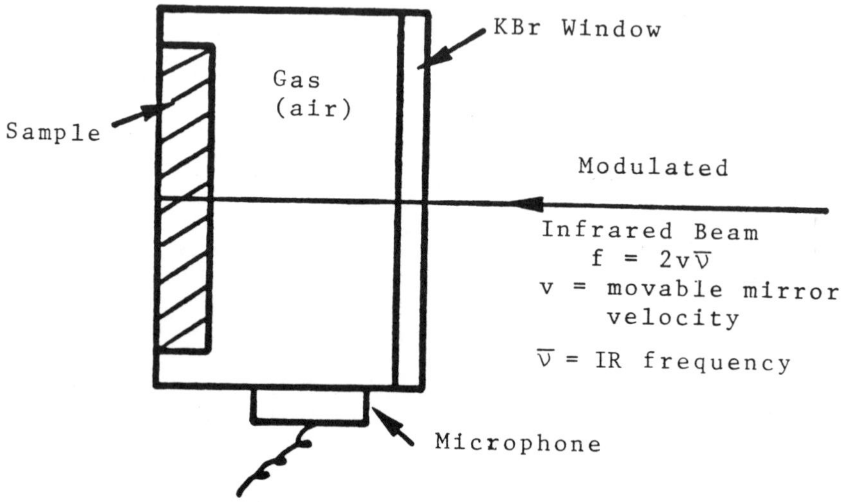

Figure 12. Schematic diagram of photoacoustic cell. Ref.2.

For quantitative work close attention must be paid to a variety of factors such as thermal diffusion length, optical path length, and the IR radiation modulation frequency. Rosencwaig and Gersho have developed a theory of the photoacoustic effect in solids [38-39]. This theory outlines six different cases for the behavior of the output signal depending on the relative values of the thermal diffusion length and the optical path length. If the thermal diffusion length exceeds the optical path length then photoacoustic saturation will occur and quantitative work becomes impossible. Furthermore, for some situations even qualitative measurements become difficult. Krishnan [37] compared ATR and PAS measurements of the same 0.7 mm thick film of poly(ethylene terephthalate) (PET). For the carbonyl band of PET, the ATR spectrum gave a peak maximum of 1715 cm^{-1} and the PAS spectrum gave 1735 cm^{-1} (see figure 13). This discrepancy was attributed to the saturation effect of PAS. Light scattering can also be a problem for PAS. It has been found to significantly alter band intensities [40].

Ellipsometry has long been used in the uv-visible region of the spectrum for probing the thickness and optical properties of surface layers and films. Sensitivities on the order of angstroms are achieved with visible light laser ellipsometry [41]. For the rapid determination of the thickness of thin oxide and other layers on metals and dielectrics, uv-visible ellipsometry is the technique of choice. The theory and practice of ellipsometry has been covered in detail in several excellent books and papers [42-44]. Ellipsometry is concerned with the change in polarization which a sample induces in a wave. The polarization of the input wave is known and controlled, and the output polarization is measured (see figure 14). From the difference between the input and output polarization states the differential phase change, delta, and the differential amplitude, psi, can be calculated. Photometry, conversely, deals with the change in intensity which a sample induces in a wave. The principles of ellipsometry apply to other spectral regions and in fact can be used with any transverse waveform [44]. Acoustical waves are one example [44]. In the last decade interest in infrared ellipsometry has increased. Although to date there have been few studies of polymer systems, the application of infrared ellipsometry to thin polymer layers on a variety of substrates seems certain to increase.

There are several reasons for doing infrared ellipsometry. First, to measure the thickness of thin films on surfaces in the range of 10-1000 nm. Second, to measure the optical constants n and k of the surface film. In any grazing angle reflection experiment the refractive index $n(\nu)$ and the absorption index $k(\nu)$ both contribute significantly to the measured spectrum. Thus one must extract both the absorption and dispersion properties of the sample to properly interpret the measured spectrum. Third, in a reflection experiment only R_p and R_s can be measured by normal photometry. With ellipsometry one can also measure delta, the differential phase change, and thus obtain phase as well as intensity information. The infrared region itself contains more chemical functional group information than is available in the uv-visible region. While one can obtain an infrared absorption spectrum of a surface species by photometry alone, ellipsometry

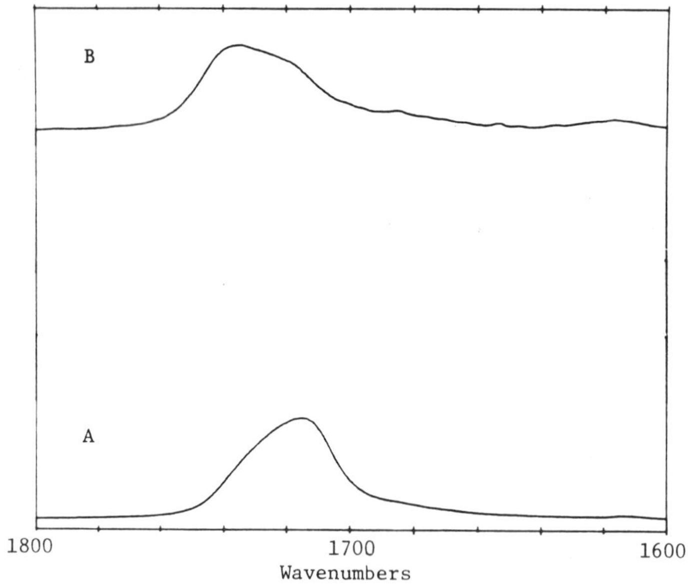

Figure 13 Spectra of polyethelene terephthalate (PET) between 1800 and 1600 cm^{-1}: (A) ATR spectrum, (B) PA spectrum Ref.: p. 552 from ref. 37.

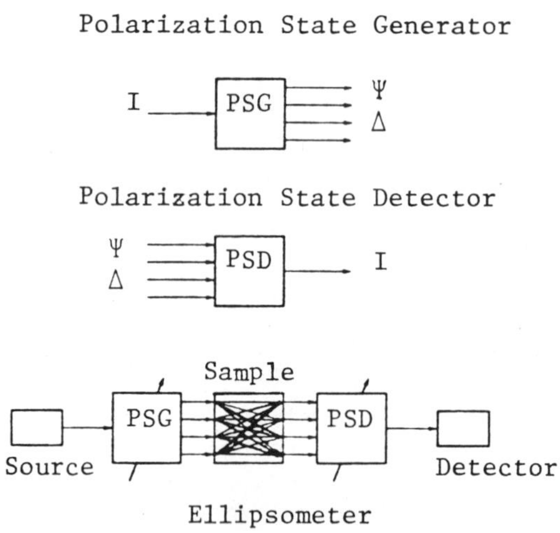

Figure 14 Ellipsometric measurement. The roles of the polarization-state generator (PSG) and detector (PSD) in transforming photon flux are shown. An ellipsometer employs both to determine by suitable optical theory the desired information about the sample S. Ref.: fig. 1 from Ref. 44.

gives both absorption and dispersion spectra and allows the direct calculation of the optical constants. Finally, the high signal-to-noise ratio and wavelength accuracy of FT-IR should be advantageous for ellipsometric purposes as has been shown by Dignam [45] and Roseler [46-47].

IV. Applications of Optical Theory

In section II classical electromagnetic theory and optics was discussed, while in section III the various infrared spectroscopic techniques were described; in section IV various applications of optical theory to infrared spectroscopy will be reviewed. The discussion of section II focussed on the different techniques and some approximate linear models for relating peak intensities to sample thickness or concentration. For many situations these equations are sufficient to obtain good quantitative results. However, for more advanced applications it is useful to be able to calculate the position, shape, and intensity of the bands one will obtain from a given infrared experiment done on a standard sample. Then similar results for unknown samples can be more readily interpreted correctly. Furthermore, some of the information from the different techniques is lost unless more advanced models are used to describe the experiments. Other applications such as ellipsometry, and polarization modulation by their nature require more sophisticated processing of the data than the basic models will allow.

Quantitative measurements in the infrared usually begin with Beer's law, and its analogs. However, even for transmission measurements several author's have shown that systematic errors in peak intensities can result if Beer's law is used without regard to reflective losses and interference effects [49-51]. Crawford et.al. [50] studied the linearity of Beer's law for benzene via transmission through a liquid cell. They showed that the dispersion of the refractive index across a strong absorption band can cause significant distortion in the measured band profile. The interference pattern created by the liquid cell contributed to this error. Jones et.al. showed similar results and by calculating the optical constants of the liquids under study they were able to correct for the optical distortion error.

Allara has pointed out that to study the interaction between different homo-polymers in blends via transmission IR spectroscopy, optical distortion effects must be taken into account [52]. For such studies the spectra of the homo-polymers is often subtracted from the blend spectra and the difference is scale expanded. Model calculations by Allara are shown in figure 15. These results show that optical effects alone can produce residual differences after subtraction which are of the same order as those expected for molecular interaction. The differences figure 15 between hypothetical mixtures and pure components are a result of the small differences in the reflectance across absorption bands from free-standing thin films of these materials.

For internal reflection spectroscopy the effective sample thickness is a function of the wavelength, refractive indices, and angle. To compare the relative intensities of bands within an ATR spectrum, one should correct the ATR spectrum for the wavelength dependence of the effective sample thickness. Conversely, since the effective thickness is a function of the angle of incidence, by varying the incident angle, it is possible in theory to probe different regions of the sample. Hobbs et.al. [53] have used this effect to probe the gradient in orientation and changing degree of crystallinity of polypropylene and poly(ethylene terephthalate) uniaxially oriented films. However, extreme care is needed to interpret the results from depth profiling studies. Again for strong absorption bands the dispersion of the refractive index will cause the effective sample thickness to vary across a band. Subtle spectral changes which arise from depth profiling experiments, may only be the result of the dispersion of the effective thickness across absorption bands. Preferably, the dispersion of the refractive index should be corrected for when doing ATR depth profiling [54]. An example of the effect of the dispersion of the refractive index on internal reflection spectra is given in figure 16.

The study of thin films on metallic surfaces by infrared reflection-absorption spectroscopy requires some care in the interpretation of band positions. The optical contribution to the measured spectra is probably higher for this technique than any other. Greenler in a study of copper oxide on copper showed that 609 cm^{-1} band of Cu_2O appeared 28 cm^{-1} wavenumbers higher in a 87° reflection spectrum than expected [12]. He accounted for this difference by using the Cu_2O optical constants of O'Keefe [55] and calculating the expected band profile for the reflection experiment using a method he previously outlined [10]. Allara showed band shifts of 10 cm^{-1} for the carbonyl band of poly(methyl methacrylate) on gold and silicon substrates [17]. In fact, Allara observed that for certain thicknesses of PMMA on silicon it was possible for the carbonyl band to have a negative absorption (see figure 17). All of his experimental observations of band shifts and shape changes were confirmed by theoretical calculations. Allara concluded that for bands with halfwidths of around 20 cm^{-1} and k values greater than 0.1, significant distortion effects could occur. For narrower peaks and bands with k less than 0.1, he concluded that reflection band shapes should be close to those of transmission.

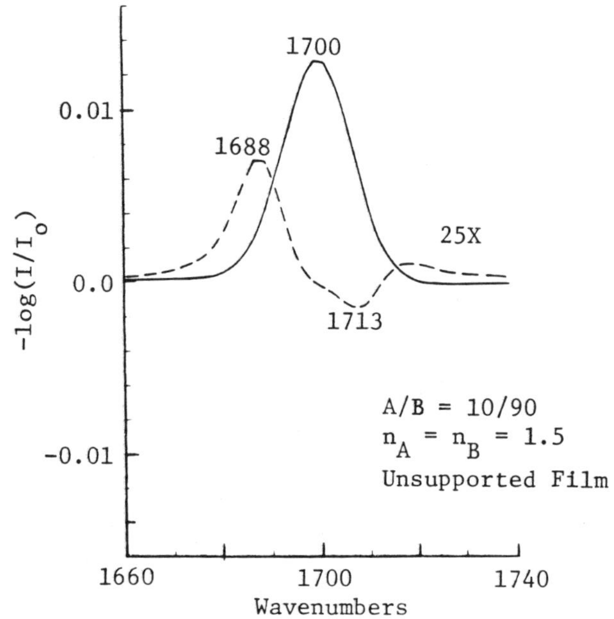

Figure 15. Absorbance spectra for a free-standing 10/90 A/B film. --- mixture spectrum; - - - difference spectrum (x25). Ref.:fig.2 from ref.52.

Several researchers have done ellipsometric measurements in the IR region. Dignam et.al. have described an IR spectroscopic ellipsometer with background noise approaching $2x10^{-5}$ absorbance units [56]. Their use of IR ellipsometry allowed them to obtain the infrared dispersion properties of surface species as well as the absorption properties (see figure 18). For surface species which exhibit optical anisotropy, Dignam [57] has shown how to extract the IR absorption and dispersion properties normal and perpendicular to the surface. The wealth of information which an infrared spectrum contains makes such surface orientation measurement very desirable. Allen and Sunderland [58] used the 10.6 μm line from a CO_2 laser to study the oxide thickness on aluminum substrates in the range of 20-200 nanometers (nm). Some of their results are shown in figure 19. They chose infrared ellipsometry because of the limitations of visible light ellipsometry for layer thicknesses approaching 200 nm, and because IR ellipsometry is not as sensitive to surface roughness of the oxide layer as uv-visible ellipsometry. The optical constants of thin silver and gold films were measured at selected infrared frequencies using infrared ellipsometry by Bashara et.al. [59].

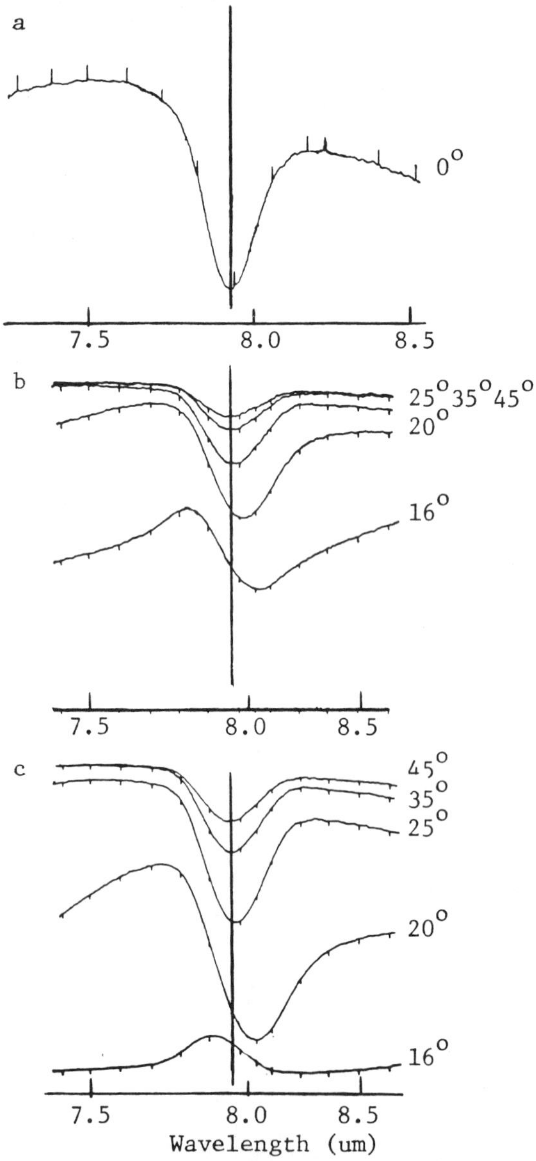

Figure 16 Spectra of band of silicone lubricant at 7.9 microns. (a) Internal reflectance spectra for unpolarized light ar various angles of incidence compared to transmission spectrum. (b) Internal reflection spectra at various angles of incidence for ⊥ polarization. (c) Internal reflection spectra at various angles of incidence for || polarization. Ref.: p. 245 from ref. 23.

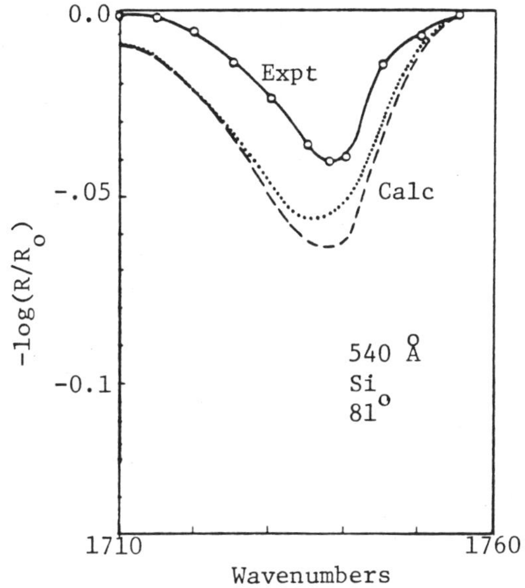

Figure 17 Band shapes for a 54 nm poly(methyl methacrylate) film on silicon at an 81° angle of incidence and p polarization: o experimental curve,---calculated curve for experimental n, ... calculated curve for Kramers-Kronig n. Ref.: fig. 5 from ref. 17.

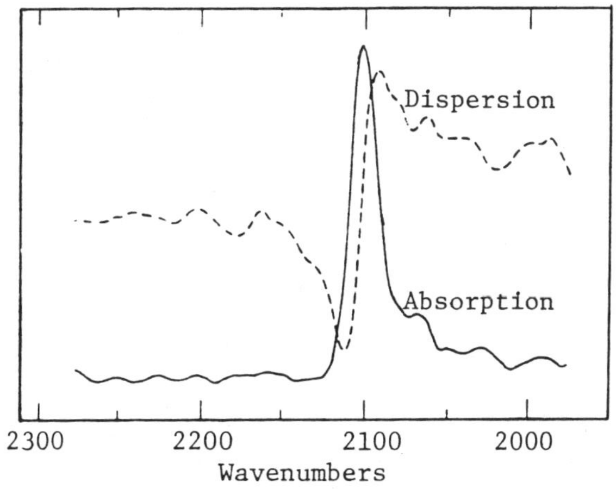

Figure 18 Absorption and dispersion spectrum for CO on a UHV deposited Cu film, for 4.2 Torr CO pressure and 293 K. Ref.: fig. 8 from ref. 56.

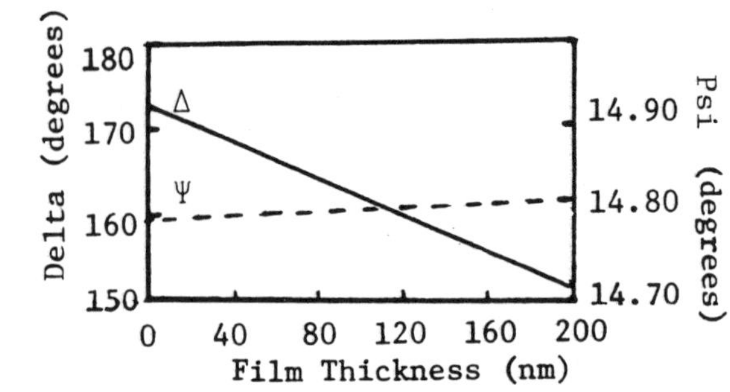

Figure 19 Computed values of psi and delta as a function of oxide thickness: circles-psi, diamonds-delta, n_{film}=1.63. Ref.: fig. 1 from ref. 58.

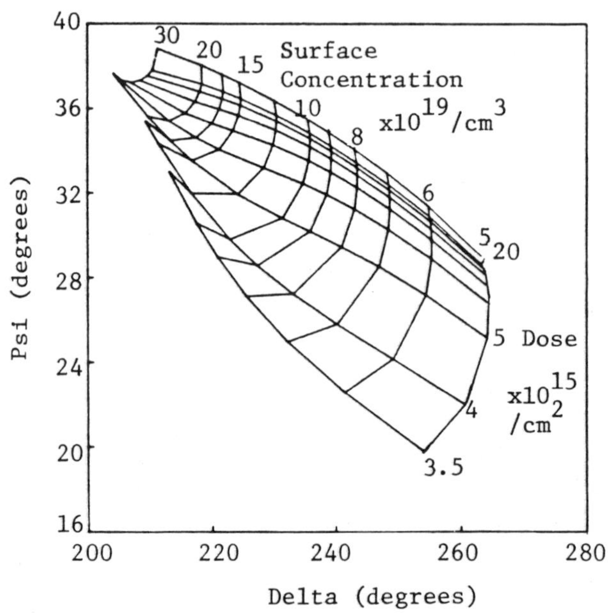

Figure 20 Psi and delta values calculated for doped layers with a gaussian impurity profile and various surface concentrations and dose values. For the calculations assumed wave length: 10.591 microns, angle of incidence 70°, sample thickness: 400 microns, substrate doping: 7.72×10^{14} atoms/cm^3. Ref.: fig. 20 from ref. 60.

The measurement of the optical constants of metals in the infrared region is of considerable interest, and ellipsometry has important advantages over reflection photometry for such determinations. Schaefer has used infrared ellipsometry to investigate ion implanted layers in silicon wafers [60]. Both the surface concentration and the dose of the dopants were measured by infrared ellipsometry (see figure 20). For the example of figure 20 a Gaussian distribution of the ions in the substrate was assumed.

V. SUMMARY

The difficulty of extracting molecular vibrational information from infrared spectroscopic measurements depends on the complexity of the sample, the number of unknown quantities, and the number of different possible sample configurations. The sample complexity can vary from homogeneous and isotropic to inhomogeneous and anisotropic. This along with the number of phases in an experiment will determine the number of spectroscopic variables. With a large number of variables it is obviously desirable to measure spectra by more than one technique. The optical configuration can also be altered by such things as changing the angle of incidence. Given a sufficient number of measurements to determine the relevant variables an appropriate optical model must be assumed, and the validity of an optical model must be determined in advance. The importance and necessity for separating molecular level information from optical artifacts has, hopefully, been demonstrated in this paper and the other papers in this book.

REFERENCES

1. S. Krimm, Fortschr. Hochpolym. Forschg. 2, 51 (1960).

2. J.L. Koenig, in "Advances in Polymer Science" vol. 54, Springer Verlag, Berlin (1983) p.87.

3. S.R. Culler, H. Ishida, J.L. Koenig, Ann. Rev. Mater. Sci., 13, 363 (1983).

4. M. Born and E. Wolf, "Principles of Optics" 6th ed., Pergamon Press, Oxford (1980).

5. R.W. Ditchburn, "Light", Interscience, New York (1953).

6. J.R. Reitz, F.J. Milford, and R.W. Christy, "Foundations of Electromagnetic Theory" 3rd ed., Addision Wesley, Reading, Mass. (1979).

7. O.Y. Ataman and H.B. Mark, Appl. Specrosc. Rev. 13, 1 (1977).

8. A. Cunningham, G.R. Davies, and I.M. Ward, Polymer 15, 743 (1974).

9. S.A. Francis and A.H. Ellison J. Opt. Soc. Am. 49, 130 (1959).

10. R.G. Greenler, J. Chem. Phys. 44, 310 (1966).

11. R.G. Greenler, R.R. Rahn, and J.P. Schwartz, J. Catalysis, 23, 42 (1971).

12. R.G. Greenler, J. Vac. Sci. Technol. 12, 1410 (1975).

13. J.D.E. McIntyre, in "Advances in Electrochemistry and Electrochemical Engineering" vol. 9, Wiley, New York (1973) p. 61.

14. H.G. Tompkins, in "Methods of Surface Analysis" vol. 1, Elsevier, Amsterdam (1975).

15. J.F. Blanke, S.E. Vincent, and J. Overend, Spectrochim. Acta 32A, 163 (1976).

16. N.J. Harrick, in "Characterization of Metal and Polymer Surfaces" vol. 2, Academic, New York (1977) p. 153.

17. D.L. Allara, A. Baca, and C.A. Pryde, Macromol. 11, 1215 (1978).

18. W.W. Wendlandt and H.G. Hecht, "Reflectance Spectroscopy" Wiley Interscience, New York (1967), chap. 2.

19. N.J. Harrick, Appl. Opt. 10, 2344 (1971).

20. J. Fahrenfort, Spectrochim. Acta 17, 698 (1961).

21. J. Fahrenfort, Spectrochim. Acta 18, 1103 (1962).

22. N.J. Harrick, J. Phys. Chem. 64, 1110 (1960).

23. N.J. Harrick, "Internal Reflection Spectroscopy" Wiley Interscience, New York (1967).

24. W.N. Hansen, in "Advances in Electrochemistry and Electrochemical Engineering" vol. 9, Wiley, New York (1973) p. 1.

25. B. Crawford, T.G. Goplen, and D. Swanson, in "Advances in Infrared and Raman Spectroscopy" vol. 4, Heyden, London (1978), chap. 2.

26. M.P. Fuller and P.G. Griffiths, Anal. Chem. 50, 1906 (1978).

27. M.P. Fuller and P.G. Griffiths, Appl. Spectrosc. 34, 533 (1980).

28. P.R. Young, B.A. Stein, and A.C. Chang, Proc. 28th Natl. SAMPE Symp. Exhib., 824 (1983).

29. R.T. Graf, J.L. Koenig, H. Ishida, Anal. Chem. 56, 773 (1984).

30. Ref. 18, chap. 3.

31. G. Kortum "Reflectance Spectroscopy", Springer-Verlag, New York (1969) chap. 4.

32. H. Hecht, J. Res. Natl. Bur. Stand. 80A, 567 (1976).

33. P. Kubelka and F. Munk, Z. Tech. Phys. 13, 593 (1931).

34. R.W. Frei and J.D. MacNeil, "Diffuse Reflectance Spectroscopy in Environmental Problem Solving" CRC Press, Cleveland (1973).

35. H. Hecht, Appl. Spectrosc. 34, 157 (1980).

36. A.G. Bell, Philos. Mag. 11, 510 (1881).

37. K. Krishnan, Appl. spectrosc. 35, 549 (1981).

38. A. Rosencwaig and A. Gersho, J. Appl. Phys. 47, 64 (1976).

39. A. Rosencwaig, "Photoacoustics and Photoacoustic Spectroscopy", Wiley, New York (1980).

40. W. Vidrine, Appl. Spectrosc. 34, 314 (1980).

41. D.E. Aspnes, Surf. Sci. 101, 84 (1980).

42. R.M.A. Azzam and N.M. Bashara, "Ellipsometry and Polarized Light" North-Holland, New York (1977).

43. R.H. Muller, in "Advances in Electrochemistry and Electrochemical Engineering", vol. 9 (1973) p.167.

44. P.S. Hauge, Surf. Sci. 96, 108 (1980).

45. M.J. Dignam and M.D. Baker, Appl. Spectrosc. 35, 186 (1981).

46. A. Roseler, Infr. Phys. 21, 349 (1981).

47. A. Roseler and W. Molgedey 24, 1 (1984).

48. R.W. Stobie, B. Rao, and M.J. Dignam, J. Opt. Soc. Am. 65, 25 (1975).

49. R.N. Jones, D. Escolar, J.P, Hawranek, P. Neelakantan, and R.P. Young, J. Mol. Struct. 19, 21 (1973).

50. T. Fujiyama, J. Herrin, and B.J. Crawford, Appl. spectrosc. 24, 9 (1970).

51. S. Maeda and P.N. Schatz, J. Chem. Phys. 35, 1617 (1961).

52. D.L. Allara, Appl. Spectrosc. 33, 358 (1979).

53. J.P. Hobbs, C.S.P. Sung, K. Krishnan, and S. Hill, Macromol. 16, 193 (1983).

54. M.J. Dignam, private communication (1984).

55. M.J. O'Keefe, J. Chem. Phys. 39, 1789 (1963).

56. R.W. Stobie, B. Rao, and M.J. Dignam, Appl. Opt. 14, 999 (1975).

57. M.J. Dignam, Polymer Preprints 25, 149 (1984).

58. T.H. Allen and R.J. Sunderland, Thin Solid Films 45, 169 (1977).

59. J.R. Adams, J.R. Zeidler, and N.M. Bashara, Opt. Comm. 15, 115 (1975).

60. R.R. Schaefer, J. Phys., Colloq. C10, 87 (1983).

CHARACTERIZATION OF POLYMERS USING POLARIZATION-MODULATION INFRARED TECHNIQUES: DYNAMIC INFRARED LINEAR DICHROISM (DIRLD) SPECTROSCOPY

Isao Noda, A.E. Dowrey and Curtis Marcott

The Procter & Gamble Company
Miami Valley Laboratories
P.O. Box 39175
Cincinnati, OH 45247

INTRODUCTION

High-frequency polarization-modulation infrared techniques [1-8] can be successfully applied to the characterization of polymers using either Fourier transform or dispersive instrumentation. These techniques include infrared reflection-absorption spectroscopy (IRRAS) [2-4], infrared linear dichroism under both static [5] and dynamic [6-8] conditions, and vibrational circular dichroism (VCD) [1] for chiral systems. Some intrinsic advantages of polarization-modulation make it especially attractive for high-sensitivity IR spectroscopy. Since VCD and IRRAS are covered thoroughly in other papers of this symposium book, our discussion will concentrate on linear dichroism of polymer films; in particular, the application of polarization modulation to dynamic infrared linear dichroism (DIRLD) spectroscopy.

DIRLD spectroscopy is a rheo-optical characterization technique based on the combination of infrared linear dichroism spectroscopy and dynamic mechanical analysis. This technique is especially suited for the study of time-dependent submolecular-level orientation responses of polymeric materials to small mechanical perturbations. Since the formalism of DIRLD closely parallels that of dynamic mechanical analysis, both will be reviewed briefly in the "Background" section. The basic instrumentation used for DIRLD will be described with examples of applications, including studies of isotactic polypropylene, an amorphous block copolymer, and linear low-density polyethylene samples. Finally, comparison of the relative advantages of dispersive and Fourier transform infrared (FT-IR) approaches will be made.

BACKGROUND

A. Polarization Modulation

Polarization-modulation is a versatile method for rapidly alternating the polarization states of electromagnetic radiation. The technique uses a linear infrared polarizer followed by a photoelastic modulator (PEM), consisting of an IR transparent material (e.g., ZnSe) coupled to a piezoelectric driver, to create externally controlled stress-birefringence (photoelastic) modulation [9]. Such modulation of the optical polarization state is especially suited for rapid dichroic measurements. Phase sensitive detection with lock-in amplifiers is then used to directly determine the small differential absorbance signal between the two polarization states in a single measurement. Because the dichroic difference signal of interest is measured as the amplitude of a periodic high-frequency "AC" signal, the spectrometer stability required to minimize low-frequency drift noise is not as crucial in a polarization-modulation experiment. In a conventional measurement, two large "DC" signals obtained sequentially after a finite amount of time are subtracted from each other or ratioed. Interferences from background gases or liquids can also be minimized with polarization-modulation because these molecules are randomly oriented and absorb both polarizations equally. Evacuating or purging the spectrometer, therefore, may become unnecessary. In addition to these general advantages, polarization-modulation provides specific advantages for certain applications.

Polarization-modulation experiments can be performed on either dispersive or Fourier transform infrared (FT-IR) spectrometers. Sensitivity advantages actually achieved using polarization-modulation vs. conventional FT-IR experiments can vary depending on the spectrometer and detector being used. A FT-IR system which is truly detector-noise limited may not realize any improvement in sensitivity with polarization modulation. If there is digitization noise, background 1/f noise, or noise due to the detector/preamplifier band width, however, utilization of the polarization-modulation technique can, indeed, result in an order-of-magnitude or more improvement in signal-to-noise ratio for the same number of scans. A major advantage of high-frequency polarization-modulation in DIRLD spectroscopy is the short time required to measure dichroic differences (less than 14 μs). This time-resolution advantage should become obvious in the following sections.

B. Infrared Dichroism

Spectroscopic measurement of infrared dichroism (i.e. a directionally preferred absorption of optical energy) has been well developed as a technique for detecting molecular orientation. The specificity of IR absorption bands to particular chemical functional groups makes this technique especially attractive for the study of stress-induced submolecular-level orientation responses of polymers. The absorption of infrared photons is caused by the interaction of the electric-field vector of the incident photon with the electric-dipole transition moment associated with a particular molecular vibration. The photon frequency must match the vibrational frequency associated with the electric-dipole transition moment in order to be absorbed. An additional requirement for absorption is that the electric-field vector of the incident photon be oscillating in a plane parallel to the electric-dipole transition moment. Photons polarized perpendicular to the dipole-transition moment cannot be absorbed. An oriented sample, therefore, often exhibits directionally preferential absorption, while a totally isotropic sample shows no dichroism. In order to measure infrared linear dichroism, light polarized in planes parallel and perpendicular to a fixed reference direction in the sample is required.

Parameters used to characterize the degree of optical anisotropy in a sample are the dichroic difference, $\Delta A = A_{\parallel} - A_{\perp}$, and the dichroic ratio, $D = A_{\parallel}/A_{\perp}$, where $A_{\parallel}$ and $A_{\perp}$ are the absorbances of light polarized parallel and perpendicular to a reference direction. For a simple unidirectional orientation, these dichroic parameters can be related to Hermans' orientation function f for the molecular chain [10,11].

$$
\begin{aligned}
f &= \frac{3\langle\cos^2\theta\rangle - 1}{2} \\
&= \frac{(D-1)(D_\infty+2}{(D+2)(D_\infty-1)} \\
&= \frac{\Delta A}{\Delta A_\infty} \qquad (1)
\end{aligned}
$$

where

$$D_\infty = 2\cot^2\alpha \qquad (2)$$

and

$$\Delta A_\infty = 3Ao(D_\infty - 1)/(D_\infty + 2) \qquad (3)$$

D_∞ and A_∞ are constants independent of orientation, α is the angle between the transition-moment direction and the chain, and $\langle \cos^2\theta \rangle$ is the average of the square of the cosine of the angle between the molecular chain axis and the stretch direction of the film. The structural factor A_o is given by

$$A_o = (A_{/\!/} + 2A_{\perp})/3 \qquad (4)$$

for a uniaxially oriented film. The polarization-modulation technique discussed above can be effectively utilized for such a measurement.

C. Dynamic Mechanical Analysis

The fundamental rheological component of DIRLD spectroscopy is dynamic mechanical analysis. Even though the basic formalism of dynamic mechanical analysis is well known [12,13], a brief review is given here to draw a comparison with DIRLD spectroscopy. In dynamic mechanical analysis, small-amplitude oscillatory strain is applied to a sample, and the resulting dynamic stress is measured as a function of temperature and deformation frequency. Both elastic and viscous components of the stress response are determined simultaneously to yield the linear viscoelastic functions of the material. The oscillatory strain amplitude used in this technique is very small, typically well below 1.0% of the total sample dimension, to assure linear viscoelastic mechanical responses. Larger strain amplitude can lead to a considerable nonlinearity.

For a time-dependent, small-amplitude, sinusoidal tensile strain applied to a sample,

$$\varepsilon(t) = \bar{\varepsilon} + \hat{\varepsilon} \sin \omega t \qquad (5)$$

the time-dependent stress response becomes

$$\sigma(t) = \bar{\sigma} + \hat{\sigma} \sin (\omega t + \delta) \qquad (6)$$

$\bar{\varepsilon}$ and $\bar{\sigma}$ are the time-independent static strain and stress often superimposed on the dynamic strain and stress. $\hat{\varepsilon}$ and $\hat{\sigma}$ are the amplitudes of the dynamic components of the strain and stress applied to a sample at frequency ω. The existence of a phase angle δ between the dynamic stress and strain is a consequence of the viscoelastic response of a polymer.

The dynamic stress response in Eq.6 can be expanded into two separate components: in-phase and 90 degrees out-of-phase with the dynamic strain.

$$\sigma(t) = \bar{\sigma} + E'\hat{\varepsilon} \sin \omega t + E''\hat{\varepsilon} \cos \omega t \qquad (7)$$

The coefficients of the in-phase and quadrature components of the time-dependent stress are the dynamic tensile storage and loss moduli, E' and E'', given by

$$E' = (\hat{\sigma}/\hat{\varepsilon}) \cos \delta \qquad (8)$$

and

$$E'' = (\hat{\sigma}/\hat{\varepsilon}) \sin \delta \qquad (9)$$

The storage modulus represents the ability of the polymeric material to elastically store the absorbed mechanical energy. The loss modulus represents the ability of the material to dissipate the absorbed energy as heat. The dissipation factor, $\tan \delta$, given by

$$\tan \delta = E''/E' \qquad (10)$$

is a convenient index of the viscoelastic state of the material. An increase in $\tan \delta$ often corresponds to an onset of a new type of dissipation of mechanical energy [12,13].

D. Dynamic Infrared Linear Dichroism (DIRLD)

The application of dynamic methods to the rheo-optical characterization of polymers has already been established for birefringence, light scattering, and x-ray diffraction [14-19]. Following the examples of other rheo-optical methods, relationships between the dynamic dichroism and strain, analogous to those between dynamic stress and strain, can be derived using a formalism similar to that used in Eqs. (5)-(10).

The time-dependent dichroic difference induced by a small-amplitude oscillatory strain (Eq. 5) is given by

$$\Delta A(t) = \Delta\bar{A} + \Delta\hat{A} \sin (\omega t + \beta) \qquad (11)$$

where $\Delta\bar{A}$ and $\Delta\hat{A} \sin(\omega t+\beta)$ are the static and dynamic components of the dichroic difference, respectively. Since the molecular orientation is a rate-dependent process, there is a phase angle β between the dynamic dichroic difference and strain. For a unidirectional orientation, a dynamic orientation function, f(t), is given by

$$f(t) = \Delta A(t)/\Delta A_\infty$$
$$= \bar{f} + \hat{f} \sin(\omega t + \beta)$$
$$= \bar{f} + N'\hat{\varepsilon} \sin \omega t + N''\hat{\varepsilon} \cos \omega t \qquad (12)$$

N' and N'' are the storage and loss components of the dynamic orientation coefficient, given by

$$N' = (\hat{f}/\hat{\varepsilon}) \cos \beta \qquad (13)$$

and

$$N'' = (\hat{f}/\hat{\varepsilon}) \sin \beta \qquad (14)$$

An orientation dissipation factor, $\tan\beta$, analogous to the mechanical dissipation factor, tan , can now be defined:

$$\tan \beta = N''/N' \qquad (15)$$

The ability to measure this orientation dissipation factor should allow us to make comparisons between macroscopic dynamic mechanical data and molecular-level information from IR spectroscopy. For example, the dynamic mechanical dissipation factor, $\tan\delta$, can be correlated with the orientation dissipation factor, $\tan\beta$, as a function of temperature and deformation frequency. Such correlations should be useful for assigning the macroscopic dynamic mechanical dispersions to changes in the microscopic response of specific functional groups in the polymer molecules. This information is crucial in elucidating the deformation and relaxation mechanisms of microstructures in polymer samples.

INSTRUMENTATION

A. Optical, Mechanical, and Thermal Components

Our DIRLD spectrometer was constructed around a modified Dynastat Dynamic Mechanical Analyzer (Imass) [20]. Its environmental chamber, regulated by the Dynatherm Temperature Controller, permits the temperature of the sample to be maintained to within ±2.0°C from -150 to 150°C. The sample deformation frequency can be varied from 0.01 to 100 Hz. Like the earlier prototype polarization-modulation spectrometer [6-8], a photoelastic modulator (PEM) is used to alternate the optical polarization state between parallel and perpendicular at 74 kHz. The instrument is capable of measuring the quadrature component of the dynamic dichroism signal as well as the component in-phase with the applied strain. This permits determination of the phase angle β between the dynamic dichroism and strain, paving the way for the assignment of dynamic mechanical dispersions in polymer films to specific functional groups.

Figure 1 shows a block diagram of the DIRLD spectrometer. The measurement technique consists of three sequential modulations of the infrared beam: chopper, polarization, and strain orientation modulations. A mechanical light chopper labels photons originating from the source at frequency ω_c, distinguishing them from background photons. The fixed wire-grid polarizer followed by the PEM causes the plane of polarization of the light to alternate rapidly between parallel and perpendicular to the strain axis of the sample at frequency $2\omega_m$. The polymer film under oscillatory deformation at frequency ω_s induces the periodic perturbation of dichroism at the same frequency.

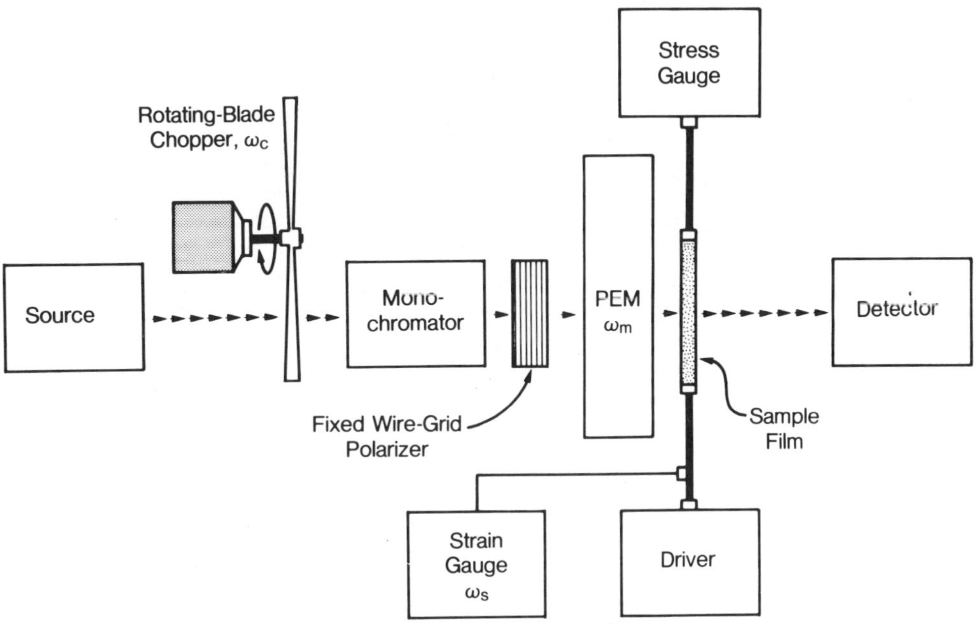

Figure 1. Block diagram of the optics for the dynamic infrared linear dichroism (DIRLD) spectrometer.

A complete optical ray diagram of the DIRLD spectrometer is shown in Figure 2. The source (L) is a high-intensity Nernst glower (Artcor) capable of black-body temperatures up to 2400°C. The focusing optics for the source (M1 and M2) are taken from a

Perkin-Elmer 521 infrared spectrometer. The light is chopped by a variable-speed light chopper (C) (Bentham Model 218) just before focusing on the entrance slit (S1) of a J-Y HR-640 two-thirds meter monochromator (f/5.7). The monochromator has interchangeable gratings (G) to optimize throughput for the spectral region of interest. There are two sets of entrance and exit slits: one continuously variable from 0 to 2.0 mm (S3 and S4); and a second set of slits fixed at 2,4 or 6 mm for lower resolution work (S1 and S2). Figure 2 shows the configuration when using the fixed slits. Light emerging from the exit slit (S2) then passes through an order-sorting filter (F), wire-grid polarizer (P), and ZnSe PEM (Hinds International). The polarizer is oriented so the transmitted light is polarized in a plane parallel to the spectrometer bed (perpendicular to the dynamic strain direction). The stress axis of the PEM is oriented at a 45 degree angle to the plane of polarization. A two-inch f/2 ZnSe lens (L1) focuses the diverging beam on the sample film (S) which is inside the environmental chamber of the Dynastat Dynamic Mechanical Analyzer.

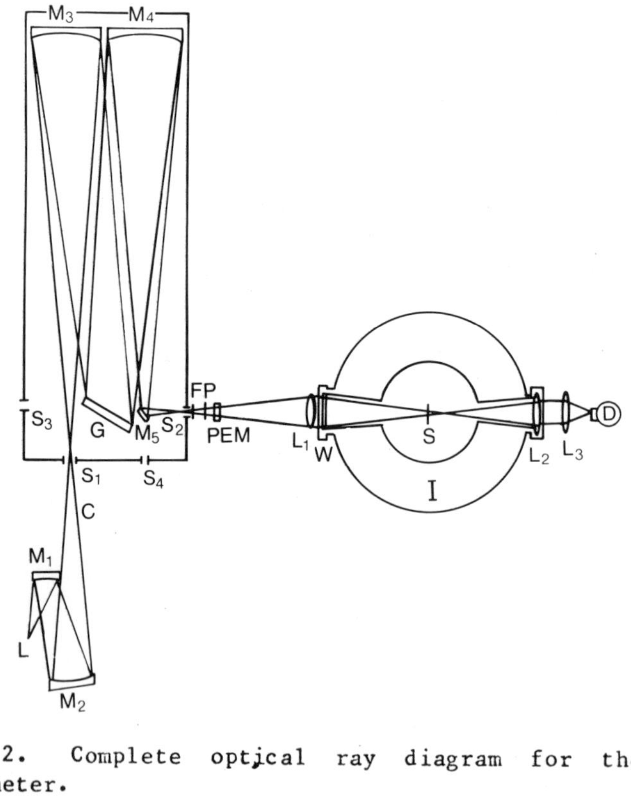

Figure 2. Complete optical ray diagram for the DIRLD spectrometer.

The double-walled environmental chamber of the Dynastat has been specifically modified for DIRLD by adding ports for the IR radiation to enter and exit the chamber. The ports have two-inch diameter openings on the outside wall and one-inch diameter openings on the inside. A solid piece of tapered Teflon lines the entrance and exit holes in the insulated region (I) between the two metal walls. This configuration matches the beam geometry while maintaining excellent temperature control. Screw caps with Teflon o-rings hold a two-inch ZnSe window (W) on the entrance port and a second two-inch f/2 ZnSe lens (L2) on the exit port. The second lens makes the diverging beam nearly parallel so it can be focused on a 1 or 2 mm detector element (D) by a third two-inch f/1 ZnSe lens (L3). Depending on the spectral region of interest, a narrow-(Type A), intermediate-(Type B), or wide-band (Type C), liquid-nitrogen-cooled Mercury-Cadmium-Telluride (MCT) photoconductive detector or an InSb photovoltaic cooled detector can be used. Our type A MCT has a long-wavelength cutoff at about 900 cm^{-1}. Our Type B MCT will go down to about 600 cm^{-1} (near the ZnSe cutoff), but is less sensitive than the Type A detector. The InSb detector is an order-of-magnitude more senstive than the MCT detectors, but it cuts off at about 1850 cm^{-1}. Each detector preamplifier has a bandpass of at least 250 kHz so that the 74 kHz polarization-modulation signal is not attenuated. All of our ZnSe optics are practically free of birefringence and are anti-reflection coated for maximum transmission at 10.6 μm. The optics after the exit slit of the monochromator are arranged in a straight line to avoid off-axis effects which can distort the polarization.

B. Signal Demodulation

A set of four lock-in amplifiers is used to demodulate the triply-modulated signal received by the detector. The configuration of the electronics is shown in Figure 3. As in the prototype configuration [6-8], the signal from the preamplifier is sent to lock-in amplifiers A and C. Lock-in amplifier C (Princeton Applied Research Co., Model 5101) demodulates the ω_c component generated by the light chopper, and its DC output is proportional to the instrument throughput or transmittance. Lock-in amplifier A (ORTEC Model 9503) demodulates the $2\omega_m$ components due to the polarization modulation. It is tuned by placing a second linear polarizer in the beam parallel to the one preceding the PEM. The second polarizer represents a totally dichroic sample and is also used to calibrate the spectrometer. The output time constant of lock-in amplifier A is set to a minimum value (0.1 msec) to pass both the static and dynamic dichroism signals. The components of this total dichroism signal can be further demodulated by lock-in amplifiers B (Ithaco 393) and D (Ithaco 391A). The signal at the output of lock-in amplifier A is dominated by the ω_c component from the light chopper, so a dual channel electronic filter

(Ithaco Model 4302) is used to selectively pass the ω_s component. Lock-in amplifier D, tuned to the chopper reference ω_c, allows us to determine the static dichroic difference spectrum. The output of lock-in amplifier D is proportional to the dichroic difference signal originating from static and residual orientations in the sample.

The dynamic strain signal from the Dynastat is used as a reference signal for lock-in amplifier B. Lock-in amplifier B is a quadrature device, so components of the dynamic dichroic difference in-phase (IP) and 90° out-of-phase (OOP) with the strain are both measured simultaneously. Lock-in amplifier B is tuned by placing a beam block (an opaque card), vibrating vertically at ω_s, in the lower half of the beam. The oscillating rigid card attached to the strain jaw of the Dynastat produces an optical signal which is exactly in phase with the strain applied to the sample. The phase of the detector signal is then synchronized with the dynamic strain.

Four DC outputs from the lock-in amplifiers (Figure 3) are digitized by a Hewlett-Packard (HP) Model 3054A Data Acquision System and HP Model 3456A Digital Voltmeter. The entire experiment, including slewing of the monochromator, is controlled by an HP Model 9836A Desktop Computer. The system is presently able to collect 154 four-and-one-half digit readings per second and is capable of multiple scans.

For noise suppression, we have used three modes of signal averaging: analog filtering, digital averaging, and multiple scanning. The output time constant of the quadrature lock-in amplifier (Figure 3, D) is selected to filter out mid-frequency range detector noise. Digital signal averaging over 200 points is carried out to minimize high-frequency (spike) noise. The co-addition of multiple scans is used to filter out low-frequency drift. Data are collected at intervals several times denser than the spectral resolution. This allows us to apply numerical noise-filtering techniques with no distortion of spectral information. Discrete data points (0.5 to 1.0 cm^{-1} increments) are connected by a fifth-order polynomial interpolation function.

RESULTS AND DISCUSSION

A. Isotactic Polypropylene

Figure 4 shows the ordinary IR absorbance spectrum of a 30 μm-thick and 20 mm-wide uniaxially stretched (4X) film of isotactic polypropylene taken at room temperature with our DIRLD spectrometer. The nominal sample length was 3.9 cm. The major band at 1168 cm^{-1} has a weak shoulder at 1153 cm^{-1}. Normal

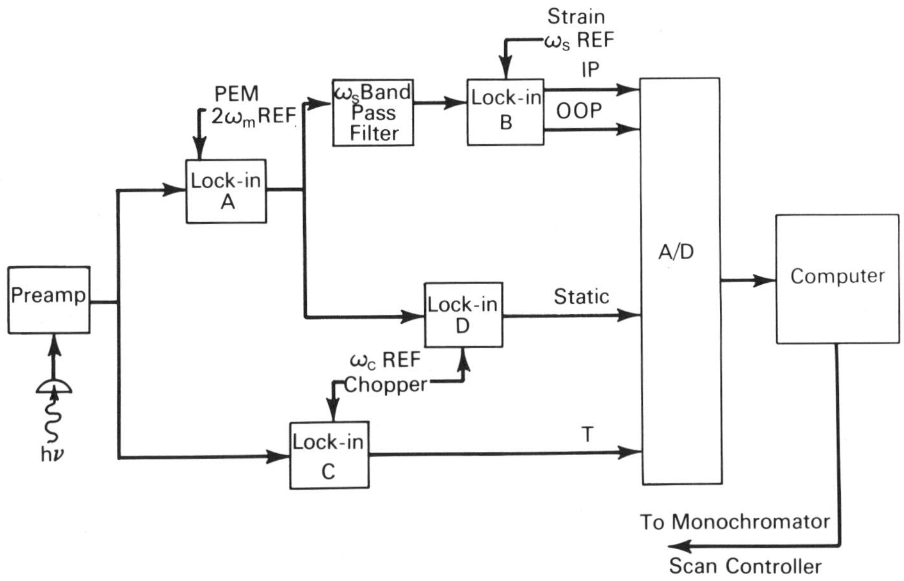

Figure 3 Block diagram of the electronics for the DIRLD spectrometer.

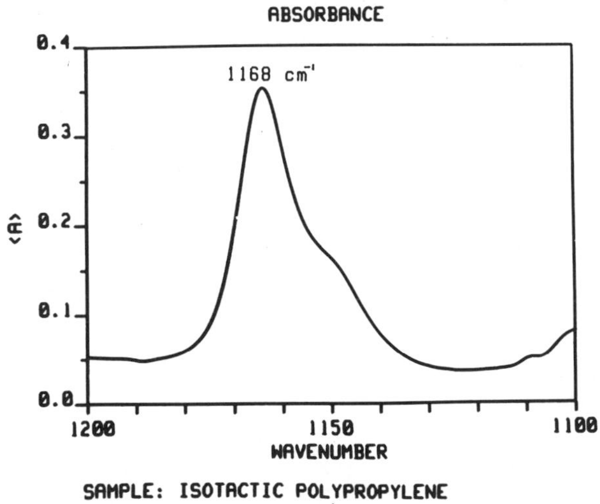

Figure 4 Transmittance spectrum of isotactic polypropylene.

coordinate calculations indicate the 1168 cm^{-1} band consists of mainly C-C skeletal stretch and CH_3 rock and is polarized parallel (A mode) to the helix axis of the crystalline polymer [21-26]. The 1153 cm^{-1} shoulder is polarized perpendicular (E mode) to the helix axis and contains significant contributions from C-CH_3 stretch and C-H bend in addition to C-C skeletal stretch and CH_2 rock [21-26]. This film has residual orientation frozen into the matrix during processing, as indicated by the nonzero static dichroism (Figure 5). The static dichroism spectrum was measured by taking the machine direction of the film as the parallel direction. There is no evidence of the 1153 cm^{-1} shoulder in the static dichroic difference spectrum. In order to obtain the dynamic dichroic difference spectra of isotactic polypropylene, the same film was dynamically deformed at a displacement amplitude of 50 μm at a strain frequency of 80 Hz. The dichroic difference induced by the oscillatory strain was detected between 1200 and 1100 cm^{-1}. Figure 6 shows the reconstructed time-resolved dynamic dichroic difference spectra (spectral resolution of 7.5 cm^{-1}) corresponding to two-and-one-half cycles of the dynamic dichroism. Each instantaneous spectrum along the time axis was plotted about 350 μsec after the previous one; enough time to repeat at least 25 cycles of the polarization modulation. A positive dynamic dichroism ($A_\parallel > A_\perp$) represents an increase in the dipole-transition moment oriented parallel to the direction of oscillatory strain. A negative dynamic dichroism ($A_\parallel < A_\perp$), on the other hand, represents an increase in the dipole-transition moment oriented perpendicular to the direction of oscillatory strain.

The dynamic dichroic difference spectra in-phase and 90° out-of-phase (quadrature) with the applied dynamic strain are plotted in Figure 7. Compared to the in-phase component, the quadrature signal is much weaker. This suggests the dynamic dichroic difference for this particular sample is almost totally in phase with the applied 80 Hz strain at room temperature. Interestingly, the band with maximum intensity at 1168 cm^{-1} in the absorbance and static dichroism spectra has now split into a negative and positive band in the dynamic dichroic difference spectra. The exact origin of this splitting is not yet clear.

Dynamic orientations of submolecular units of polymers under oscillatory strain should be functions of the strain amplitude. For a small-amplitude strain, we assumed a linear relationship between the dynamic dichroic difference and strain (Eqs. 11 and 12). In order to study the effect of strain amplitude on the dynamic dichroism spectrum, a series of dynamic spectra of the same isotactic polypropylene film deformed at different amplitudes of dynamic displacement were obtained (Figure 8). The size of the dynamic dichroic signal increases with the increase in dynamic displacement for both the positive and negative bands. The magni-

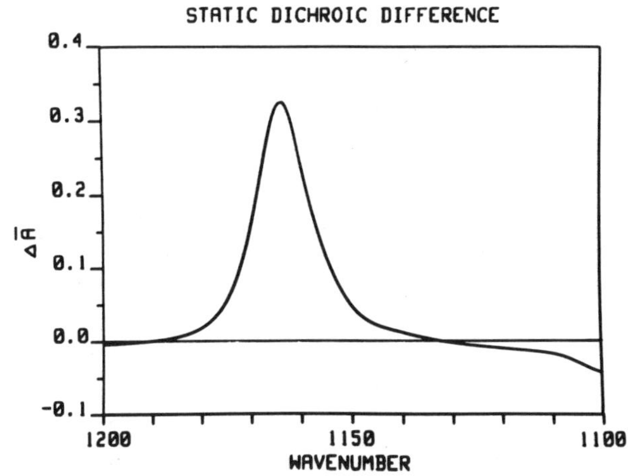

Figure 5 Static dichroic difference spectrum of isotactic polypropylene.

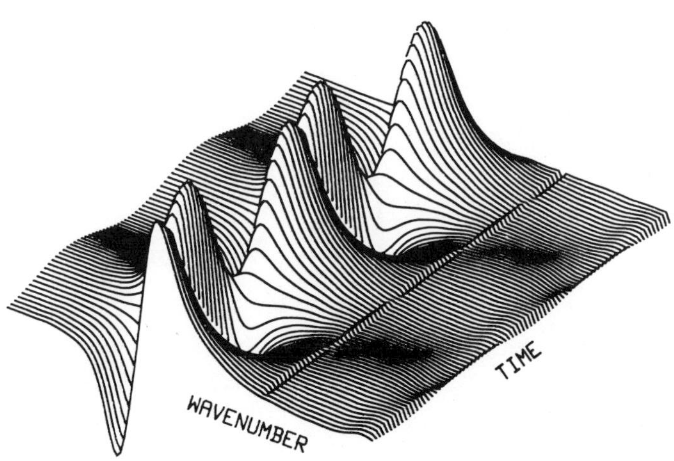

Figure 6 Time-resolved dichroic difference spectrum of iso-polypropylene. Two-and-one-half cycles of the 80 Hz oscillatory strain are shown between 1200 and 1100 cm^{-1}.

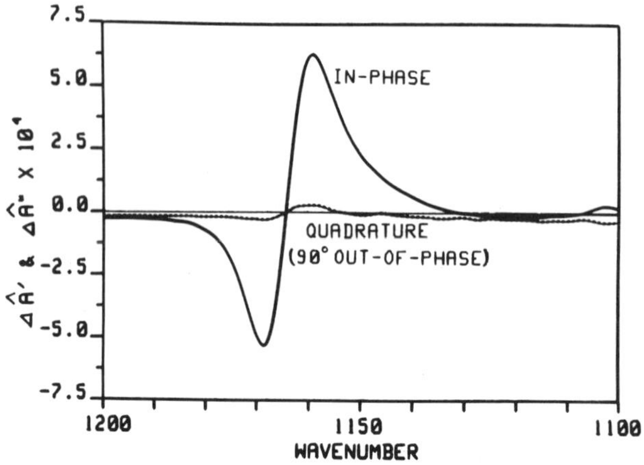

Figure 7 In-phase and quadrature components of the dynamic dichroic difference spectrum of isotactic polypropylene.

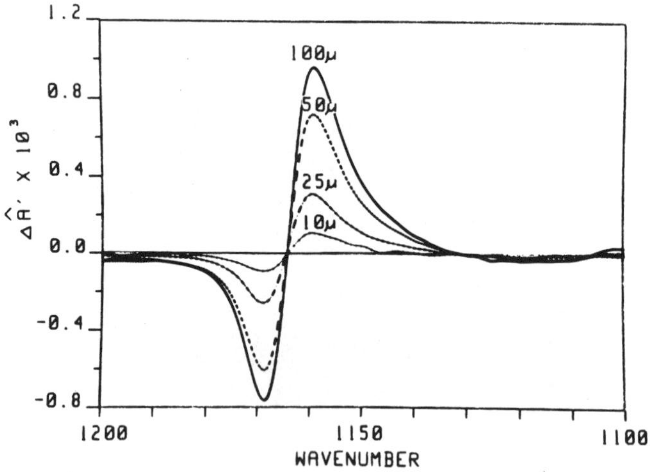

Figure 8 In-phase component of the dynamic dichroic difference spectrum of isotactic polypropylene as a function of dynamic-strain amplitude.

tudes of each peak are plotted as functions of displacement amplitude in Figure 9. At very small displacement amplitudes, the magnitude of the dynamic dichroic difference is proportional to the displacement amplitude. Under higher displacement amplitudes, however, a considerable deviation from linearity was observed. For our sample size (c.a. 3.9 cm) a breakdown in linearity was observed at a strain level less than 0.5%. This result strongly suggests that the application of linear theory to dynamic infrared rheo-optical studies obtained at strain amplitudes above 1% may not be justified. The nonlinear dichroic response due to large-amplitude strain will very likely contain a substantial level of higher harmonic components, which may complicate the interpretation.

In order to avoid such nonlinear rheo-optical responses, the strain amplitude must be kept to a very small level. At such small deformation amplitudes the dichroic signals also become small. Extraordinary spectroscopic sensitivity, therefore, is required for meaningful infrared rheo-optical studies of polymers.

B. Styrene-Butadiene-Styrene Triblock Copolymer

For an example of a DIRLD study of amorphous polymers, we have chosen a styrene-butadiene-styrene triblock copolymer (Kraton D1102, Shell Chemical Co.). This polymer is known to form micro-phase-separated domains consisting of polystyrene and butadiene blocks which result in multiple glass-to-rubber transition temperatures (T_g) corresponding to each constituent phase. A thin sample film was cast from toluene solution and annealed between heated plates. We have studied the dynamic orientation of this block copolymer sample around the T_g of polybutadiene near -80°C under 80 Hz dynamic strain. The sample was supported on a thin Teflon film which was IR transparent between 1400 and 1500 cm^{-1}. Absorption bands in the CH-deformation region are observed at 1492 cm^{-1} (polystyrene) and around 1450 cm^{-1} (polystyrene and polybutadiene).

The in-phase component of the dynamic dichroic difference spectrum (recorded with a spectral resolution of 12 cm^{-1}) shows a negative band around 1450 cm^{-1} which monotonically decreases in absolute value as the temperature is raised from -100 to -60°C (see Figure 10). The quadrature component of the dynamic dichroic difference spectrum (see Figure 11), on the other hand, goes through a negative peak around the T_g of the polybutadiene phase. Well above or below this T_g, no significant signal intensity in the quadrature component is observed, i.e., the dynamic orientation of the submolecular unit corresponding to this band is totally in phase with the applied dynamic strain. However, considerable out-of-phase perpendicular orientation is observed around the

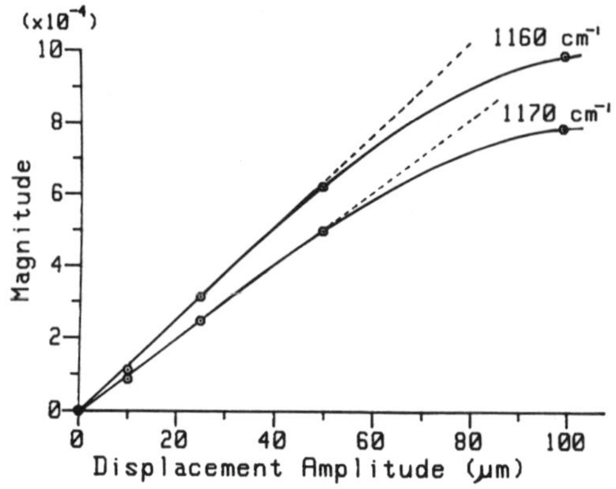

Figure 9 Magnitude of the 1170 and 1160 cm^{-1} bands as a function of dynamic strain amplitude for isotactic polypropylene.

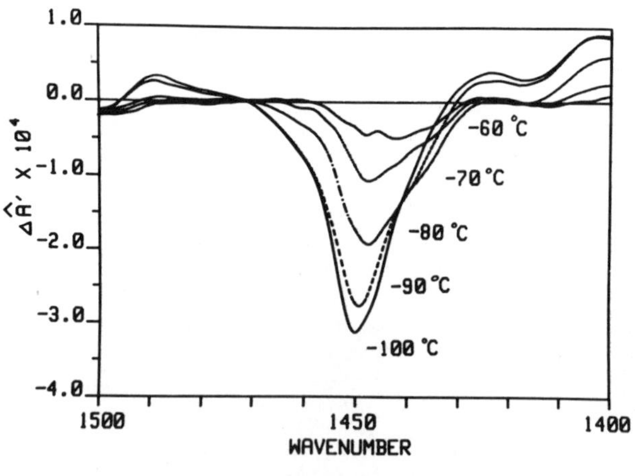

Figure 10 In-phase component of the dichroic difference spectra of Kraton D 1102 at -100, -90, -80, -70, and -60°C.

T_g. It is interesting to note that similar behavior is observed in the dynamic stress response, where the in-phase component decreases monotonically and the quadrature component shows a maximum around the T_g. The temperature dependence of the DIRLD spectra, however, is far more specific to the orientation and mobility of particular submolecular units compared to the macroscopic stress response.

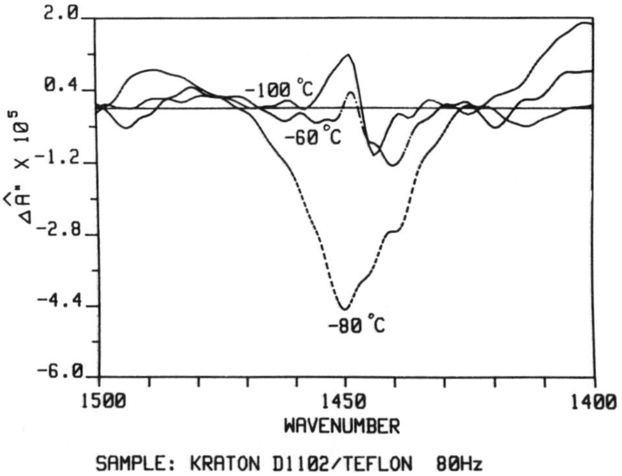

Figure 11. Quadrature component of the dynamic dichroic difference spectra of Kraton D1102 at -100, -80, and -60°C.

C. Linear Low-Density Polyethylene

As shown in the section on block copolymers, temperature studies using DIRLD can reveal correspondence between mechanical and molecular orientational relaxation phenomena. We have performed a similar temperature study on a semicrystalline polymer using DIRLD spectroscopy. The CH-deformation region between 1500 and 1400 cm^{-1} of linear low-density polyethylene was studied. A 37.5 µm-thick film (Mobil Chemical Co.) was cut to dimensions 6 cm x 2 cm and mounted between the jaws of the Dynastat. The nominal sample length between the two jaws was 4.1 cm.

The sample was studied at various temperatures while being deformed with an oscillatory strain. A strain amplitude of 50 µm and strain frequency of 80 Hz were used throughout the experiment. While data were collected, the temperature was controlled to within ±2.0°C by liquid nitorogen boiloff heated to the specified temperature. A low-level static stress (approximately 0.1 MPa) was imposed on the sample to keep it from flopping.

IR spectra were scanned from 1480 to 1450 cm^{-1} every 0.5 cm^{-1} at a spectral resolution of 4 cm^{-1}. Two hundred data collections (taking a total time of approximately 2 sec) were made on the in-phase and quadrature (90 degree out-of-phase) components of the dynamic dichroic difference signal at each wavenumber position. A 12.5 sec time constant was used for the lock-in amplifier to filter out mid-range noise. In addition, a total of four scans were coadded, a five-point smoothing function applied, and the spectra interpolated using a fifth order polynomial function before plotting. The absorbance and static dichroism spectra were collected simultaneously. The total time spent collecting data at each temperature was about 25 min.

The isochronal dynamic mechanical spectra of the oven-annealed LLDPE collected on the Autovibron (Imass) operated at 110 Hz are shown in Figure 12. Both the loss modulus and the dissipation factor tan δ show distinct peaks around -120°C and -20°C which correspond to the onset of mechanical γ- and β-relaxation processes, respectively [12,13,27]. Our discussion will concentrate on the peak around -20°C (β-relaxation), which is believed to be caused by interlamellar grain-boundary shearing motions [28-32]. Although the details of the mechanical relaxation phenomena of polyethylenes are not completely understood, considerable progress has been made in understanding the origin of relaxation processes by using various rheo-optical techniques [29-33].

Absorbance and static dichroic difference spectra of the LLDPE sample collected at 20, -65, and -150°C are shown in Figures 13 and 14. Some small frequency and intensity differences due to the temperature changes are observed among the spectra. The 1473 cm^{-1} band has been assigned to the CH_2 scissors in the crystalline component. It completely disappears upon heating above the melting point of polyethylene. The 1463 cm^{-1} band is a superposition of a CH_2 scissors of the amorphous component and a crystalline component. The small shoulder at 1457 cm^{-1} is due to the CH_3 antisymmetric bending vibration. The static dichroic difference spectra indicate the sample film had considerable residual orientation. The dipole-transition moment associated with the 1463 cm^{-1} band is aligned parallel to the dynamic stretch direction while the transition moment associated with the 1473 cm^{-1} band is oriented perpendicular.

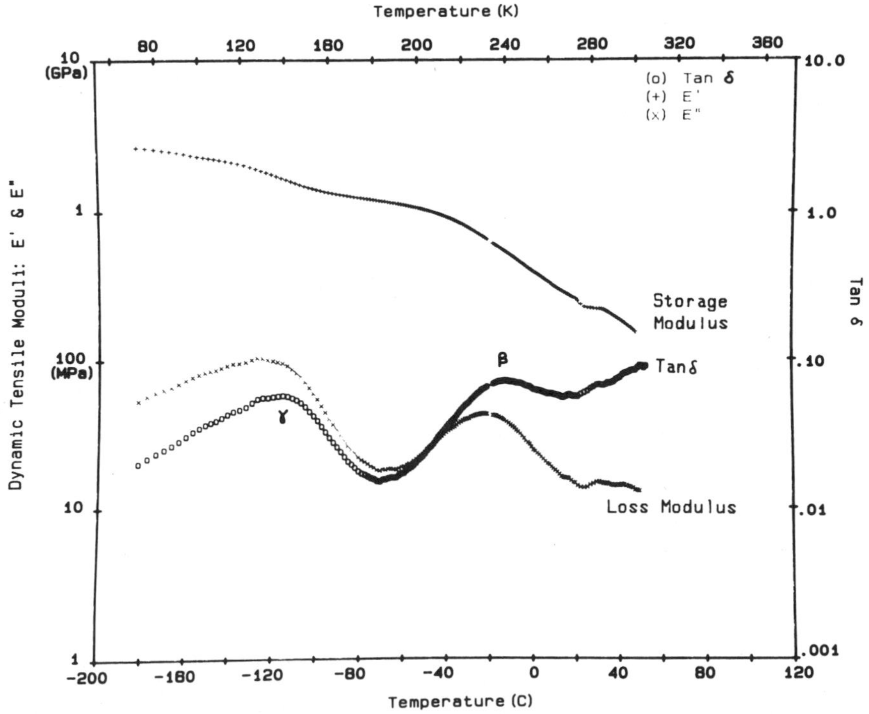

Figure 12. Dynamic mechanical spectra of linear low-density polyethylene (LLDPE).

Figures 15-18 show the in-phase and quadrature components of the dynamic dichroic difference spectra of LLDPE at +20, 0, -10, -25, and -45°C. The in-phase component decreases in absolute value as the temperature is lowered from +20 to -25°C. The 1463 cm^{-1} band shifts down approximately 1 cm^{-1} between the -10 and -45°C in-phase spectrum, suggesting one component of the doublet has partially vanished and the other is being formed around -25°C, the temperature corresponding to the peak in the mechanical relaxation. The 1473 cm^{-1} band in the in-phase spectrum does not shift. The absolute value of this band becomes small first, then

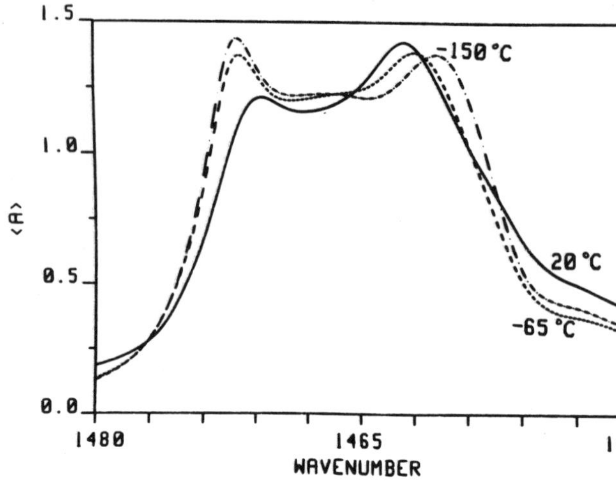

Figure 13 Absorbance spectra of LLDPE at 20, -65, and -150℃.

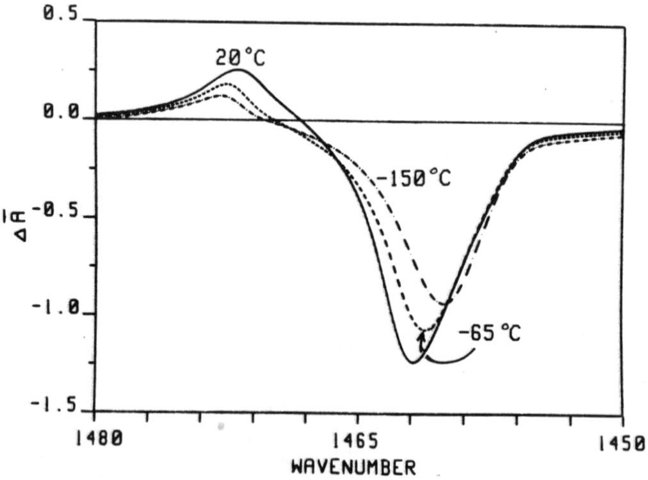

Figure 14 Static dichroic difference spectra of LLDPE at 20, -65, and -150℃.

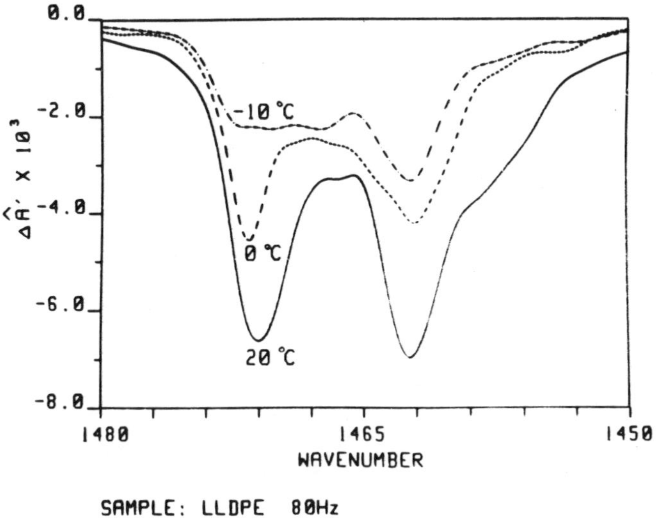

Figure 15 In-phase component of the dynamic dichroic difference spectra of LLDPE at 20, 0, and -10℃.

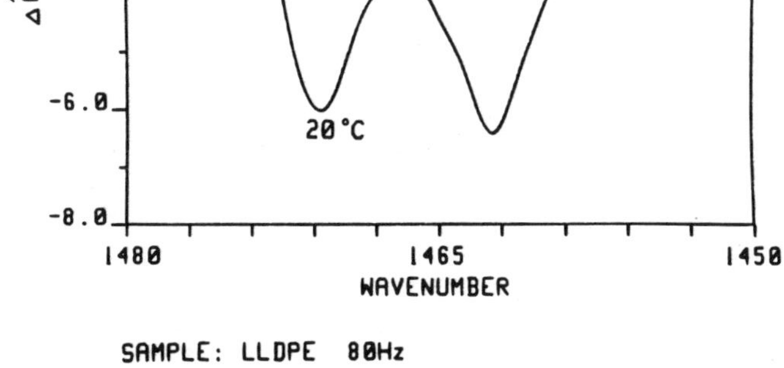

Figure 16 Quadrature component of the dynamic dichroic difference spectra of LLDPE at 20, 0, and -10℃.

increases again around the temperature range of the β relaxation. The 1473 cm^{-1} band is smallest at $-10^{o}C$ compared to $-25^{o}C$ for the 1463 cm^{-1} band. In the quadrature spectrum, both the 1473 and 1463 cm^{-1} bands decrease in absolute value from +20 to $-10^{o}C$. From -10 to $-45^{o}C$ a sign change from negative to positive is observed in both bands upon passage through the β relaxation. At $-45^{o}C$ the 1463 cm^{-1} band is not shifted as it was in the in-phase spectrum. A methyl deformation at 1457 cm^{-1} is now observable in some of the quadrature spectra. Many weak bands appear which remain unassigned.

The abrupt change in the sign (i.e., the overall orientation direction) of the dynamic dichroism spectrum is a very important observation. Such a change in sign of dynamic rheo-optical response has been observed for the dynamic birefringence of spherulitic high-density polyethylene in the vicinity of the β relaxation [31]. This anomalous rheo-optical behavior can be explained in terms of a progressive change in the contributions from two competing orientation responses with directions perpendicular to each other. As the sample is cooled, gradual counter-balance of the negative dynamic dichroism arising from intra-lamellar relaxation (i.e., dynamic orientation of crystal grains within the orienting lamellae) by the dynamic dichroism associated with interlamellar relaxation (i.e., dynamic orientation of crystal) takes place. Dynamic x-ray diffraction results [31] also strongly support this hypothesis. Our observation of the dynamic dichroism around the β relaxation is the first direct evidence of submolecular-level orientation responses strongly coupled with the low-level perturbation of supermolecular morphology of semi-crystalline polymers.

FT-IR vs. Dispersive Approach

Dynamic infrared linear dichroism [6-8] is an example of a technique that, for now, is done with highest sensitivity on a dispersive spectrometer. The oscillatory strain frequencies typically used in dynamic mechanical analysis are in the same range or slower than the interferometer modulation frequencies (100 Hz to 5 kHz). Although there exist several examples of the successful application of double-modulation FT-IR [1-5], the modulation frequencies need to be at least one order of magnitude higher than the interferogram frequencies. To obtain FT-IR spectra of polymer films under oscillatory deformation at these slower modulation frequencies, therefore, requires the use of time-resolved techniques [34-41]. FT-IR time-resolved spectroscopy (TRS), however, involves time-consuming data collection and file-sorting procedures.

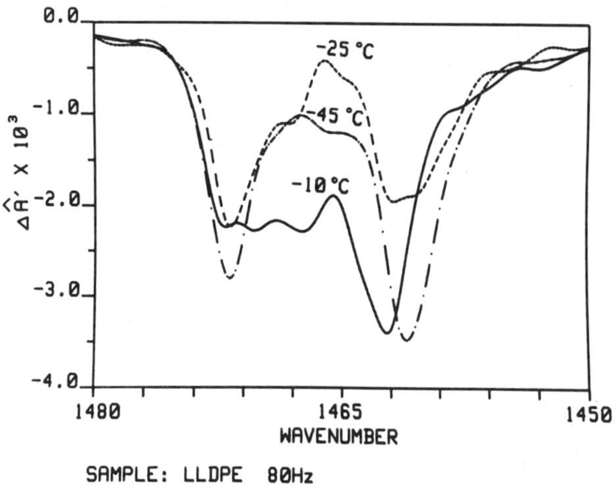

Figure 17 In-phase component of the dynamic dichroic difference spectra of LLDPE at -10, -25, and -45°C.

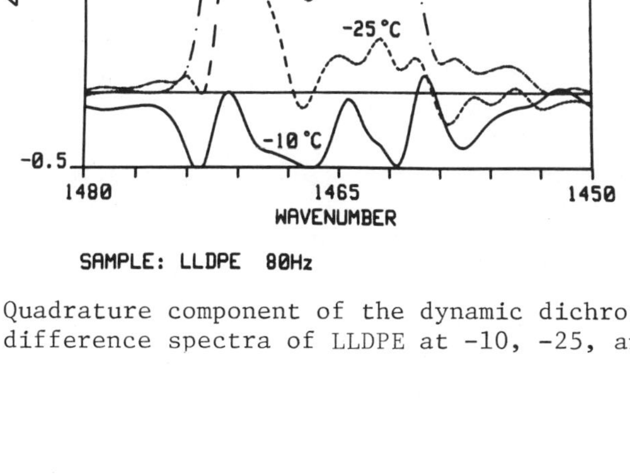

Figure 18 Quadrature component of the dynamic dichroic difference spectra of LLDPE at -10, -25, and -45°C.

Using our triple-modulation dispersive spectrometer equipped with a Type A MCT detector, it is possible to scan a 100 cm^{-1} region with 8 cm^{-1} resolution in 35 min with a noise level of 5 x 10^{-6} absorbance units (e.g., Figure 7). Moreover, the transmission spectrum, static dichroism spectrum, as well as both the in-phase and quadrature components of the dynamic dichroism are collected simultaneously in our technique. Assuming a 0.1% noise level for a single time-resolved FT-IR scan (close to the theoretical limit for a 16-bit analog-to-digital converter), forty thousand interferogram coadditions would be required to achieve a noise level equivalent to that of the dispersive instrument. Since a single time-resolved interferogram scan may take over ten minutes to acquire [36,37], considerable improvement in the FT-IR technique will be required before it becomes competitive in signal-to-noise ratio with DIRLD spectra obtained on our dispersive instrumentation. Certainly, as the number of resolution elements in a given spectrum increases, the advantage of using FT-IR also increases; this is an undeniable incentive for developing FT-IR dynamic dichroism technology.

The ability to determine the dichroism of a polymer sample in a single measurement, rather than comparing two separate measurements, is another advantage of our dispersive instrument. Progress in this area is being made by combining polarization modulation and time-resolved FT-IR spectroscopy [38], but signal-to-noise ratios fall short of the current dispersive approach. Although dynamic infrared dichroism signals can be made larger by increasing the oscillatory strain amplitude, as demonstrated earlier, it is desirable to keep this amplitude relatively small (well below 1%) to minimize the nonlinear viscoelastic effects.

REFERENCES

1. L.A.Nafie and D.W.Vidrine, in "Fourier Transform Infrared Spectroscopy, Techniques Using Fourier Transform Interferometry", Vol. 3, J.R.Ferraro, L.J.Basile, Eds., Academic, New York (1982).

2. A.E.Dowrey and C.Marcott, Appl. Spectrosc., 36, 414 (1982).

3. W.G.Golden and D.D.Saperstein, J. Electr. Spect. Relat. Phenom., 30, 43 (1983).

4. W.G.Golden, D.D.Saperstein, M.W.Severson and J.Overend, J. Phys. Chem., 88, 574 (1984).

5. C.Marcott, Appl. Spectrosc., 38, 442 (1984).

6. I.Noda, A.E.Dowrey and C.Marcott, J. Polym. Sci. Polym. Lett. Ed., 21, 99 (1983).

7. I.Noda, A.E.Dowrey and C.Marcott, Polym. Prepr., 24, 122 (1983).

8. I.Noda, A.E.Dowrey and C.Marcott, Proc. 1985 Internatl. Conf. Fourier and Computerized Infrared Spectroscopy, SPIE, 553, 56 (1985).

9. J.C.Cheng, L.A.Nafie, S.D.Allen and A.I.Braunstein, Appl. Opt., 15, 1960 (1976).

10. J.J.Hermans, P.H.Hermans, D.Vermaas and A.Weidinger, Rec. Trav. Chim. Pays-Bas., 65, 427 (1946).

11. R.J.Samuels,"Structured Polymer Properties", Wiley, New York (1974).

12. N.G.McCrum, B.E.Read and G.Williams,"Anelastic and Dielectric Effects in Polymeric Solids", Wiley, New York (1967).

13. J.D.Ferry,"Viscoelastic Properties of Polymers", 3rd Ed., Wiley, New York (1980).

14. R.S.Stein, Polym. Eng. Sci., 9, 320 (1969).

15. R.S.Stein, S.Onogi and D.A.Keedy, J. Polym. Sci., 57, 801 (1962).

16. R.Yamada, C.Hayashi, S.Onogi and M.Horio, J. Polym. Sci. Part C, 5, 123 (1964).

17. H.Kawai, T.Itoh, D.Keedy and R.S.Stein, J. Polym. Sci. Polym. Lett. Ed., 2, 1075 (1964).

18. G.L.Wilkes, J. Macromol. Sci. Revs. Macromol. Chem., C10, 149 (1974).

19. T.P.Lodge, J.W.Miller and J.L.Schrag, J. Polym. Sci. Polym. Phys. Ed., 20, 1409 (1982).

20. S.S.Sternstein, Adv. Chem. Ser., 203, 123 (1983).

21. T.Miyazawa, Y.Ideguchi and K.Fukushima, J. Chem. Phys., 38, 2709 (1963).

22. T.Miyazawa, J. Polym. Sci. Part C, 7, 59 (1964).

23. R.G.Snyder and J.H.Schactschneider, Spectrochim. Acta, 20, 853 (1964).

24. H.Tadokoro, M.Kobayashi, M.Ukita, K.Yasufuku, S.Murahashi and T.Torii, J. Chem. Phys., 42, 1432 (1965).

25. P.C.Painter, M.M.Coleman and J.L.Koenig, "The Theory of Vibrational Spectroscopy and Its Application to Polymeric Materials", Wiley, New York (1982).

26. B.Jasse and J.L.Koenig, J. Macromol. Sci. Rev. Macromol. Chem., C17, 61 (1979).

27. T.Hashimoto, R.E.Prud'homme and R.S.Stein, J. Polym. Sci. Polym. Phys. Ed., 11, 709 (1973).

28. S.Suehiro, T.Yamada, H.Inagaki, T.Kyu, S.Nomura and H.Kawai, J. Polym. Sci. Polym. Phys. Ed., 17, 763 (1979).

29. R.J.Cembrola, T.Kyu, S.Suehiro, H.Kawai and R.S.Stein, J. Polym. Sci. Polym. Phys. Ed., 20, 1279 (1982).

30. R.J.Cembrola, T.Kyu, R.S.Stein, S.Suehiro and H.Kawai, J. Polym. Sci. Polym. Phys. Ed., 21, 329 (1983).

31. T.Kyu, M.Yamada, S.Suehiro and H.Kawai, Polym. J., 12, 809 (1980).

32. H.Kawai, S.Suehiro, T.Kyu and A.Shimomura, Polym. Eng. Rev., 3, 109 (1983).

33. M.Takayanagi, Memoirs Facul. Eng. Kyushu Univ., 23, 41 (1963).

34. W.G.Fateley and J.L.Koenig, J. Polym. Sci. Polym. Lett. Ed., 20, 445 (1982).

35. D.J.Burchell, J.E.Lasch, R.J.Farris and S.L.Hsu, Polymer, 23, 965 (1982).

36. W.M.Grim, J.A.Graham, R.M.Hammaker and W.G.Fateley, Amer. Lab., 16, 22 (1984).

37. J.A.Graham, W.M.Grim and W.G.Fateley, J. Mol. Struct., 113, 311 (1984).

38. J.E.Lasch, E.Dobrovolny, S.E.Molis and S.L.Hsu, Proc. ACS Div. Polym. Mater. Sci. Eng., 50, 182 (1984).

39. S.L.Hsu and D.J.Burchell, Org. Coat. Plastics Chem., 44, 635 (1981); and Polym. Preprints, 22, 305 (1981).

40. D.J.Burchell, J.E.Lasch, E.Dobrovolny, N.Page, J.Domian, R.J.Farris and S.L.Hsu, Appl. Spectrosc., 38, 343 (1984).

41. J.E.Lasch, D.J.Burchell, T.Masaoka and S.L.Hsu, Appl. Spectrosc., 38, 351 (1984).

A COMPARISON OF SPECTRAL SUBTRACTION AND POLARIZATION MODULATION SPECTROSCOPY FOR USE IN DEFORMATION STUDIES OF POLYMERS

J.E. Lasch, E. Dobrovolny, S.E. Molis and S.L. Hsu*

Department of Polymer Science and Engineering
University of Massachusetts
Amherst, MS 01003
* To whom correspondence should be addressed

ABSTRACT

The applicability of the spectral subtraction and polarization modulation techniques to analyze small segmental orientation changes in mechanical-vibrational spectroscopy of polymers has been evaluated. Unless special circumstances are considered, comparison of the two techniques does not provide an overwhelming advantage for modulation that would favor it over the simpler subtraction technique.

INTRODUCTION

The overall mechanical properties of polymeric materials in terms of stress-strain behavior are controlled by the deformation mechanisms operating on the microstructural level. Therefore, a considerable amount of research effort has been devoted to clarifying the type and rate of various molecular motions. Generally, these studies are not directed at measuring the mechanical values of the properties but rather at correlating molecular slippage, conformational change, extension of amorphous segments, and rotation of crystallites with the properties obtained.

Essentially, the experiments can be divided into two types, static or dynamic. The former type deals with the assessment of the resultant structural changes. The latter type deals mainly with the rate of change. The static experiments are generally not difficult and the considerable amount of data obtained can be interpreted and compared to the various deformation models proposed. On the other hand, the dynamics of microstructural changes have not been studied as extensively as in the first area. Actually, only a few physical techniques are applicable to follow the rate of change. Dynamic x-ray diffraction scattering, birefringence, and light scattering either used singularly or in conjunction have been successful in measuring the rate of the elongation and disruption of the spherulites, rotation and reorganization of the crystallites, and the extension and slippage of amorphous chains [1-9]. In a typical experiment, small angle light scattering (SALS) was used to follow the spherulitic elongation [2]. Wide angle x-ray scattering (WAXS) was used to measure the crystal orientation, and then the amorphous orientation was calculated by subtracting the crystalline contribution from the total birefringence change.

In contrast, the selectivity of infrared spectroscopy allows the direct and simultaneous measurement of the changes of each component, not only in chemically homogeneous materials, but in heterogeneous ones as well, e.g. copolymers and polymer blends. The infrared spectroscopic data can distinguish not only chemically distinct chains, but different conformations, various types of crystalline chain packing, and the strength of hydrogen bonds [10-12]. Furthermore, when the angle of transition dipole moment with respect to the chain axis is known, the orientation of each component can be calculated by determining the infrared dichroism with polarized infrared measurement [13,14]. Therefore, this technique measures one of the most sensitive molecular parameters, segmental orientation, to the degree of macroscopic deformation, i.e. sample strain.

In polymer materials, the orientation of chain segments is the most commonly measured quantity in mechanical-vibration experiments and it is determined by measuring the absorbance of infrared radiation polarized parallel and perpendicular to the stretching direction of a polymer film. The measurement is normally accomplished by taking two separate spectra (one of each polarization) and calculating their difference or ratio. In these studies, the amount of noise in the obtained spectra is an experimental limitation. For the irreversible extension experiments, which are conducted by the rapid scan technique, the time resolution is controlled by two factors: the number of scans required to yield a signal-to-noise ratio (SNR) in each spectrum and the time required to switch between parallel and perpendicular polarizations. In the case of cyclic deformation experiments, to achieve extremely high time resolution low strain amplitude must be maintained to ensure the reproducible stress-strain behavior in the sample [15-18]. The resulting small spectral changes lead to reduced SNR in the data.

Instead of using the generally accepted subtraction technique described above, polarization modulation is a differential method which can be used to obtain the difference spectrum directly in a single measurement. Polarization modulation is accomplished by alternately passing two differently polarized light beams through the sample. The detected signal is processed by a phase sensitive amplifier synchronized with the changing polarization, and converted to a signal which represents the intensity difference between the two polarizations. The main advantages of this technique are improved photometric accuracy and SNR and the convenience and improved time resolution which result from not having to collect two separate spectra. The purpose of our present study is to evaluate the advantages of polarization modulation relative to subtraction for the measurement of polymer orientation.

Nafie et al. have published several papers describing the measurement of infrared circular dichroism by combining the polarization modulation with a Fourier transform infrared study [19-22]. Their results indicate that the Fourier transform infrared technique shows a definite advantage over the dispersive infrared technique. The obtained circular dichroic spectra are of higher quality in both aspects, the spectral resolution and SNR [19,22]. Dowrey and Marcott have compared subtraction and modulation measurements of linear dichroism and they observed an improved SNR by modulation as well as elimination of certain artifacts associated with the subtraction spectra [23,24]. Generally, the observed sensitivity increase for the modulation scheme has been attributed to a reduced dynamic range in the experiment [19,25,26]. However, no further elaboration is given. Therefore, the increase in sensitivity cannot be predicted from the experimental conditions. In this study, the improvement in sensitivity has been examined from a theoretical viewpoint in conjunction with experimental measurements. A direct comparison of a polarization modulation technique with the subtraction technique is also presented here.

EXPERIMENTAL

A Hinds International Series II Photoelastic Modulator with a ZnSe crystal was used. The ZnSe optical element at 750 cm^{-1} has a low transmission value of 50%; and the lower limit for which half wave retardance can be produced is 1250 cm^{-1}. The ZnSe crystal is 5 cm in diameter and a mask 1.6 cm in diameter was used in our experiment to ensure that the retardance was uniform over the infrared beam. The crystal is driven at 74kHz (ω_d), alternating the polarization at 148 kHz ($2\omega_d$). The modulator control unit allows the wavelength of half wave retardance to be set and provides a reference signal at ω_d and $2\omega_d$.

A mount was constructed to hold the modulator in the beam with the modulation principal axes rotated 45° from sample parallel and perpendicular directions.

A Harrick quad diamond polarizer, mounted before the sample compartment, can be placed into the IR beam and set at any angle or can be removed from the beam. When a polarizer was required behind the modulator, a Perkin Elmer wire grid polarizer was mounted on the modulator.

A Princeton Applied Research model 124A lock-in amplifier was used as a phase sensitive amplifier (mixer). A model 116 preamplifier was used in the direct input mode and the input high pass filter was set to 40 kHz. The lock-in amplifier has a specific frequency response up to 200 kHz, presumably adequate for the input channel, but in reality the mixer was required to have a higher limit. The in-phase mixer output signal for a 150 kHz sine wave input is a $\sin t$ or $\sin^2 t$ waveform with a fundamental frequency of 300 kHz plus higher frequency harmonics. The 90° out-of-phase mixer output signal has even stronger high frequency harmonics. The limited frequency response causes a phase shift between the fundamental and harmonics, producing an uninterpretable mixer output signal. This prevents the reference phase from being adjusted while observing the mixer output with an oscilloscope. Nonetheless, the mixer provides a reasonable signal proportional to the amplitude of the $2\omega_d$ component of the input signal. The signal must be taken directly from the mixer output because the following output stages (with an approximately 1 msec time constant) would attenuate the $\omega_{\bar{\nu}}$ components (100 Hz to 10 kHz) severely. Care must be taken not to overload the input preamplifier stage which has a 1.6 volt clipping level. The maximum IR detector signal is 10 Volts but this level is attenuated by all the optical elements used in the modulation experiment.

The detector signal from the Nicolet Fourier transform infrared spectrometer was taken from the detector 1 input on the beam path control board (Figure 1) and mounted to the lock-in input. The lock-in output is returned to the detector 2 input on the beam-path control board. This allows a selection of the AC or DC interferogram by selecting detector 2 or 1 through the spectrometer operating software, respectively. The signal is removed and reinserted before the band pass filters of the spectrometer as seen in the block diagram (Figure 2).

RESULTS AND DISCUSSION

The general concept of modulating a signal to improve the quality of measurement is quite old. The intensity of a light beam is normally measured by passing it onto a detector which converts it into an electrical signal that is amplified. In the amplifier and in some detectors, there is a noise component with an amplitude inversely proportional to frequency [27]. By chopping the light beam, the signal is shifted into the higher frequency region in which it is amplified with less noise. In addition, alternating current amplification is less prone to drift and gain errors than direct current amplification.

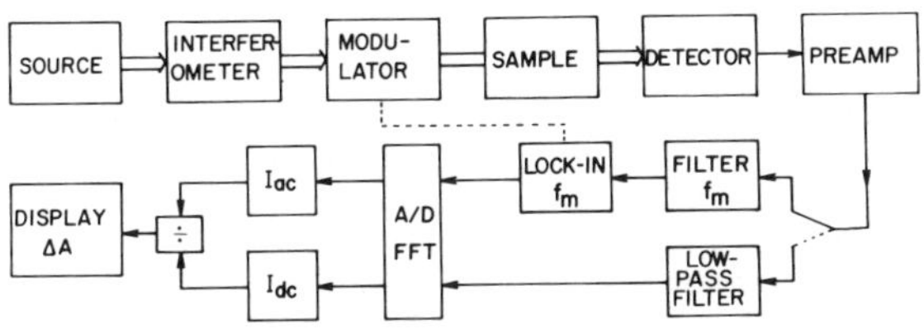

Figure 1. Fourier transform infrared signal path modifications for the polarization modulation experiment.

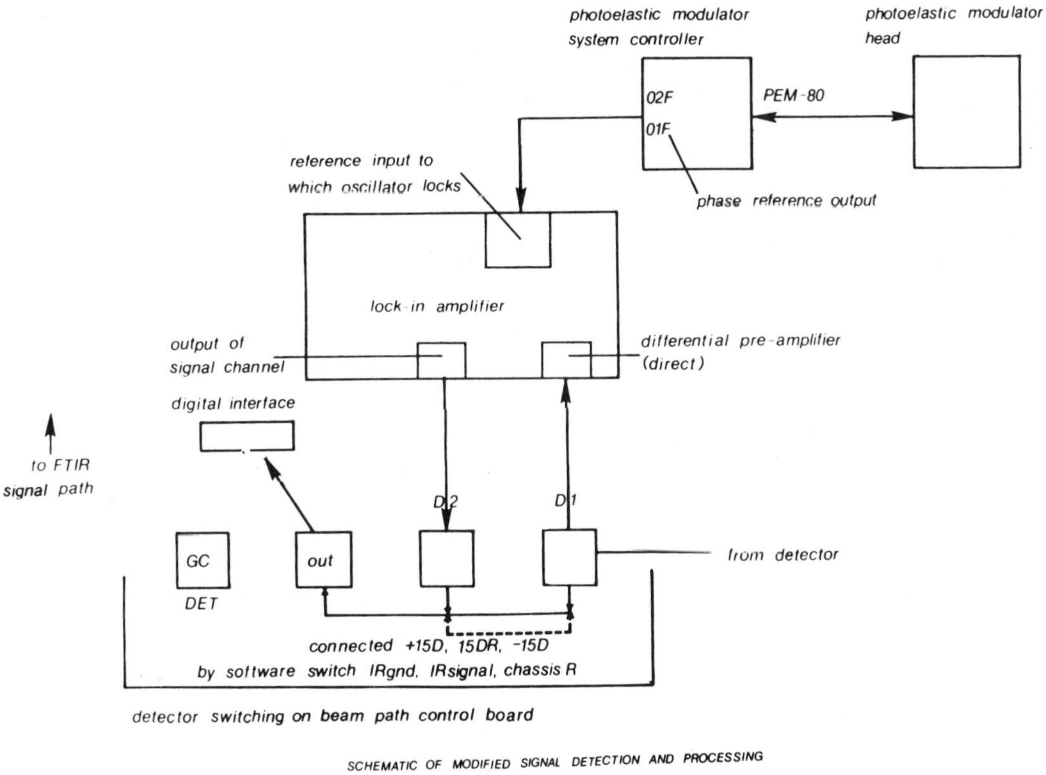

Figure 2. Block diagram of polarization modulation experiment.

This concept was broadened to include double beam ratio spectrometers and other forms of differential measurements, where the absolute light intensity is no longer measured. Only the difference of two intensities is determined in these schemes. The two signals are alternately passed onto a single detector and a phase sensitive amplifier is used to produce the different signal. The process of switching between the two signals produces the same benefits for the noise reduction as does chopping the signal. Furthermore, the direct measurement of the difference signal requires a much smaller dynamic range than would be required to obtain the difference of the two separately-measured signals with the same accuracy.

In addition to the advantage of the dynamic range reduction stated above, the modulation scheme eliminates several problems associated with the subtraction scheme in which the two spectra are measured sequentially in two different geometries. In each mechanical-vibrational experiment, errors may be caused by the change in the sample orientation between the measurement of the two spectra. Also, the time needed to switch between parallel and perpendicular polarization can lengthen the interval between successive spectra, degrading the time resolution. Additionally, either the sample or, normally, the polarizer is rotated between successive measurements. Both experimental arrangements generate artifacts in the subtraction spectrum if the sample orientation or geometry and the intensities of the two polarized beams are not uniform over the area illuminated by the infrared beam. One such artifact that has been observed is film interference fringes which do not cancel out in the ratio or difference spectra [23]. The modulation scheme is less vulnerable to these effects mentioned above, because the two polarization conditions are quickly repeated and averaged many times in a short time period. In some instances these additional effects can be more important than the SNR and dynamic range advantages realized by the use of polarization modulation.

The dynamic range and signal-to-noise considerations involved in difference (subtraction) and differential (modulation) measurements should be considered carefully. Since the analytical expressions describing noise in the two techniques cannot easily be described, procedures for numerical evaluation of the dynamic range will be discussed. Although the analysis is complex, causing the results to be inapplicable to other samples, considerable insight on modulation spectroscopy can still be gained by examining them.

Generally speaking, the mercury cadmium telluride (MCT) detector is the major noise source in our Fourier transform infrared spectrometer. Its noise level is independent of the light level falling on the detector [27]. The detector noise which is contained in the measured interferogram signal is converted to a wavenumber distribution by the Fourier transformation. The single beam distribution can be determined by subtracting two single beam spectra. As seen in Figure 3, this noise distribution is almost achromatic, i.e. little change is observed from 4000 to 400 cm^{-1}. The absorbance noise, however, is not achromatic and the effects that shape the absorbance noise distribution can be seen as follows.

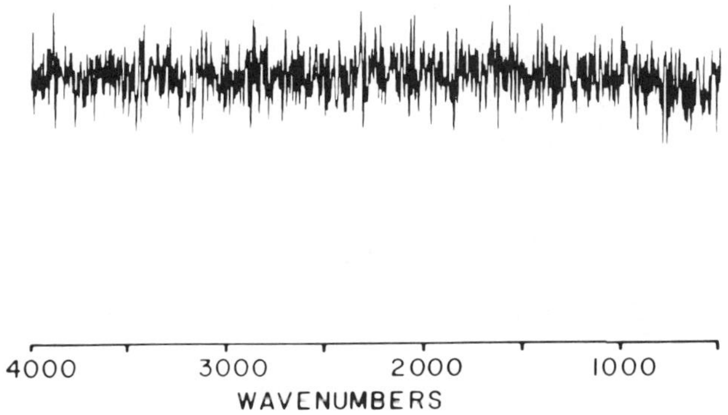

Figure 3. Spectra (32 scans at 4 cm^{-1} resolution) showing the wavenumber dependence of the single beam noise.

The single beam noise, N_s, can be converted into absorbance noise, n_A, using Eq. (1) which follows from the definition of the absorbance, A, and its derivative, dA/dI;

$$n_A = -\frac{1}{I_o \ln 10} 10^A N_s \qquad (1)$$

$$n_A = k\, 10^A N_s$$

$$A = -\log(I/I_o) \qquad (2)$$

$$dA = -\frac{1}{I_o \ln 10} 10^A dI \qquad (3)$$

Both the noise in the absorbance spectrum (Eq. (1)) and that in the transmission spectrum depend on the reciprocal of I_o, which depends on instrument throughput. Therefore, in an open beam 100% line (the ratio of two background spectra taken under identical conditions), the noise level is highest where I_o is smallest.

The noise level of a spectrum is commonly determined by examining a spectral region free from peaks. The deviation from the average in this region is taken to be the noise level. The knowledge of the true noise level is particularly important in evaluating the small differences obtained with our mechanical-vibrational experiments, because in many cases spectral features of polymers have unfamiliar band shapes. The noise level near a peak may be considerably higher than other regions due to the 10^A dependence shown in Eq. (1). Figure 4 shows the parallel spectrum and the difference of two polarizations, $A_p - A_s$, measured by modulation and subtraction techniques. The actual noise observed for each technique is greatest at the peaks or the least transmittance of the infrared spectrum. The actual signal-to-noise ratio applicable in determining the orientation change can be determined by subtraction of two spectra obtained under identical conditions. The noise may also be calculated if both $I_o(\nu)$ and $A(\nu)$ are taken into account.

For the subtraction method, the noise power in the difference spectrum is an additive sum of the noise in the two original spectra. Therefore, adding the two separate noise sources of the same level gives $\sqrt{2}$ times the original noise level. For subtraction, the absorbance noise is given by

$$n_{sub} = \sqrt{2}\, K(10^{A_p} + 10^{A_s})N_s \qquad (4)$$

where A_p and A_s are the parallel and perpendicular absorbances and N_p and N_s, the parallel and perpendicular noise levels, are assumed to be equal. The absorbance signal is $S_{sub}=\Delta A=A_p-A_s$.

The advantage of the modulation technique is that the difference can be measured directly [19]. If the sample is characterized by two polarized components $I_p(\bar{\nu})$ and $I_s(\bar{\nu})$, where $\bar{\nu}$ is the frequency of the infrared radiation, when a sinusoidal modula-

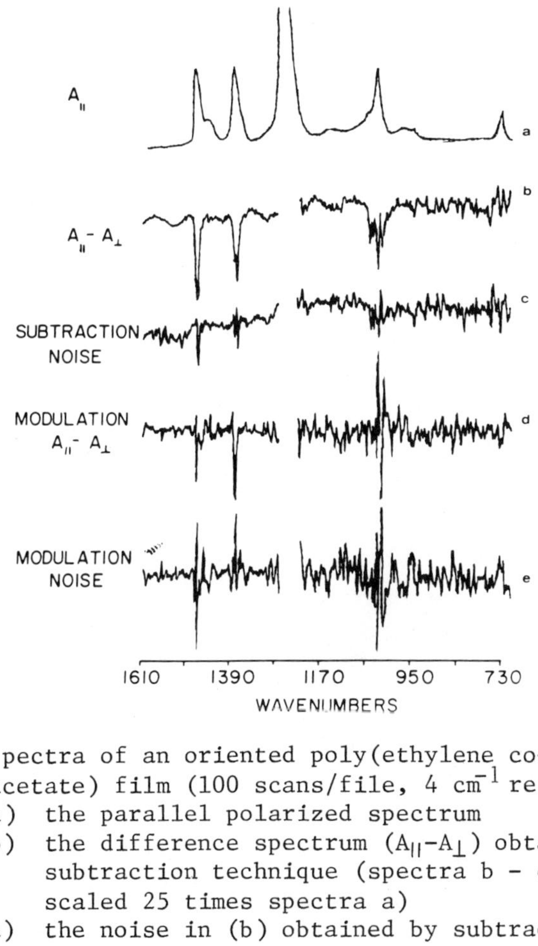

Figure 4 Spectra of an oriented poly(ethylene co-vinyl acetate) film (100 scans/file, 4 cm^{-1} resolution)

a) the parallel polarized spectrum
b) the difference spectrum ($A_{\parallel}-A_{\perp}$) obtained by the subtraction technique (spectra b - e are scaled 25 times spectra a)
c) the noise in (b) obtained by subtraction two difference spectra
d) the difference spectrum ($A_{\parallel}-A_{\perp}$) obtained by the modulation technique
e) the noise in (d) obtained by subtracting two difference spectra

tion is applied to these components at a frequency ω_m, the total intensity reaching the detector as a function of time and frequency is

$$I(\bar{\nu},t) = I_{dc}(\bar{\nu}) + I_{ac}(\bar{\nu})\sin(2\pi\omega_m t) \quad (5)$$

where the average intensity, I_{dc}, and the difference, I_{ac}, are given by

$$I_{dc}(\bar{\nu}) = 0.5\left[I_p(\bar{\nu}) + I_s(\bar{\nu})\right] \quad (6)$$

and

$$I_{ac}(\bar{\nu}) = 0.5\left[I_p(\bar{\nu}) - I_s(\bar{\nu})\right] \quad (7)$$

The difference intensity, I_{ac}, which contains the desired structural information, can be expressed in terms of A, the absorbance. The expressions are then

$$I_{dc}(\bar{\nu}) = 0.5\, I_o(\bar{\nu})\left[10^{-A_p(\bar{\nu})} + 10^{-A_s(\bar{\nu})}\right] \quad (8)$$

$$I_{ac}(\bar{\nu}) = 0.5\, I_o(\bar{\nu})\left[10^{-A_p(\bar{\nu})} - 10^{A_s(\bar{\nu})}\right] \quad (9)$$

In our experiments, the change in the segmental orientation can best be expressed as

$$I_{ac}/I_{dc} = (10^{A_p} - 10^{-A_s})/(10^{-A_p} + 10^{-A_s}) \quad (10)$$

$$= \tan h\left[0.5\ (\ln 10)\ \Delta A_{11\perp}\right]$$

where $\Delta A_{11\perp}$ is $A_p - A_s$. For small values of ΔA, the hyperbolic tangent may be replaced by its argument with little error.

For the modulation method, the noise expression, N_{mod} follows from the derivative of the ratio;

$$R = I_{ac}/I_{dc} = x/y \quad (11)$$

and

$$\frac{dR}{dI} = \frac{\partial R}{\partial x}\frac{\partial x}{\partial I} + \frac{\partial R}{\partial y}\frac{\partial y}{\partial I} \quad . \tag{12}$$

Thus the noise n_{mod} is given by

$$n_{mod} = \frac{I_{ac}}{I_{dc}}\left(\frac{N_{ac}}{I_{ac}} - \frac{N_{dc}}{I_{dc}}\right) \tag{13}$$

where N_{ac} and N_{dc} are the noise in the single beam AC and DC spectra. Eq. (13) can be simplified if $N_{ac}=N_{dc}$ and $I_{ac}<<I_{dc}$;

$$n_{mod} = \frac{N_{ac}}{I_{dc}} \tag{14}$$

Using the modulation signal, S_{mod}, given by

$$S_{mod} = \frac{I_{ac}}{I_{dc}}$$

the modulation signal-to-noise ratio, SNR_{mod}, is

$$SNR_{mod} = \frac{I_{ac}}{N_{ac}} = \frac{0.5\ I_o(10^{-A_p} - 10^{-A_s})}{N_{ac}}$$

Meanwhile, the subtraction signal-to-noise ratio, SNR_{sub}, is

$$SNR_{sub} = \frac{I_o \ln 10}{\sqrt{2}\,N_s}\ (A_p - A_s)/(10^{A_p} + 10^{A_s}) \quad .$$

Since both expressions contain I_o and the noise level, these two factors can be brought over from the right hand side of the equations and the two results compared by plotting the quantity (SNR x Noise/I_o) for various values of A_s and A_p (Figure 5). As seen in that figure, the subtraction and modulation values are close for small values of (A_p-A_s). At larger values, the approximation used in the SNR expressions and that in Eq. (10) lose their accuracy. Thus the difference seen for larger values can be considered insignificant. These plots demonstrate that the SNRs for the two techniques are equivalent; there is no inherent signal-to-noise gain in using the modulation technique - any improvement must come from a reduction in the dynamic range requirements.

The dynamic range measurement is simply the ratio of the largest to smallest signal that can be measured. The spectral dynamic range is the ratio of the largest spectral feature to the root mean square noise level [27]. If the spectral dynamic range is greater than the measurement dynamic range, digitization noise (the difference between the measured and actual value) will be added to the spectrum. The spectral dynamic ranges of A_p, A_s, and ΔA can easily be determined from their absorbance spectra. However, in Fourier transform infrared spectroscopy, it is not the absorbance dynamic range or even the single beam dynamic range that is important but the dynamic range of the interferogram.

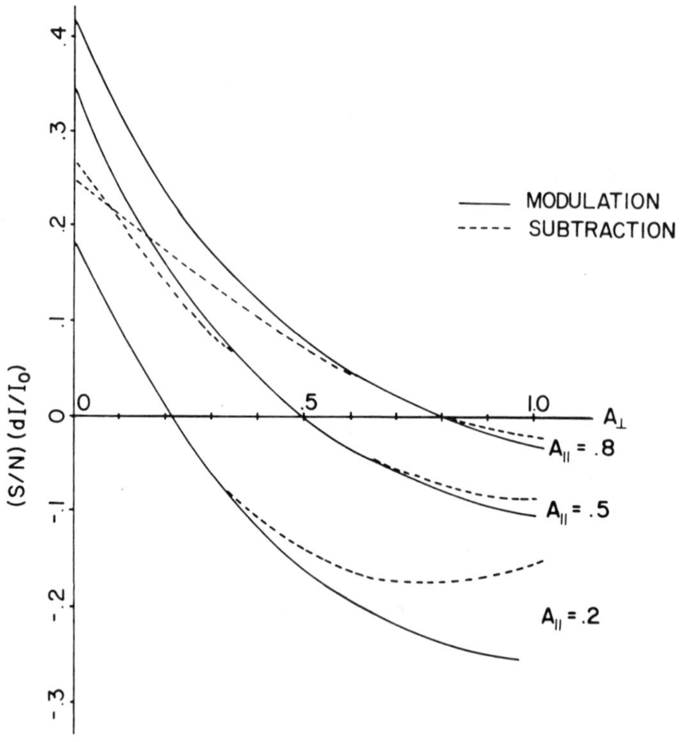

Figure 5. Calculated signal to noise ratios for the subtraction and the modulation techniques.

The interferogram which is expressed by the Fourier transform

$$V(\delta) = \int_{\bar{\nu}_1}^{\bar{\nu}_2} B(\bar{\nu}) \cos\left[2\pi\bar{\nu}\delta - \theta(\bar{\nu})\right] d\bar{\nu} \qquad (18)$$

where V, B, δ, and θ are the interferogram, intensity at the detector, retardation, and the phase factor respectively. In this expression, both the phase, $\theta(\bar{\nu})$ and $B(\bar{\nu})$ over the entire spectral region are required. The consideration may be simplified by setting the phase to zero for all wavenumbers. The maximum value of the interferogram occurs at $\delta=0$ and is given by the integral of $B(\bar{\nu})$ from $\bar{\nu}_1$ to $\bar{\nu}_2$.

$$V(\delta)_{max} = \int_{\bar{\nu}_1}^{\bar{\nu}_2} B(\bar{\nu})\, d\bar{\nu} \qquad (19)$$

By assuming a flat $B(\bar{\nu})$, the integral is proportional to the number of resolution elements, N, given by $(\bar{\nu}_2-\bar{\nu}_1)/\Delta\nu$ where $\Delta\bar{\nu}$ is the spectral resolution. The smallest contribution to the interferogram comes from the single resolution element of minimum intensity:

$$V(\delta)_{min} = B_{min}(\bar{\nu}) \cos(2\pi\bar{\nu}\delta) \qquad (20)$$

If the cosine is set equal to one, the estimated interferogram dynamic range becomes N times the spectral dynamic range. This estimate provides an upper limit for the interferogram dynamic range. In fact, a factor of N times the spectral dynamic range has been suggested as a more realistic estimate [27,28]. For a 4 cm^{-1} resolution spectrum from 4400 to 400 cm^{-1} the N is about 30.

As an example, if A_p is 1.00000, A_s is 1.00011, and ΔA is 1.1×10^{-4} then the requirement for 10:1 dynamic range in ΔA or 10^5:1 in A_p or A_s, results in an interferogram dynamic range of 300 to 1 for modulation and 3×10^6 to 1 for subtraction. A 15 bit A/D converter has a dynamic range of 3.3×10^4 to 1 and could accurately sample the modulation interferogram. The dynamic range of subtraction, 3×10^6 to 1, cannot be handled directly. However, it is unlikely that the 10^5 to 1 absorbance signal-to-noise ratio could be achieved in a single scan. This shortcoming is normally corrected by utilizing the fact that the signal-to-noise and the measurement dynamic range can both be improved by coadding scans [27]. In the optimum case, the dynamic range can be expected to increase with the square root of the number of scans [27]. Therefore, if 10^4 scans with a dynamic range of $3x10^4$ to 1 were coadded, the dynamic range could be improved 100 times to $3x10^6$ to 1. Thus, it is theoretically possible to use subtraction to directly measure a very small signal. However, in actual practice, these results are not always realized. Theory predicts that spectral signal-to-noise should decrease with the square root of the number of scans, i.e. noise is proportional to $N^{-0.5}$. This is observed initially, but the improvement in the signal to noise ratio after a large number of scans begins to level off, since other possible source of noise, including digitization noise or round off errors in the numerical computation process, can become relatively important.

The peak-to-peak noise level in the absorbance curve measured in the region of 3000 to 2000 cm^{-1} obtained by modulation and subtraction techniques are shown in Figures 6 and 7. The noise plotted versus $N^{-0.5}$ in Figure 6 shows the decrease in noise to be essentially linear up to 1000 scans. However, it should be noted that the method by which the data was plotted in Figure 6 is inconvenient since the data is highly compressed for large numbers of scans. In Figure 7 the data are plotted as the noise x $N^{0.5}$ versus N, and therefore are free from the vertical compression shown in the previous figure. In the second method of plotting the data, any uncertainty in measuring the noise is increased by $N^{0.5}$. In either case, the two plots show 1) that the noise is still continuing to decrease for both methods at 1000 scans, 2) the modulation noise is three to four times greater than the subtraction noise, and 3) a noise level of less than $2x10^{-4}$ absorbance units was obtained at 1000 scans by subtraction. The difference spectra (A_p-A_s) of a highly oriented poly(p-phenylene benzobisthiazole) film shown in Figure 8 confirms the above conclusions.

The most significant result obtained in our study is the higher noise level obtained by using the modulation technique. This is principally caused by the optical losses of the modulator crystal and the reduction in beam aperture. The observed difference spectrum between the two polarizations decreases at lower wavenumbers as the modulation efficiency falls off. At the same time the noise also increases at lower frequency as the modulator transmission decreases. Our results confirm that modulation fails to yield any signal-to-noise advantage over the subtraction technique and that in fact it is observed to operate at a disadvantage under most applicable experimental conditions.

In an earlier study of Fourier transform infrared polarization modulation, Marcott et al. reported that modulation produced better results than the subtraction technique. But equivalent results for modulation and subtraction were obtained in a second experiment [24]. We feel the relative merits of the two techniques not only depend on the sensitivity and noise level of the detector but also on the overall throughput of the spectrometer. If the Fourier transform infrared spectrometer throughput and detector area are properly matched, it may be possible to increase the aperture opening to compensate for the optical losses of the modulator [23], thus possibly allowing modulation to out-perform subtraction.

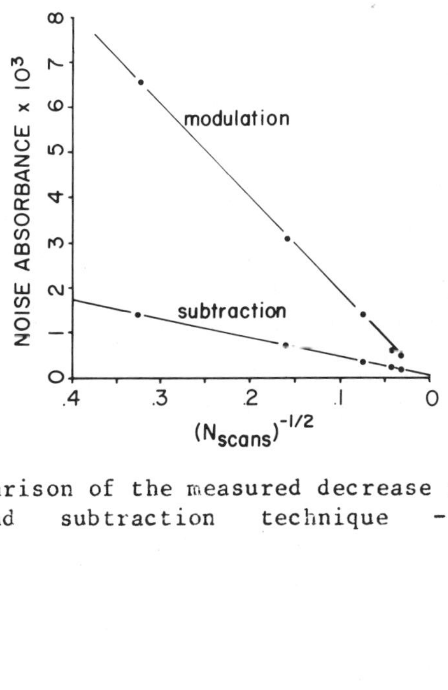

Figure 6. Comparison of the measured decrease in noise for the modulation and subtraction technique - noise versus $(N_{scans})^{-1/2}$.

To perform infrared mechanical-vibrational measurements of polymer films undergoing small amplitude oscillatory strains, Noda, Dowrey, and Marcott have employed a triple modulation scheme in conjunction with a dispersive instrument [29,30]. The three modulations used were 1) chopping the infrared beam, 2) sample strain, and 3) polarization modulation. For a polyethylene film undergoing a 0.1% oscillatory strain (100 Hz), a sensitivity of 1.3×10^{-4} absorbance units was measured for the 1465 cm^{-1} band. This experiment truly increased significantly the sensitivity limit of polarization modulation experiments. In most other experiments, the advantages of the polarization modulation technique may be divided into two types: intrinsic - relating to a reduction in dynamic range and improvement in SNR due to differential as opposed to difference measurement, and practical - which result from not having to measure two separate spectra and not having to rotate the sample or polarizer. The practical advantages include relaxing the requirement that the sample be identical when the parallel and perpendicular spectra are measured, which can be extremely difficult if the sample undergoes rapid irreversible changes. The polarization modulation technique also reduces the likelihood of having artifacts associated with non-uniform samples, infrared beam profiles, and polarizer imperfections.

The practical advantages can be quite significant in infrared mechanical-vibrational experiments in which time between acquiring succeeding spectra is an important parameter and in which the sample is often changing continuously. The practical disadvantages should also be considered. These include a significant attenuation of the infrared beam, a low frequency limit of about 800 cm^{-1}, as well as the cost, complexity, and an additional source of experimental complication.

Our examination of the intrinsic advantages has shown that the dynamic range reduction of modulation is not that important. For subtraction, the measurement dynamic range is higher than the spectral dynamic range for measurement down to $2x10^{-4}$ absorbance units. This was proven by the continued decrease in noise with the square root of the number of scans. Therefore, measurement of spectral features down to the noise level should be possible. For different experimental conditions in which the measurement of the interferogram is limited in dynamic range, modulation can significantly improve the sensitivity limit.

We intended to incorporate the polarization modulation technique into the time resolved spectroscopic technique developed earlier [16-18]. However, the samples used often exhibit a static dichroism that is at least 10^{-2} absorbance units and often higher. The high SNR required to measure differences of this magnitude are difficult to obtain by any method. In a triple modulation experiment, Noda et al. demodulated the strain dependence to eliminate the static dichroism and can directly measure the dynamic dichroism. With Fourier transform instruments, the strain frequencies fall at or below the Fourier frequencies and therefore cannot be demodulated.

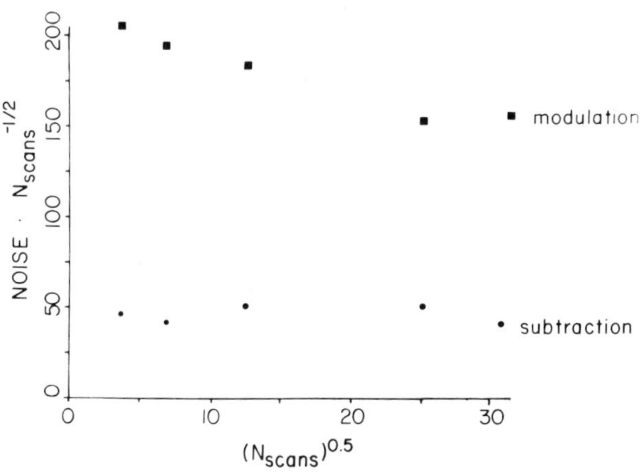

Figure 7 Comparison of the measured decrease in noise for the modulation and subtraction techniques - noise·$(N_{scans})^{\frac{1}{2}}$ versus $(N_{scans})^{\frac{1}{2}}$.

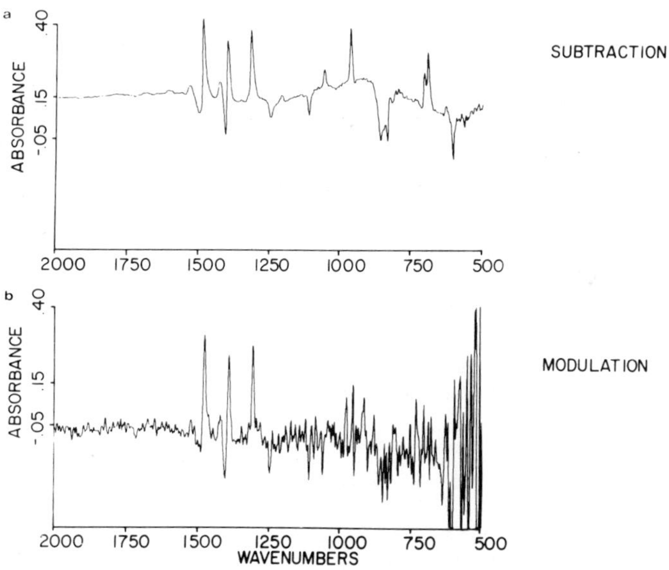

Figure 8 The difference spectra (A -A) for a highly oriented poly(p-phenylene benzobisthiazole) film which shows the difference in noise levels and spectral features (10 scans at 4 cm^{-1} resolution)
a) the subtraction technique
b) the modulation technique.

Before combining two such complex experiments, each must be well understood or the results will be totally unreliable. Because of the stringent repeatability requirements of TRS and the inefficiency in data collection of the present TRS implementation, the practical modulation advantages are quite significant even if the intrinsic advantages do not come into play.

CONCLUSION

Comparison of modulation and subtraction does not provide an overwhelming advantage for modulation that would favor it over the much simpler and more straightforward subtraction technique. Direct comparison of the two techniques for two polymer samples showed subtraction produced better results due to the significant optical losses in the modulation experiment. The polarized spectra obtained for both techniques for a highly oriented polymer film are identical except for the noise levels.

The modulation technique should be considered when the optical losses can be compensated by opening up the aperture, as can be done on an IBM-98 Fourier transform infrared spectrometer for example, or when a high dynamic range interferogram is encountered (a high interferometer throughput or a small area and/or high D detector), or when the practical advantages warrant it (TRS dichroic experiments).

REFERENCES

1. R.S.Stein, Accounts Chem. Res., 5, 121 (1972).
2. R.S.Stein, J. Polym. Sci., C15, 185 (1966).
3. D.P.Pope and A.Keller, J. Polym. Sci. Polym. Phys. Ed., 13, 533 (1975).
4. A.Peterlin, J. Mat. Sci., 6, 490 (1971).
5. K.Fujita, S.Suehiro, S.Nomura and H.Kawai, Polym. J., 14, 545 (1982).
6. R.Yamada and R.S.Stein, J. Appl. Phys., 36, 3005 (1965).
7. Y.Kobayashi, S.Okajima and A.Narita, J. Appl. Polym. Sci., 11, 2515 (1967).
8. Y.Uemura and R.S.Stein, J. Polym. Sci. A2, 10, 1691 (1972).

9. T.Asada, J.Sasada and S.Onogi, Polym. J., 3, 350 (1972).

10. H.Ishihara, I.Kimura, K.Saito and H.Ono, J. Macromol. Sci.-Phys., B10, 591 (1974).

11. H.W.Siesler, Polym. Bull., 9, 557 (1983).

12. ibid., 9, 382 (1983).

13. R.Zbinden,"Infrared Spectroscopy of High Polymers", Academic, New York (1964).

14. R.J.Samuels, Die Makromol. Chem., Suppl., 4, 241 (1981).

15. D.J.Burchell, J.E.Lasch, R.J.Farris and S.L.Hsu, Polymer, 23, 965 (1982).

16. J.E.Lasch, T.Masaoka, D.J.Burchell and S.L.Hsu, Polym. Bull., 10, 51 (1983).

17. J.E.Lasch, D.J.Burchell, T.Masaoka and S.L.Hsu, Appl. Spectrosc., 38, 351 (1984).

18. W.G.Fately and J.L.Koenig, J. Polym. Sci. . Polym. Lett. Ed., 20, 445 (1982).

19. L.A.Nafie and D.W.Vidrine, in "Fourier Transform Infrared Spectroscopy", Vol.3, J.R.Ferraro and L.J.Basile, Eds., Academic, New York (1982).

20. L.A.Nafie and M.Diem, J. Appl. Spectrosc., 33, 130 (1979).

21. E.D.Lipp, C.G.Zimba,L.A.Nafie and D.W.Vidrine, J. Appl. Spectrsc., 36, 496 (1982).

22. E.D.Lipp, C.G.Zimba and L.A.Nafie, Chem. Phys. Lett., 90, 1 (1982).

23. A.E.Dowrey and C.Marcott, Appl. Spectrosc., 36, 414 (1982).

24. C.Marcott, Appl. Spectrosc., 38, 442 (1984).

25. W.G.Golden, D.S.Dunn and J.Overend, J. Catalysis, 71, 395 (1981).

26. M.F.Russel, M.Billardon and J.P.Badoz, Appl. Optics, 11, 2375 (1972).

27. P.R.Griffiths,"Chemical Infrared Fourier Transform Spectroscopy", Vol.43, John Wiley and Sons, New York (1975).

28. C.H.Perry, R.Geick and E.F.Young, Appl. Optics, 5, 1171 (1966).

29. I.Noda, A.E.Dowrey and C.Marcott, J. Polym. Sci. Polym. Lett. Ed., 21, 99 (1983).

30. ibid., International Union of Pure and Applied Chemists, 28th Macromolecular Symposium, 1982.

FOURIER TRANSFORM INFRARED VIBRATIONAL CIRCULAR DICHROISM IN THE CARBONYL STRETCHING REGION OF POLYPEPTIDES AND URETHANE AMINO ACID DERIVATIVES

Laurence A. Nafie, Elmer D. Lipp, Anita Chernovitz
and Germana Paterlini

Department of Chemistry
Syracuse University
Syracuse, New York 13244-1200

ABSTRACT

Fourier transform vibrational circular dichroism (FT-IR VCD) and ordinary FT-IR absorption spectra are presented for the polypeptide, poly(ε-CBZ-L-lysine), and the urethane amino acid derivative, N-t-Boc-L-alanine, in the carbonyl stretching region. The assignment of the spectra are discussed and two basic mechanisms for VCD intensity are described and applied to the VCD spectra presented. The stereosensitivity of VCD spectra to molecular structure in these and other molecules is illustrated and the need for the establishment of quantitative structural relationships is stressed.

INTRODUCTION

Vibrational circular dichroism (VCD) is a relatively new form of molecular spectroscopy which combines the absolute stereospecificity of optical activity with the structural sensitivity of vibrational spectroscopy. A number of reviews of the subject, written from a variety of perspectives, have appeared in recent years [1]. VCD can be defined in an operational sense as the difference in the absorption of a sample between left and right circularly polarized radiation in the infrared, vibrational region of the spectrum. To support VCD, a sample must be chiral. In most instances this is satisfied at the molecular level where the constituent molecules are non-superimposable relative to their mirror image forms.

Since its discovery just over a decade ago, VCD has advanced both experimentally and theoretically [1]. Instrumentation has improved from early measurements restricted to the hydrogen stretching region to include coverage throughout the mid-infrared region down to 600 cm^{-1}. In addition, a double modulation Fourier transform technique has been devised to permit the measurement of VCD with Fourier transform infrared (FT-IR) spectrometers [2]. FT-IR VCD spectra exhibit a combined resolution and signal quality that is unsurpassed by dispersive methods [3] and has also made possible the application of Fourier self-deconvolution (FSD) to VCD spectra, further enhancing the spectral resolution of these spectra [4].

Early theoretical descriptions of VCD focused on two models, the degenerate coupled oscillator (DCO) model [5] and the fixed partial (FPC) model [6]. In both models, one is primarily concerned with the details of the nuclear motion to establish the expected VCD intensity and a closer examination of the FPC model reveals its form to be that of the DCO model extended to all pairs of nuclear displacements in a normal vibrational mode. Progress toward more complete descriptions has involved attempts to model more explicitly the contribution of the electronic charge density to VCD intensity. A fundamental difficulty encountered in this effort has been the need to exceed the limitations of the Born-Oppenheimer approximation to achieve a non-zero electronic contribution to the magnetic dipole transition moment. Models receiving the most attention in this area have been the localized molecular orbital model [7], the charge flow model [8], the dynamic polarization model [9] and the bond moment model [10]. Recently we have developed an approach based on a vibronic coupling expansion which is free of the restrictions imposed by the models mentioned above [11]. A central concept emerging from the more sophisticated model is that of electronic current density induced by nuclear velocities during vibrational motion.

The aim of this paper is two-fold. The first is to present FT-IR VCD spectra in the carbonyl stretching region of a polypeptide and a urethane amono acid derivative to demonstrate the feasibility and sensitivity of the experimental method. The second is to interpret these spectra on the basis of two elementary, conceptual mechanisms for generating VCD intensity in molecules. The establishment of the presence of these intensity mechanisms in the carbonyl stretching spectra then provides a basis for gaining a clearer understanding of the absolute stereosensitivity of VCD to molecular structure and vibrational motion.

Several papers dealing with VCD in polypeptides have appeared in the literature [12] and recently we reported the first example of FT-IR VCD in a polypeptide, together with its Fourier self-deconvolution [13]. The results for polypeptides presented in this paper are an extension of this recent work. The VCD spectra in urethane amino acids has not received previous attention in the literature.

EXPERIMENTAL

The FT-IR VCD spectra are measured using a Nicolet 7199 FT-IR spectrometer equipped for VCD measurement as described previously [3,4,13]. Briefly, an optical filter operating between 2000 cm^{-1} and 1200 cm^{-1}, a BaF_2 wire grid polarizer, a zinc-selenide modulator, a HgCdTe detector and a phase sensitive lock-in amplifier are the only required accessories for VCD measurement. The VCD spectra were obtained from 12,288 AC-scans and 800 DC-scans requiring 2.5 h of collection time. The final result is the difference between the raw VCD spectra of the sample in its enantometric counterpart. The noise spectra are derived from the difference of two sequential blocks of raw VCD spectra of 6144 AC-scans.

Poly(ε-CBZ-lysine) samples were purchased from Sigma Chemical Co. and the samples of N-t-Boc-L-alanine were obtained from Chemical Dynamics, Inc. Samples were dissolved in reagent grade chloroform-d_o or d_1 and used without further purification.

CONCEPTUAL MODELS

As described above, the intent of the present paper is to provide an appreciation of the origin and sensitivity of VCD spectra. In keeping with this goal we will not attempt a quantitative analysis of the spectrum to be presented below, but rather we will provide the basis for interpreting the origin of spectra in terms of two conceptual models.

The first model is based on the coupling of two vibrational electric transition moments. This model is concerned with nuclear motion and hence derives from the conceptual bases of the coupled oscillator model, as well as the fixed partial charge model [5,6]. The model describes the VCD intensity expected from two electric transition moments that are coupled to one another vibrationally and skewed dissymmetrically in space with respect to one another. The coupling of the two transitions leads to two new transitions representing in-phase and out-of-phase combinations of the original transitions. If the original transition moments are degenerate, the mixing will be complete. If they are non-degenerate, with distinct unperturbed vibrational frequencies, the mixing will be partial; i.e. only part of the character of the transition will be a phased combination with respect to the other transition, and each mode will retain some of its original unperturbed character. A pictorial representation of this model is provided below where oscillators μ_1 and μ_2 are skewed with respect to one another by the angle ϕ . For simplicity it is assumed that μ_2 lies in a plane parallel to μ_1. If this is not true, an additional skew angle is necessary [14]. The VCD expected from the coupling

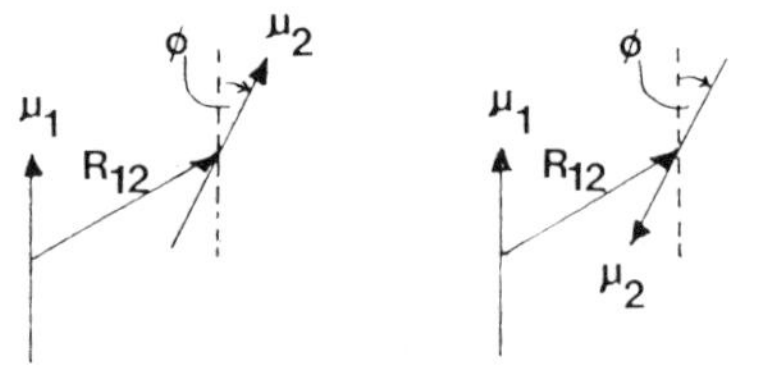

of two transitions as depicted above is a bisignate couplet having equal positive and negative VCD intensities given by the following expression for the rotational strength, R.

$$R^{\pm} = + \frac{\pi}{2C} R_{-12} \cdot \mu_1 \times \mu_2$$

where the upper sign refers to the in-phase component and the lower sign to the out-of-phase component. As drawn above, the in-phase vibration would yield negative VCD intensity and the out-of-phase component, positive intensity. The opposite sense of the angle, ϕ, would reverse these signs, as found in the enantiometric, mirror-image structure. The sense of the observed VCD couplet in an actual spectrum also depends upon the relative frequency locations of the in-phase and out-of-phase components. For a truly degenerate pair of oscillators, the transitions are split by their mutual radiative coupling. For a non-degenerate pair of oscillators, the unperturbed frequencies of the oscillators, as well as vibrational coupling factors, also influence the relative frequency location. Finally, we note that the closer two non-degenerate oscillators are to one another before coupling, the stronger their mutual interaction and VCD intensity due to the coupled oscillator mechanism. In the limit of large frequency separation, the two transitions no longer effectively couple and the coupled oscillator VCD vanishes. Several clear examples of the coupled oscillator mechanism have appeared in the literature.

The second conceptual model is based on vibrationally induced current density that moves in an arc or a ring [15]. Referred to as the ring current mechanism, this source of VCD intensity has recently been associated with large monosignate VCD intensity observed in the hydrogen stretching region of certain broad classes of molecules, such as the amino acids and their transition metal complexes . Usually the ring, or ring fragment in which the current flows is formed by an intramolecular hydrogen bond, transition metal bond or a similar intramolecular electronic association. Current is induced in the ring by the stretching motion of an oscillator that is either external to the ring or contained within the ring. In the latter case, the ring must be puckered to prevent the circular dichroism intensity from vanishing. The induced ring current corresponds to the motion of electron density that is distinct from the nuclear motion, and results in a large contribution to the electronic part of the magnetic dipole transition moment that is not compensated by a similar but oppositely signed nuclear contribution. Pictorially, the ring current mechanism can be illustrated by the diagrams below.

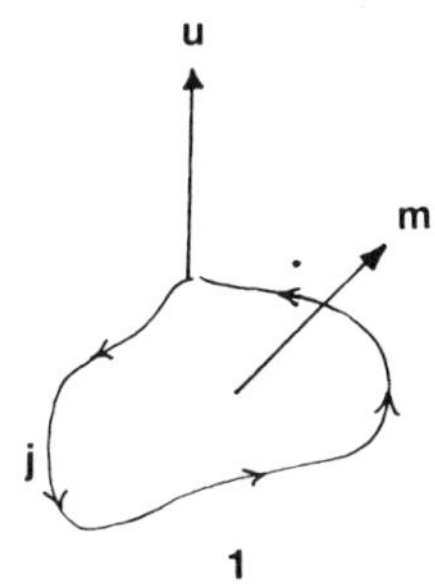

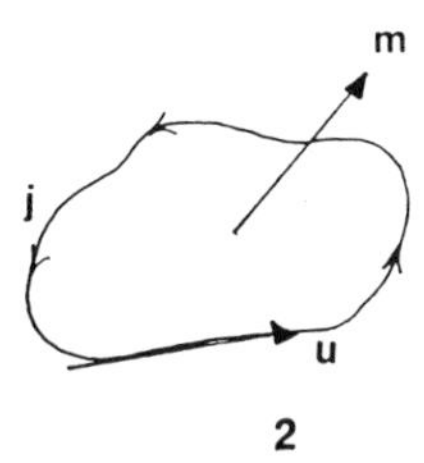

In diagram 1, the oscillator μ is external to the intramolecular ring current; the magnetic moment m is normal to the average plane of the ring current, being portional to the integration of $r \times j$ around the ring. In diagram 2, the oscillator is contained in the ring. VCD intensity of one sign then arises from the general intensity expression for the rotational strength

$$R = \mathrm{Im}(\mu \cdot m) \qquad (2)$$

where Im is the imaginary-part-of. Thus, the sign of the observed VCD from this mechanism is diagnostic of the stereochemical relationship between the inducing dipole, μ, and the ring geometry and sense of ring current giving rise to the magnetic moment, m.

It follows that for a given spectral region, if only one vibrational transition is present that is activating the ring current mechanism, the VCD intensity in that region will be biased to the sign of the VCD associated with the ring current mechanism. Conversely, VCD intensity that is associated with the nuclear motion according to the coupled oscillator mechanism, or a complete set of coupling oscillators in the FPC model, will be conservative (equal positive and negative intensity) in nature. As a result, a mere inspection of the VCD associated with given region of the spectrum that is not vibrationally coupled to a nearby region will be indicative of the relative extent to which these two basic VCD mechanisms are contributing to the spectrum. The spectra presented in this paper will be explained qualitatively in terms of these two basic mechanisms.

RESULTS AND DISCUSSION

The absorbance and VCD spectra of the polypeptide poly(ε-CBZ-L-lysine) in $CHCl_3$ solution are presented in Figure 1 for the region 1800 to 1500 cm^{-1}. The primary structure of this polymer can be represented as:

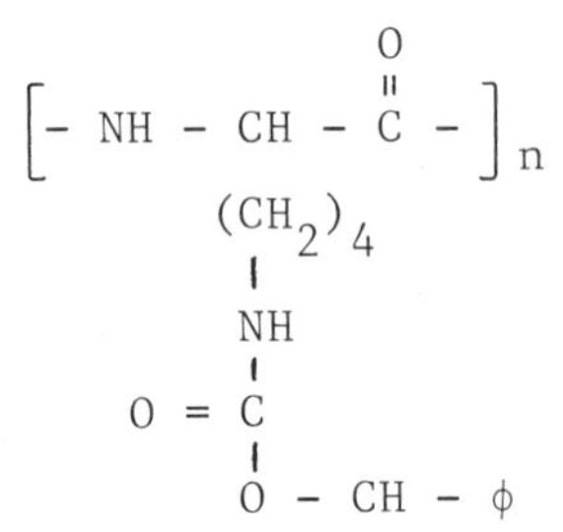

The broad absorption band at about 1710 cm^{-1} is due to the carbonyl stretching of the CBZ group, while the narrow absorption band at 1650 cm^{-1} is due to the amide I vibration of peptide backbone. The latter is due primarily to the amide carbonyl stretching vibration. The absorption bands between 1550 and 1500 cm^{-1} are associated with the amide II vibrational region; due to strong solvent absorption in this region little VCD is observable. The band at 1710 cm^{-1} gives rise to no discernible VCD intensity whereas the amide I band can be associated with a large bisignate VCD effect.

It is known that poly(ε-CBZ-L-lysine) assumes a right-handed α-helical structure in $CHCl_3$ solution [16]. Previously, we have demonstrated that this polypeptide exhibits a bisignate VCD spectrum in the amide I region using a conventional dispersive scanning VCD instrument [12b]. Close agreement is found between the earlier spectrum and the result in Figure 1. Close agreement also exists between our previous measurement of the amide I FT-IR VCD spectrum of poly(γ-benzyl-L-glutamate) [13,17]. Our results of VCD measurements to date of -helical polypeptide in the amide I region indicates that the VCD arises from the polypeptide backbone structure and is relatively insensitive to varia-

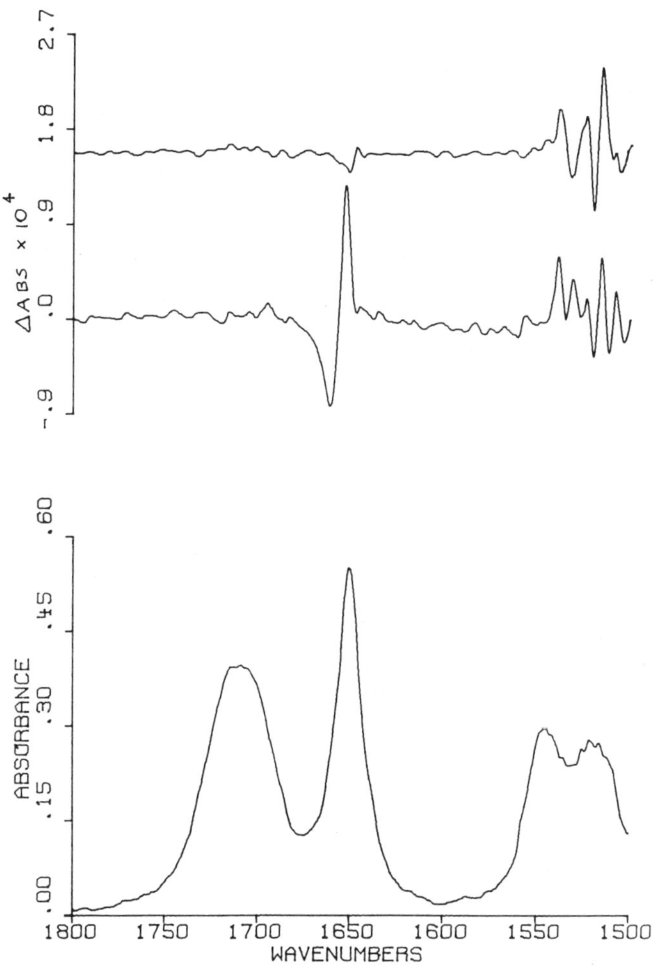

Figure 1 Absorbance (bottom), VCD (middle) and VCD noise estimate (top for poly(ε-CBZ-L-lysine) under the following conditions: spectral bandwidth, 4 cm^{-1}; concentration, 3 mg/ml in $CHCL_3$ solution; path-length 0.500 mm. The noise estimate is one-half the difference between two consecutive blocks of 6144 AC scans and represents the noise level in the displaced VCD spectrum.

tions in the amino acid side chain.

The qualitative explanation of the VCD in Figure 1 is that it arises from coupling of carbonyl stretching motion in the α-helix of the polypeptide. The VCD spectrum is not heavily biased although it appears to favor negative VCD intensity by a small degree. The major source of intensity may therefore be ascribed to the nuclear motion coupling mechanism of VCD, which is conservative with respect to overall VCD intensity. We have implicitly assumed that the various amide vibrational modes are independent of one another and that no other type of vibration is involved in the composition of the observed amide I band. The VCD couplet therefore arises from the phased vibrational motion of adjacent carbonyl groups that are skewed dissymmetrically with respect to one another by virtue of the α-helical secondary structure of the polypeptide. A more detailed quantitative analysis of the VCD in the amide I region of polypeptides has been described in another publication [13] for poly(γ-benzyl-L-glutamate). The small amount of negative VCD bias most likely arises from electronic current density that either propagates in a helical path along the hydrogen bonded network of the peptide units, or that travels in intramolecular rings in local regions of the polypeptide

Turning to the example of a urethane derivative of an amino acid, we consider the FT-IR VCD and absorption spectra of N-t-BOC-L-alanine in $CDCl_3$ solution in the region between 1800 and 1400 cm^{-1}. The structure of this molecule can be written as

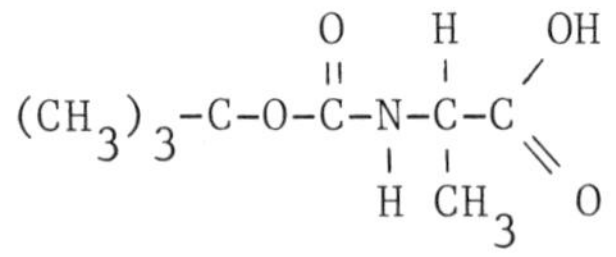

where a dashed line is drawn between the urethane carbonyl oxygen and the hydrogen of the carboxylic acid group to denote an intramolecular hydrogen bond that forms in this molecule in chloroform solution. Although other structures may also be present in solution that involve either intermolecular hydrogen bonding or the absence of hydrogen bonding, the structure indicated above, known as the C_7 conformation bacause of its seven membered ring, is of particular interest to us for the present discussion.

It is clear that a strong bisignate VCD can be associated with the absorption at 1710 cm^{-1}, and that other bands have little if any VCD by comparison. Closer inspection of the main absorption band reveals the onset of splitting into at least two component bands. Previous workers in fact have assigned a number of vibrational modes in this region [18]. They have assigned a weak band at 1760 cm^{-1} to the free acid carbonyl stretching mode and three bands between 1730 and 1690 cm^{-1} as due to strongly hydrogen bonded acid carbonyl stretching, free urethane carbonyl stretching, and weakly hydrogen bonded urethane carbonyl stretching. A final band at 1660 cm^{-1} has been assigned to strongly hydrogen bonded urethane carbonyl stretching. According to these assignments, the main absorption peak in Figure 2 is comprised of three vibrational modes. In this context, strong hydrogen bonding refers to <u>inter</u>molecular hydrogen bonding where interaction bond distances can assume optimum values, whereas weak hydrogen bonding refers to <u>intra</u>molecular hydrogen bonding as in the C_7 conformation. Namely, at higher frequency, the strongly hydrogen bonded acid carbonyl stretching mode mixes with the weakly hydrogen bonded urethane carbonyl. The free urethane band, although present, is presumed not to contribute for reasons to be made clear below. The nature of the interaction that gives rise to the observed positive-negative VCD couplet is that of the skewed transition dipole mechanism of in-phase and out-of-phase vibrations described in the previous section. The skewed orientation arises from the relative orientation of the two carbonyl groups in the C_7 conformer. Similar bisignate carbonyl groups are skewed in orientation relative to one another [19]. We note that the juxtaposition in frequency of the two modes giving rise to the couplet in Figure 2 requires <u>inter</u>molecular hydrogen bonding for one group, the acid carbonyl, and <u>intra</u>molecular hydrogen bonding for the other, the urethane carbonyl.

A second feature of the VCD in Figure 2 is its bias to negative VCD intensity. Although the bias is not strong, it is a persistent effect in all VCD spectra of urethane amino acid derivatives in the carbonyl stretching region. The bias is associated with the stretching of the urethane carbonyl group which lies within the C_7 intramolecular hydrogen bonded ring. This indicates that the lower frequency VCD band is predominantly the urethane carbonyl mode whereas the higher frequency number is due to the acid carbonyl, although a considerable amount of mixing has occurred.

The explanation of the VCD spectrum in Figure 2 as arising from the presence of the coupled oscillator effect as the principal mechanism and the existence of a ring current associated with the urethane carbonyl stretching vibration as a secondary mechanism, is reinforced by consideration of the VCD and absorption

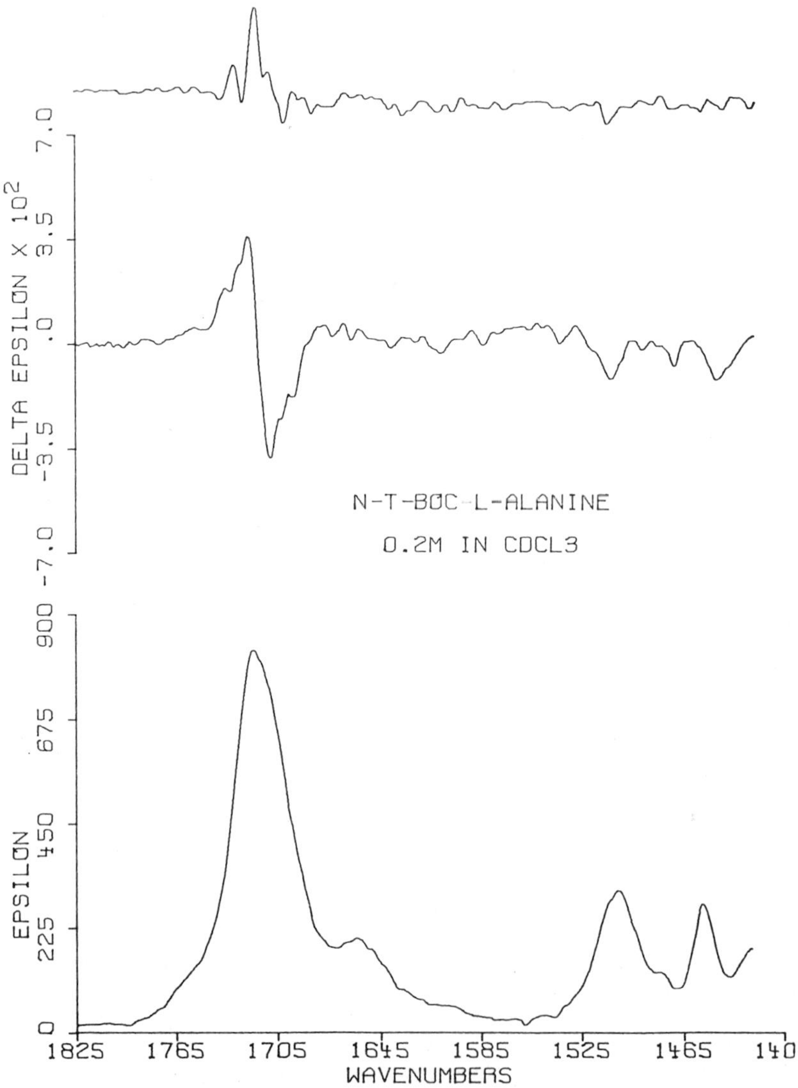

Figure 2 Absorbance (bottom) and VCD (middle) for N-t-Boc-L-alanine in 0.2M in $CDCl_3$ solution in units of the absorption coefficient, $\varepsilon(\Delta\varepsilon)$, Lcm^{-1} mol^{-1}. The VCD noise estimate (top) was obtained as described in Figure 1. The spectral resolution was 4 cm^{-1}.

spectra of N-t-BOC-L-alanine in dilute $CHCl_3$ solution as provided in Figure 3. The absorption spectrum shows the absence of a strongly (intermolecular) hydrogen bonded urethane carbonyl band at 1660 cm^{-1}, as expected at low concentration, a decrease in intensity on high frequency side of the main absorption band at 1730 cm^{-1} corresponding to the loss of the strongly hydrogen bonded acid carbonyl band and the growth of the free acid carbonyl band near 1760 cm^{-1}.

The VCD spectrum at low concentrations is much weaker than the higher concentration VCD spectrum. This is explained by the loss of VCD intensity due to the coupled oscillator, nulear motion, mechanism which requires the coupling of nuclear motion in different parts of the molecule. The loss of intermolecular hydrogen bonding upon dilution drastically reduces one of the two unperturbed transitions contributing to the coupled oscillator VCD effect. Nevertheless, the other transition corresponding to the C_7 urethane carbonyl group is largly unaffected by the loss of intermolecular hydrogen bonding. If anything, this band gains intensity due to the loss of the strongly hydrogen bonded urethane band. As a result, even though most of the coupled oscillator part of the VCD spectrum has been lost, the ring current contribution still persists and can be seen rather clearly in Figure 3, still associated with the urethane carbonyl band at 1710 cm^{-1}. Although the negative VCD bias was more difficult to discern in the presence of a large coupled oscillator effect in Figure 2, it is exposed without interference in Figure 3.

CONCLUSIONS

The overall result emerging from our study of VCD in the carbonyl region is the considerable stereoselectivity to molecular conformation realized in this region. For molecules possessing more than one carbonyl group, the coupled oscillator, coupled nuclear motion mechanism can contribute if the carbonyl transitions involved are sufficiently degenerate in frequency. The bisignate VCD spectra in Figures 1 and 2 are specific examples of this coupling. In addition, for molecules with only one carbonyl group, a biased, monosignate VCD spectrum may arise if the carbonyl stretching motion can induce an oscillating ring current in phase with its own stretching motion. In most of the cases we have studied to date, the ring current is supported by an intramolecular ring possessing considerable delocalizable electron density. The development of quantitative stereo-specific relationships between VCD spectra and the details of molecular conformation in the carbonyl stretching region will bring to fruition what appears to be a most promising area of spectroscopic investigation.

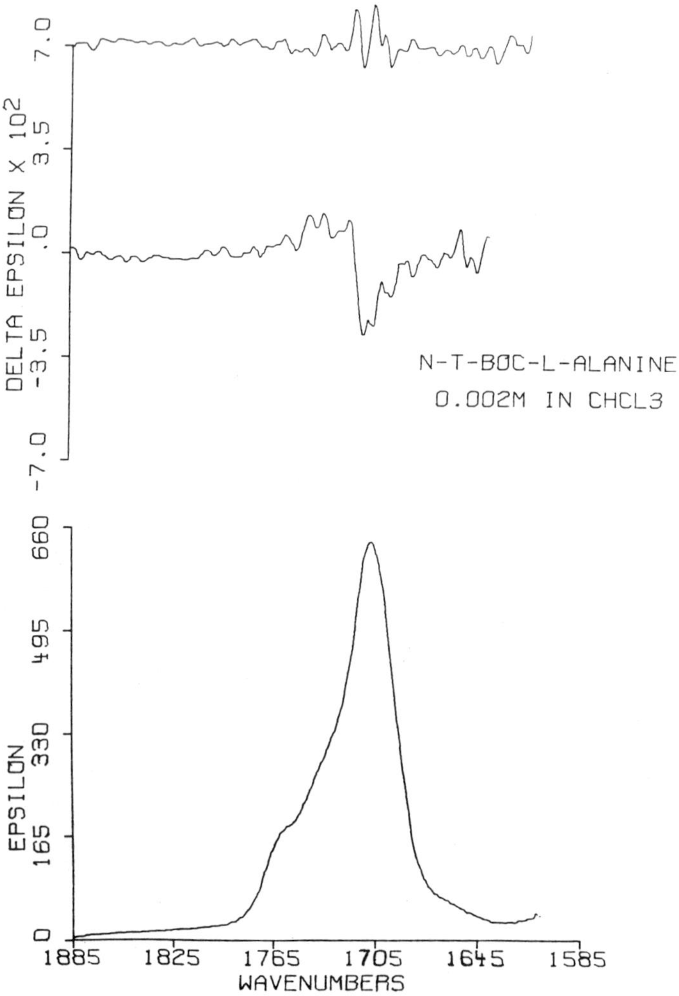

Figure 3 Absorbance (bottom), VCD (middle) and VCD noise estimate (top) for N-t-Boc-L-alanine in 0.002M $CHCl_3$ solution obtained under the same condititons as Figure 2.

ACKNOWLEDGEMENT

The results presented in this article were obtained with support from the National Institutes of Health (GM-23567) and the National Science Foundation (CHE 83-02416).

REFERENCES

1. a)L.A.Nafie and M.Diem, Acc. Chem. Res., 12, 296 (1976).

b) P.J.Stephens and R.Clark, in "Optical Activity and Chiral Discrimination", S.F.Mason, Ed., D.Reidel, Dordrecht (1979) p.263.

c) T.A.Keiderling, Appl. Spectrosc. Rev., 17, 189 (1981).

d) L.A.Nafie in "Vibrational Spectra and Structure", Vol. 10, J.R.Durig, Ed., Elsevier, Amsterdam (1981) p.153.

e) L.A.Nafie in "Advances in Infrared and Raman Spectroscopy", Vol.11, R.J.H.Clark and R.E.Hester, Eds., Wiley-Heyden, Chichester (1984) p.49.

f) P.L.Polavarapu in "Vibrational Spectra and Structure", Vol.13, J.R.Durig, Ed., Elsevier, Amsterdam (1984) p.103.

2. a) L.A.Nafie and M.Diem, Appl. Spectrosc., 33, 130 (1979).

b) L.A.Nafie, M.Diem and D.W.Vidrine, J. Am. Chem. Soc., 101, 496 (1979).

c) L.A.Nafie, E.D.Lipp and C.G.Zimba in "Proceedings of the 1981 Interfational Conference on Fourier Transform Infrared Spectroscopy", H.Sakai, Ed., SPIE, Bellingham (1981) p.457.

d) L.A.Nafie and D.W.Vidrine, in "Fourier Transform Infrared Spectroscopy", Vol.3, J.R.Ferraro and L.J.Basile, Eds., Academic, New York (1982) p.83.

3. a) E.D.Lipp, C.G.Zimba and L.A.Nafie, Chem. Phys. Lett., 90, 1 (1982).

b) E.D.Lipp and L.A.Nafie, Appl. Spectrosc., 38, 20 (1984).

4. E.D.Lipp and L.A.Nafie, Appl. Spectrsc., 38, 774(1984).

5. a) G.Holzwarth and I.Chabay, J. Chem. Phys., 57, 1632 (1972).

b) T.R.Faulkner, Ph.D.Thesis, University of Minnesota, Minnesota (1976).

c) H.Sugeta, C.Marcott, T.R.Faulkner, J.Overend and A.Moscowitz, Chem. Phys. Lett., 40, 397 (1976).

6. a) J.A.Shellman, J. Chem. Phys., 58, 2882 (1973).

b) Ibid., 60, 343 (1974).

c) C.W.Deutsche and A.Moscowitz, J. Chem. Phys., 49, 3257 (1969).

d) Ibid., 53, 2530 (1970).

7. a) L.A.Nafie and T.H.Walnut, Chem. Phys. Lett., 49, 441 (1977).

b) T.H.Walnut and L.A.Nafie, J. Chem. Phys., 67, 1501 (1977).

c) L.A.Nafie and P.L.Polavarapu, J. Chem. Phys., 75, 2935 (1981).

d) P.L.Polavarapu and L.A.Nafie, J. Chem. Phys., 75, 2945 (1981).

8. S.Abbate, L.Laux, J.Overend and A.Moscowitz, J. Chem. Phys., 75, 3161 (1981).

9. C.J.Barnett, A.F.Drake, R.Kuroda and S.F.Mason, Mo. Phys., 41, 455 (1980).

10. a) L.D.Barron in "Optical Activity and Chiral Discrimination", S.F.Mason, Ed., D.Reidel, Dordrecht (1979) p.219.

b) P.L.Polavarapu, Mol. Phys., 49, 645 (1983).

11. a) L.A.Nafie and T.B.Freedman, J. Chem. Phys., 78, 7108 (1983).

b) L.A.Nafie, J. Chem. Phys., 79, 4950 (1983).

12. a) R.D.Singh and T.A.Keiderling, Biopolymers, 20, 237 (1981).

b) B.B.Lal and L.A.Nafie, Biopolymers, 21, 2161 (1982).

c) A.C.Sen and T.A.Keiderling, Biopolymers, 23, 1519 (1984).

d) Ibid., 23, 1533 (1984).

13. E.D.Lipp and L.A.Nafie, Biopolymers, 24, 799 (1985).

14. I.Tinoco, Rad. Res., 20, 133 (1963).

15. a) L.A.Nafie, M.R.Oboodi and T.B.Freedman, J. Am. Chem. Soc., 105, 7499 (1983).

b) M.R.Oboodi, B.B.Lal, D.A.Young, T.B.Freedman and L.A.Nafie, J. Am. Chem. Soc., 107, 6205 (1985).

16. G.R.Bird and E.R.Blout, J. Am. Chem. Soc., 81, 2499 (1959).

17. L.A.Nafie, E.D.Lipp, A.Farrell and M.G.Paterlini, Polymer Preprints, 25, 145 (1984).

18. E.Benedetti, B.DiBlasio, V.Pavone, C.Pedone, C.Toniolo and G.M.Borova, Biopolymers, 20, 1635 (1981).

19. V.Narayanan and T.A.Keiderling, J. Am. Chem. Soc., 105, 6406 (1983).

APPLICATION OF FT-IR MICROSAMPLING TECHNIQUES TO SOME POLYMER SYSTEMS

K. Krishnan

Bio-Rad Digilab Division
237 Putnam Avenue, Cambridge, MA 02139

ABSTRACT

An universal microsampling accessory that can be used to record the FT-IR transmission or reflection spectra of samples as small as 20 μm x 20 μm in linear dimensions is described. This accessory consists of an all-reflecting infrared microscope coupled with a small area, high sensitivity mercury-cadmium-telluride detector. The use of this accessory for the study of imperfections in polymers, single filament polymer fibers, and the characterization of different layers in a multilayer polymers will be illustrated from their transmission spectra. A diamond anvil cell can be used in conjunction with the microsampling accesory to produce high quality FT-IR transmission spectra from thick polymer samples. Using a polarizer with the accessory, dichroic spectra of single filament polymer fibers can be obtained. In the reflection mode, the use of the accessory for the study of the cratering in paint panels will be described.

INTRODUCTION

Fourier transform infrared (FT-IR) spectroscopy with its high optical throughput advantage is ideally suited for the study of microsamples. The prior infrared microsampling studies had been carried out by placing a suitable small aperture over the sample area of interest and recording the infrared transmission spectrum [1-4]. Using this sample preparation technique in conjunction with an 8x beam condenser, Cournoyer et al. [2] have obtained the FT-IR spectrum of 90 picograms of triphenylphosphate in cellulose acetate. A nichrome wire source and a deuterated triglycine sulfate (DTGS) detector were used and 96,000 scans at 8 cm^{-1} resolution were coadded (corresponding to a measurement time of around fourty hours) to produce this FT-IR transmission spectrum. Lacy [4] used a liquid-nitrogen cooled mercury-cadmium-telluride (MCT) detector and was able to record the FT-IR spectra of similar samples by coadding 2,000 scans. However, this method still involved time-consuming sample preparation.

Krishnan and Kuehl [5] have desribed two different micro-sampling accessories specifically designed for FT-IR instruments. One accessory could be used to obtain the microtransmission spectra and the other for microreflection spectra. They have described the use of such accessories for epitaxial thickness measurements on silicon from the reflection technique. In this paper, an universal microsampling accessory that can be used to record the transmission or reflection FT-IR spectra of micro-samples will be described. Some applications of this accessory to study various polymers will be illustrated.

EXPERIMENTAL

The schematic optical diagrams of the universal microsampling accessory are shown in Figures 1a and 1b.

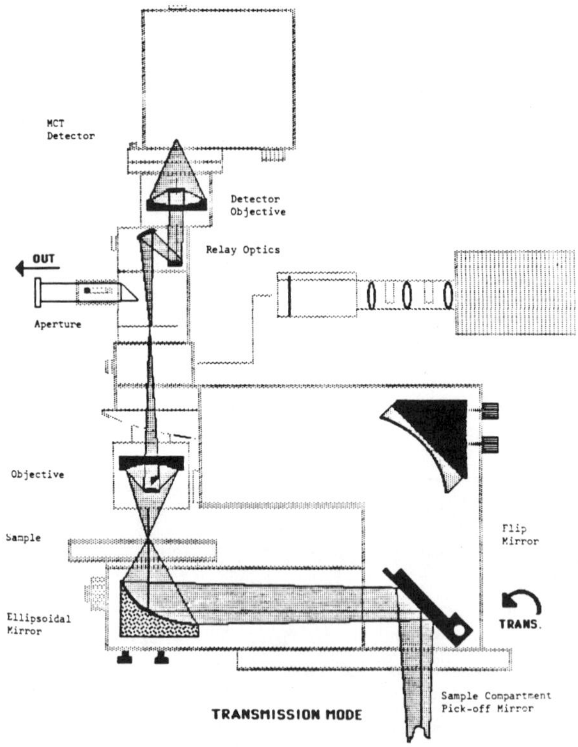

Figure 1a. Schematic optical diagrams of the universal microsampling accessory in the transmission mode.

The accessory consists of an all-reflecting infrared microscope coupled to a small area, high sensitivity MCT detector. The sample is placed horizontally on a standard microscope X-Y stage. In the transmission mode (Figure 1a), the flipping mirror M1 is moved into the optical path so that the infrared radiation from the FT-IR instrument is focussed on the sample from below. The

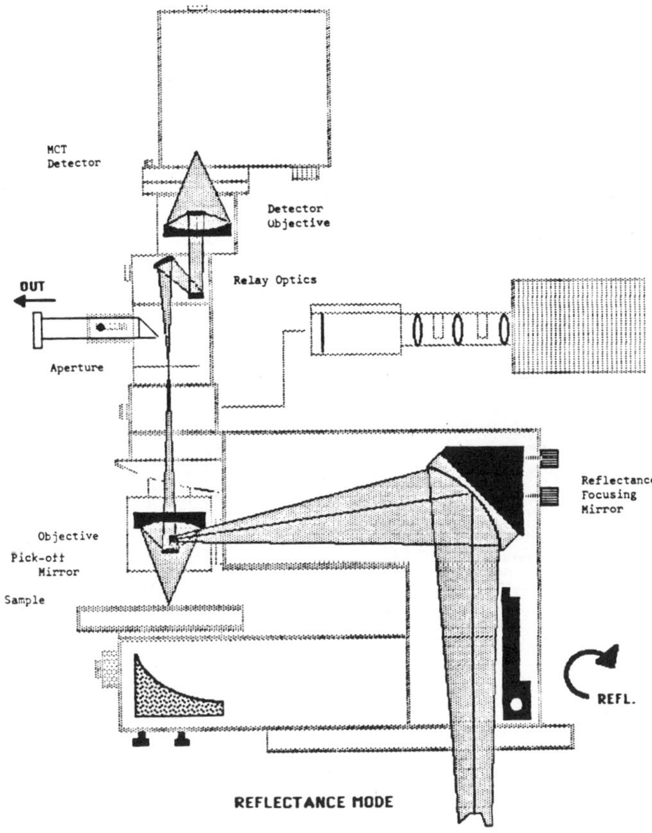

Figure 1b. Schematic optical diagrams of the universal microsampling accessory in the reflection mode.

radiation transmitted through the sample is collected by a Cassergranian objective and an image of the sample, magnified 32 times, is produced within the barrel of the microscope. A standard microscope visible illumination system that is collinear with the infrared optical path can be used for the visual examination of the image of the sample. A 10x eyepiece is included in the visible illumination system and thus the image is examined at

320 magnification. A variable aperture, circular or rectangular, is placed in the image plane and can be used to isolate the sample area of interest. This technique of aperturing the image of the sample eliminates the need for the time-consuming pinhole masking of the sample itself. The radiation transmitted through the variable aperture is collected and focused on a small area, high-sensitivity MCT detector by a Cassegranian condenser. The noise produced by an MCT detector is proportional to its linear dimensions. The use of the small area detector instead of the conventional 1mmx1mm or 2mmx2mm apparatus results in spectra of enhanced signal-to-noise ratios. The on-axis optics employed in the microscope insures distortion-free imaging of the radiation transmitted by the sample on the small area detector.

In the reflection mode (Figure 1b), the flipping mirror M1 is moved outside the optical path and the infrared beam of the FT-IR instrument is focussed on the sample from above through a small mirror inside the Cassegranian objetive. The radiation reflected by the sample is collected and focussed on the detector in exactly the same fashion as in the transmission mode. Thus, by moving the flipping mirror M1 in and out of the optical path, the accessory can be converted from the transmission to the reflection mode.

Most of the spectra reported in this article were recorded using a 0.25μmx0.25μm narrow range (4000-700 cm^{-1}) MCT detetor in the microsampling accessory. The accessory was mounted on a Digilab FTS-90 FT-IR spectrometer and a moving mirror velocity of 0.64 cm/sec was employed for the measurements. Most of the reported spectra were recorded at 4 cm^{-1} spectral resolution coadding 256 scans corresponding to measurement times of around two minutes per sample.

RESULTS AND DISCUSSION

Sensitivity

Figure 2a shows the transmission spectrum of polystyrene from a 10 μm x 10 μm area of the sample (aperture at the image plane set to 320 μm x 320 μm). The background spectrum was recorded using a 100 μm x 100 μm effective sampling area (aperture at the image plane set to 3200 μm x 3200 μm). A wide range (4000-450 cm^{-1}) MCT detector was used and 1000 scans at 8 cm^{-1} were coadded (measurement time of four minutes) to produce the spectrum. One can see that the baseline in the spectrum falls off at the long wavelengths (lower frequencies). This is due to the diffraction effects at the smaller aperture. Figure 2b shows the transmission spectrum of polystyrene recorded using a narrow range MCT detector, with the background spectrum also recorded through a 10 μm x 10 μm area. Now the diffraction effects in the sample and reference spectra are balanced out and the spectrum shows a flat baseline. the excellent spectral quality in Figure 2b shows that using the microsampling accessory, spectra from even such microscopic polymer samples (corresponding to less than five nanograms of the sample) can be recorded in a few minutes. Figure 2b also illustrates the fact that one can obtain the high quality infrared spectra of samples whose linear dimensions are comparable to the wavelength.

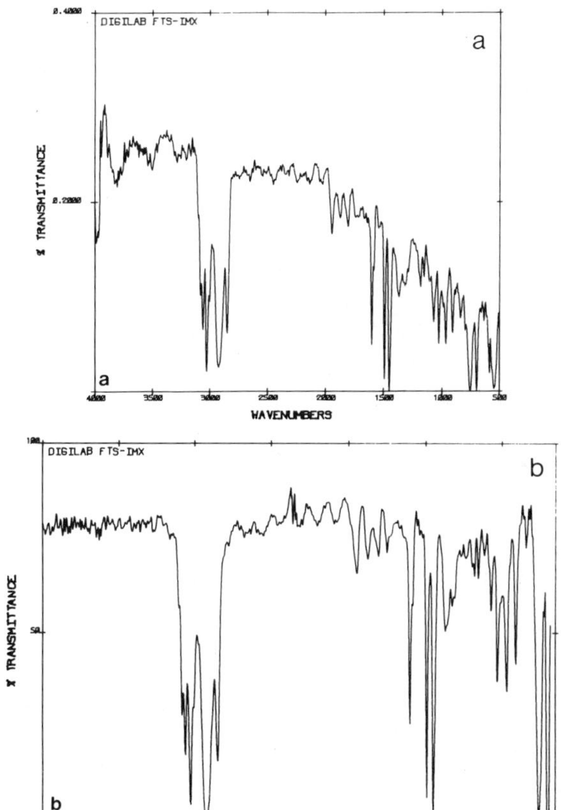

Figure 2. Transmission spectrum of polystyrene film from a 10 μm x 10 μm sampling area. Background spectrum recorded through a (a) 100 μm x 100 μm sampling area and (b) 10 μm x 10 μm sampling area. Note that the diffraction effects are balanced out in figure 2b.

Imperfections in Polymers

Many polymer films exhibit imperfections such as gel spots or haze. Using the visible illumination system of the microsampling accessory, the variable aperture could be used to isolate only the imperfection in the polymer and its transmission spectrum could be recorded. The transmission spectrum of the normal polymer film could be recorded through the same aperture and the subtraction between the two spectra can yield the spectrum of the imperfection. Figure 3 shows the results of such a study on a haze spot in polyethylene. The spectrum from 2000 to 450 cm^{-1} through the haze spot (100 μm x 100 μm) is shown at the top, spectrum of normal polyethylene in the middle and the difference at the bottom. These spectra were all recorded using a wide range MCT detector, coadding 256 scans at 8 cm^{-1} resolution (one minute measurement time). Figure 3 illustrates the ease with which such difficult problems could be studied.

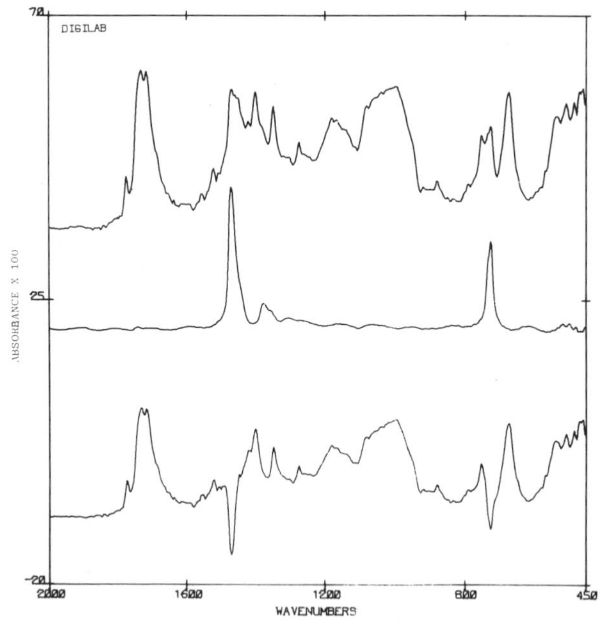

Figure 3. FT-IR transmission through a 100 μm x 100 μm sampling area of (top) a haze spot in polyethylene; (middle) normal polyethylene and (bottom) the difference spectrum.

Multilayer Polymers

Different layers in a multilayer polymer could easily be characterized by the use of the microsampling accessory. If the sample is thick, a very thin (approximately 5 to 10 μm) cross-section of it could be microtomed. Very thin multilayer polymers could first be embedded in epoxy matrix and then microtomed across their cross-section. This cross-section could then be supported on a KBr plate on the microscope X-Y stage, and the different layers in the cross-section could be isolated using the variable aperture and their spectra recorded. Figures 4A and 4b show the transmission spectra of two layers in a multilayer polymer. A 50 m wide, 100 m long sampling area was used in each layer and the spectra were recorded at 4 cm^{-1} resolution coadding 64 scans (30 seconds measurement time).

Polymer Fibers

The microsampling accessory can be used to obtain the FT-IR spectra of single polymer filaments. Figure 5 (top) shows the spectrum of a single poly(ethylene terephthalate) (PET) fiber. The diameter of the fiber was 15 m and the aperture was set to sample at 10 m x 150 m area of the fiber. The spectrum was recorded at 8 cm resolution coadding 512 scans (two minute measurement time). As can be seen from the figure, the absorption bands around 1200 cm are quite broad and saturated indicating that the 15 m sample thickness (the diameter of the fiber) is too large. In such cases, one can press the fiber into a very thin film between the two faces of a diamond anvil cell to obtain a high quality spectrum. (For details of the diamond anvil cell technique see, for example, Krishnan and Ferraro [6].) Figure 5 (bottom) shows such a spectrum of the same PET fiber recorded through the diamond anvil cell using the microsampling accessory. The fiber was pressed between two one millimeter face diamonds and one of the diamonds containing the pressed fiber was placed on the X-Y stage to record the spectrum. the same measurement conditions as used for recording Figure 5 (top) were used and a 50 m x 50 m area of the pressed sample was examined.

Dichroic Spectra

The microsampling accessory can be used to obtain the FT-IR transmisson spectra of polymer films or single fibers by inserting a polarizer in the optical path of the microscope above the variable aperture. Two background spectra could be recorded for the two mutually orthogonal orientations of the polarizer and two more such spectra could be recorded with the sample in place. The sample and background spectra for each polarizer orientation could then be ratioed to produce the two dichroic spectra of the sample.

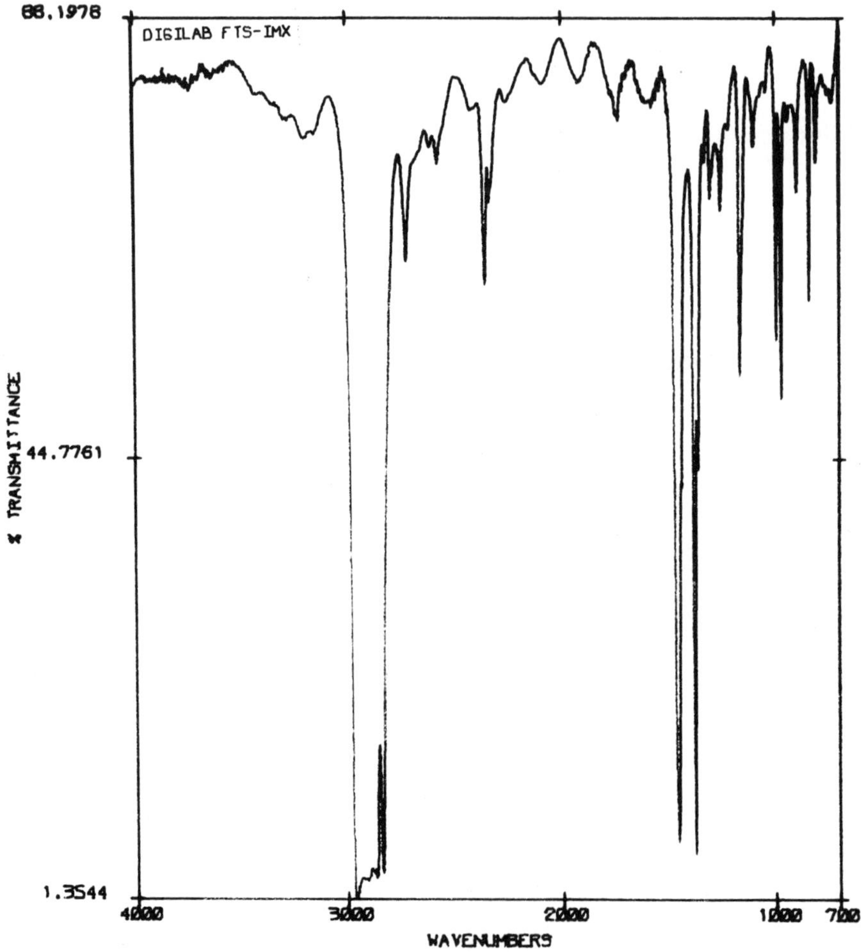

Figure 4 a, b FT-IR transmission spectra from 50 μm x 100 μm sampling areas of two layers of a multi-layer polymer film.

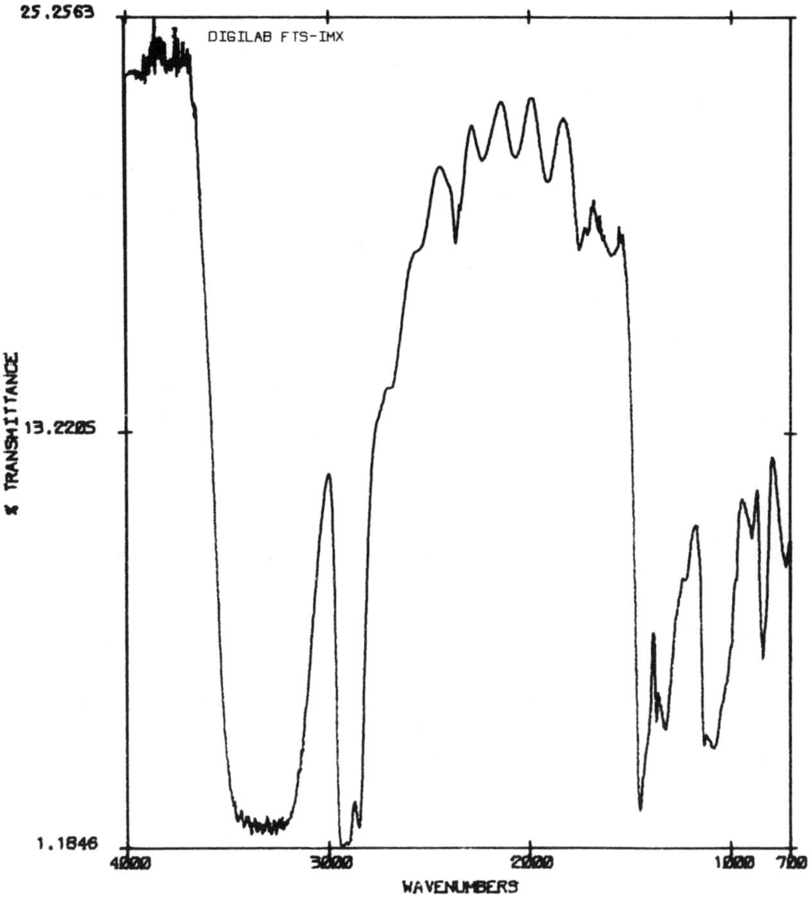

25.2563
DIGILAB FTS-IMX
% TRANSMITTANCE
13.2205
1.1846
4000
3000
2000
1000
700
WAVENUMBERS

Figure 4b.

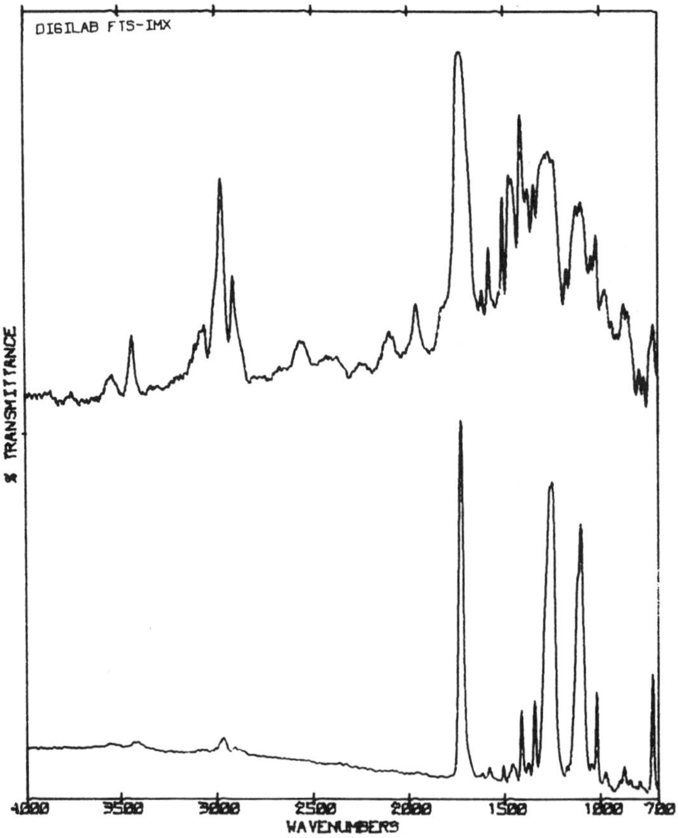

Figure 5. Top: transmission spectrum of a single 15 μm diameter PET fiber. Bottom: spectrum of the same fiber after pressing between two diamond faces in a diamond anvil cell.

Figures 6a (k_x) and 6b (k_y) show the dichroic spectra of the single 15 μm diameter PET fiber. These spectra were recorded at 4 cm^{-1} for measurement times of two minutes per sample. One can easily see the dichroic behaviour of the different bands in the spectra, particularly the weaker bands around 900 cm^{-1} and around 1350 cm^{-1}. The stronger bands around 1200 cm^{-1} show saturation and their dichroic ratios could not be determined. When the polarized transmission spectra of the same fiber were recorded in the diamond cell after pressing them into a thin film, in order to obtain the dichroic ratios of the stronger bands, it was found that none of the PET bands exhibited any dichroism. Obviously, the diamond anvil cell that applies fairly high pressure on the fiber destroys the orientation in the fiber. In any case, spectra shown in Figure 6 indicate that some information on the orientation of different functional groups in fibers such as PET could be obtained by this technique.

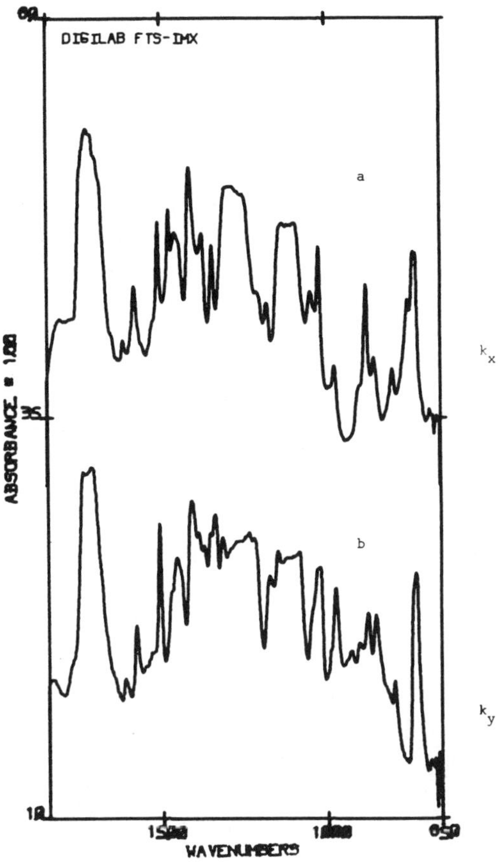

Figure 6. Dichroic transmission spectra of a single 15μm diameter PET fiber (a) the k_x spectrum and (b) the k_y spectrum.

Microreflectance Spectra

In the reflection mode, the microsampling accessory can be used to easily obtain the infrared spectra of polymer films on metal surfaces. Figure 7 shows the microreflectance spectrum of a polymer contaminant (100 μm x 100 μm) on a gold-coated electrical contact. This spectrum was recorded in thirty seconds at 8 cm^{-1} resolution. This technique can also be used to obtain the FT-IR reflectance spectra of imperfections such as craters on paint panels. Figure 8a shows the reflection spectrum recorded at 8 cm^{-1} resolution, 256 scans, from a 100 μm x 100 μm area of a crater in a green paint panel. Figure 8b shows the spectrum from a good area of the paint. One can see the significant differences between the two spectra. Figure 8c shows the difference spectrum, crater-paint. Figure 8a, 8b and 8c illustrate the ease with which problems of this type could be studied.

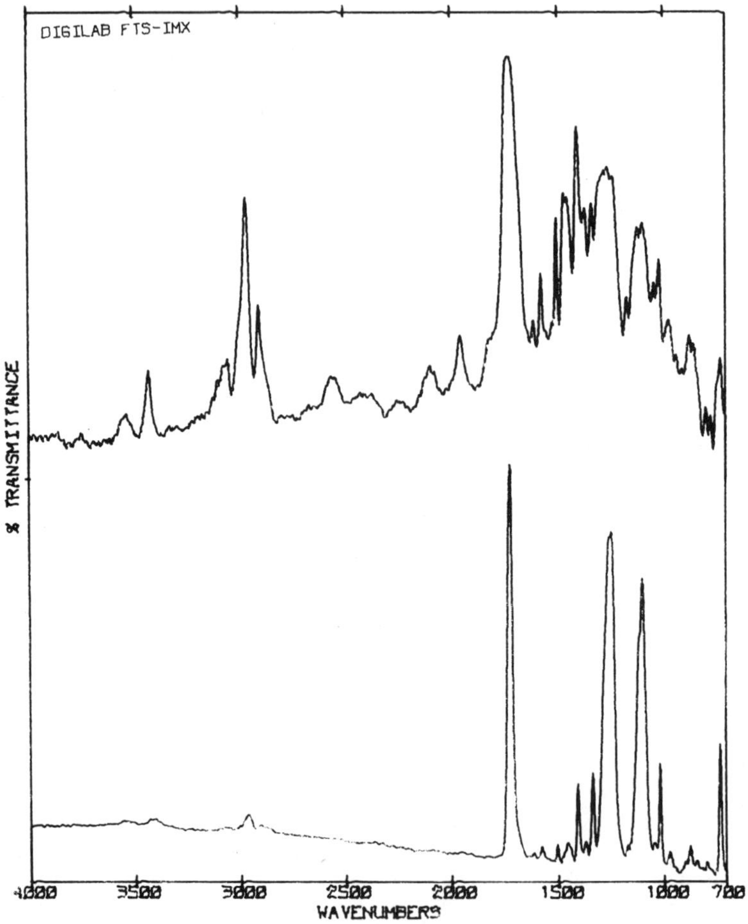

Figure 7 FT-IR microreflectance spectrum from a polymer contaminant on a gold-coated electrical contact. The spectrum was recorded from a 100 μm x 100 μm area of the contaminant.

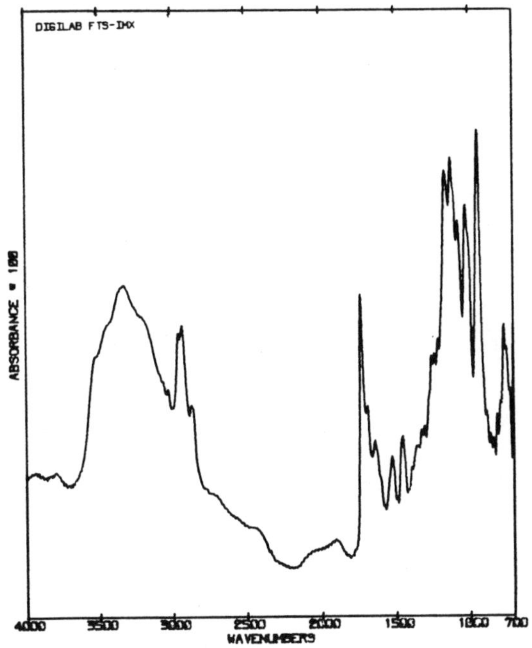

Figure 8a. Microreflectance spectrum from 100 μmx 100 μm area of crater in a green paint panel.

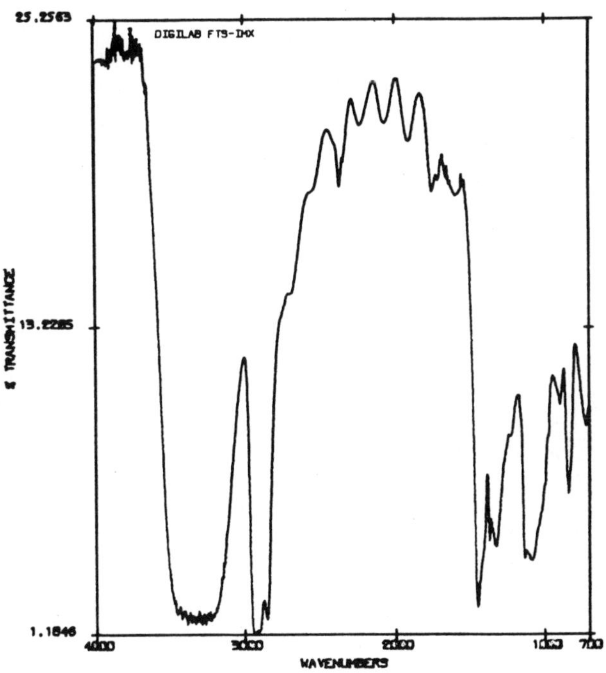

Figure 8b. Microreflectance spectrum from 100 μm x 100 μm area of good paint area.

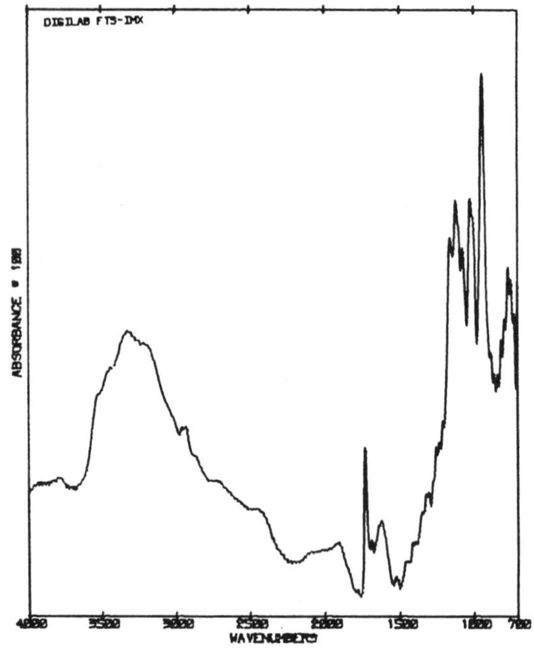

Figure 8c. The difference spectrum of Figures 8a and 8b, (crater - paint).

SUMMARY

It has been shown that the universal FT-IR microsampling technique can be used to characterize a variety of polymer systems. Samples such as microcontaminants in polymers, single polymer fibers etc. which previously were very difficult to handle by FT-IR spectroscopy can be routinely analyzed. Dichroic spetra can be obtained from the microscopic polymer samples to obtain information regarding molecular orientation. The microreflectance technique is especially suited for the study of polymers on metal substrates.

REFERENCES

1. D.H.Anderson and T.E.Wilson, Anal. Chem., 47, 2482 (1975).

2. R.Cournoyer, J.C.Shearer and D.H.Anderson, Anal. Chem., 49, 2275 (1979).

3. M.F.Lacy, Contamination Control Seminar, Anaheim, Calfornia, Oct. 16-19 (1979).

4. Ibid., "Proceedings of the 28th Annual Technical Meeting of the Institute of Environmental Sciences" (1983).

5. K.Krishnan and D.Kuehl, in "Semiconductor Processing ASTM Special Technical Publication 850", D.C.Gupta, Ed., ASTM, Philadelphia (1984) p.325.

6. K.Krishnan and J.R.Ferraro, in "Fourier Transform Infrared Spectroscopy", Vol. 3, J.R.Ferraro and L.J.Basile, Eds., Academic, New York (1982) p.149.

IR-PAS STUDIES: SIGNAL-TO-NOISE ENHANCEMENT AND DEPTH PROFILE ANALYSIS

Richard W. Duerst and P. Mahmoodi

Central Research Laboratories
3M Company
St. Paul, MN 55144

Marilyn D. Duerst

Chemistry Department
University of Wisconsin-River Falls
River Falls, WI 54022

ABSTRACT

The IR-PAS technique is hampered in many instances in its applications by the low signal-to-noise (S/N) ratio, which is generally due to environmental noise present around the equipment. A chamber is described which acoustically isolates the detector, resulting in a S/N ratio enhancement of more than a factor of three. Preliminary studies involving different mirror velocities hint that a phase delay may exist between a given interferometer modulation frequency and the corresponding acoustical wave, a phenomenon which may be applicable to depth profile studies.

INTRODUCTION

Increased utilization of the infrared photoacoustic spectroscopic (IR-PAS) technique for the characterization and identification of a wide variety of samples [1-7] is due to a number of inherent advantages, including the fact that the PAS technique requires essentially no sample preparation and thus avoids those practical problems associated with mulls and the KBr pellet method. In IR-PAS, the signal is generated by the sample, rather than the signal resulting from lack of absorption by the sample. In addition, IR-PAS is inherently capable of sample mapping.

SIGNAL-TO-NOISE STUDIES

The IR-PAS technique is hampered in many instances in its applications by the low signal-to-noise ratio (S/N). An attempt to overcome this disadvantage involved identification of the sources of noise that contribute to the poor S/N ratio, such as structurally borne noise and airborne noise.

Attenuation of environmental noise by acoustical isolation of the detector was sought by designing a chamber to house the photoacoustic detector, constructed of 3/4" plywood and 3/4" fiberglass mat, measuring 12.5"x8.5"x8.0" (see Figure 1). Wire leads were passed through a hole in the chamber and the hole was then filled with about 3/4" of fiberglass wool; about 3/4" of clay was applied to the outside part of the hole. A 1" diameter hole was drilled in the chamber to admit the infrared radiation. For the fiberglass wool used, the noise reduction coefficient (N.R.C.), defined as

$$\frac{\alpha 250\ \text{Hz} + \alpha 500\ \text{Hz} + \alpha 1000\ \text{Hz} + \alpha 2000\ \text{Hz}}{4}$$

with α as the fraction of energy absorbed at the specified frequency, has a value of about 0.7.

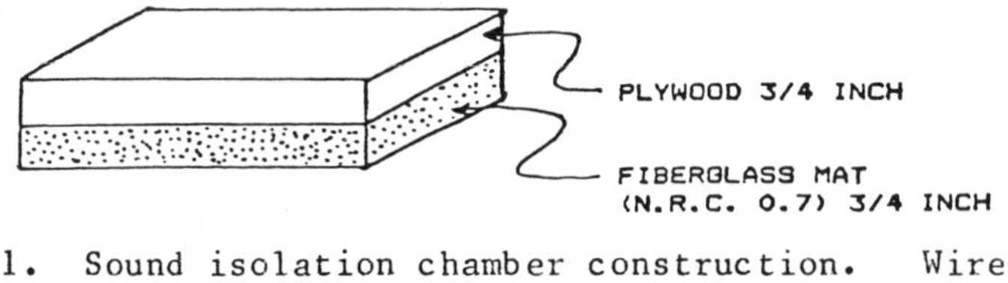

Figure 1. Sound isolation chamber construction. Wire leads through hole fitted with fiberglass wool and clay. 1 Inch hole drilled to admit radiation.

The current configuration employs an EG & G Princeton Applied Research Model 6003 photoacoustic sample cell interfaced to a Nicolet 7199 FT-IR spectrometer.

The degree of attenuation of room noise by the chamber was estimated with the use of a Bruel and Kjaer Type 20 Frequency Analyzer and an HP 3582A Spectrum Analyzer with a B & K No. 4145

1" diameter microphone. This study was undertaken in order to eatablish whether or not specific frequencies should be attenuated. Figure 2 illustrates the decibel level for room noise as a function of frequency for measurements made both inside and outside the isolation chamber. The frequencies measured range from 0 to 2500 Hz. Results show a broad spectral distribution of room noise.

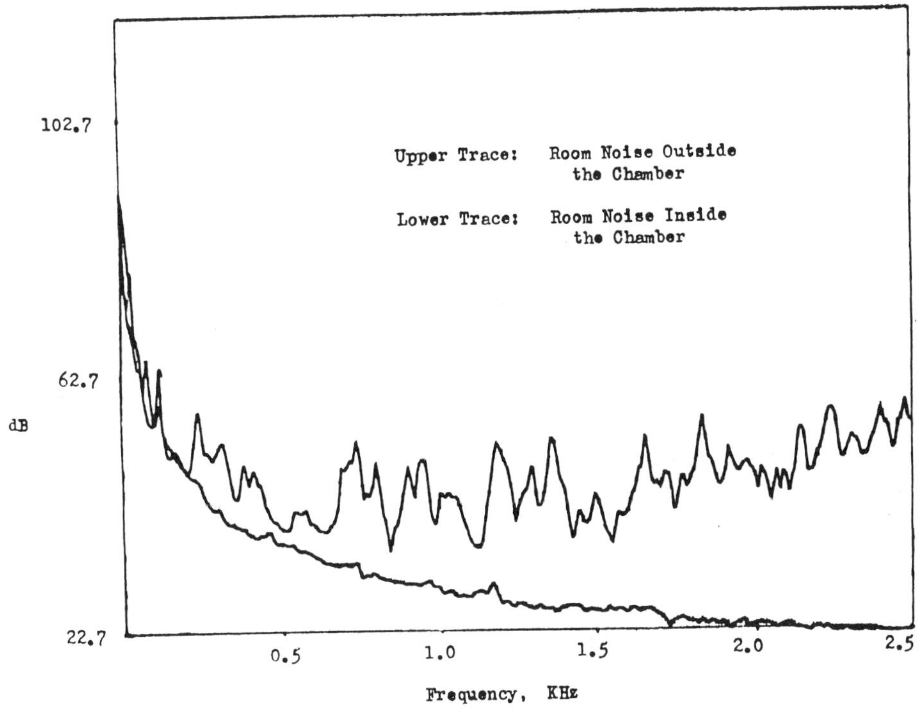

Figure 2. Spectrum analysis of room noise inside and outside chamber.

An additional set of data was used to estimate the S/N ratio enhancement due to the attenuation of room noise by the chamber over a frequency rang from 0 to 1000 Hz. The results are illustrated in Figure 3. If we were to average over the frequency range from 50 to 600 Hz, we would anticipate an enhancement factor of

3.1. For higher mirror velocities, a higher enhancement factor is expected.

In order to demonstrate that use of the chamber involves the attenuation of atmospheric noise rather than structurally borne noise, a preliminary study of a polypropylene sample was undertaken. One and one hundred scans were averaged both with and without the chamber. Two 10,000 scan averages were used to establish a baseline. Spectra without the chamber were taken with the sample cell placed upon the cover of the chamber (also made of plywood and fiberglass mat). At a resolution of about 4 cm^{-1}, the overall S/N ratio enhancement was estimated to be about four to one for a mirror velocity of 0.074 cm/s. (See published spectra in reference 8).

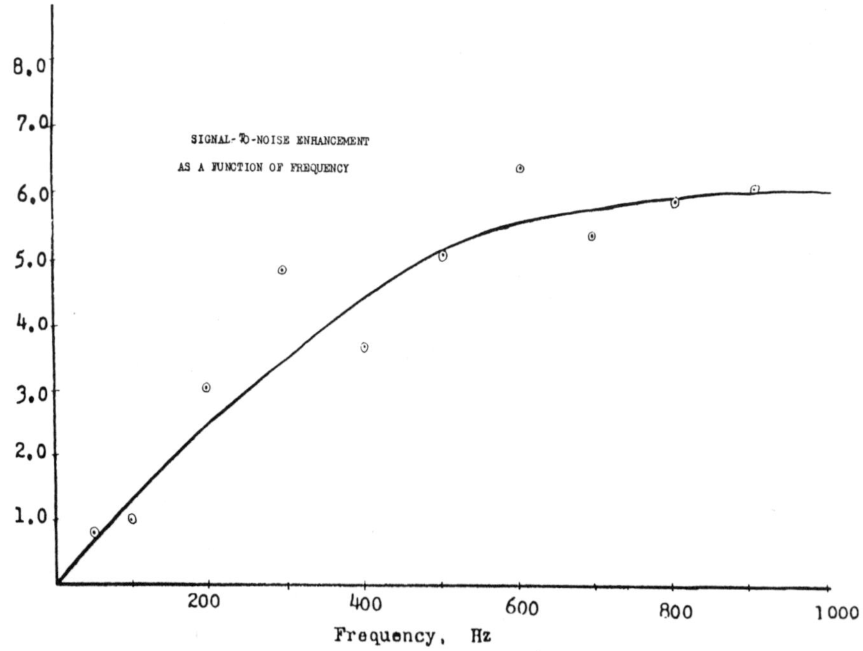

Figure 3. Signal-to-noise enhancement as a function of frequency.

A previously published method [3] to establish the noise level is the "100% line test", which involves ratioing two carbon lampblack (Fisher Scientific Co.) spectra at 8 cm^{-1} resolution taken with 128 scans at a velocity setting of 0.055 cm/s. The sample cell was purged with helium for about five minutes before spectra were taken. A peak-to-peak noise amplitude of approximately 8% T was obtained without the chamber (but, again, resting on the cover); the peak-to-peak noise level dropped to about 2% T with the chamber.

To alleviate the arbitrary nature of the noise level estimation, an algorithm was developed using the digitized spectra in order to determine mathematically the root mean square noise level, N_{RMS}. N_{RMS} is defined as

$$N_{RMS} = \sqrt{\frac{\Sigma(\%T\ (n\ scans) - \%T\ (10{,}000\ scans))^2}{P}}$$

i.e. the square root of the sum of the squares of the difference between the %T for n scans and the %T for 10,000 scans (both ratioed against a different 10,000 scans), divided by the number of points, P, (in this case, 105 points) summed between 1900 and 2100 cm^{-1}. The algorithm was subjected to the constraint that the sum of the difference should be equal to zero, i.e.

$$\sum_{1900\ cm^{-1}}^{2100\ cm^{-1}} (\%T\ (n\ scans) - \%T\ (10{,}000\ scans)) = 0$$

A S/N ratio may thus be determined by representing the ratioed carbon signal, S, as 100% T, and the noise as N_{RMS}. The results obtained with a slow mirror velocity (0.074 cm/s) are as follows (with the peak-to-peak centerburst voltage at about 7 V):

Table 1

	S/N Ratio		
No. of Scans	1	10	100
No Chamber	1.0:1	3.3:1	7.2:1
With Chamber	2.9:1	12.2:1	21.9:1
Enhancement	2.9	3.7	3.0

These results suggest that the effect of atmospheric room noise, which appears to be a major contributer to the poor S/N ratio previously obtained with the IR-PAS technique, may be markedly deminished through the utilization of this chamber.

Current improvements in the design or use of the chamber, such as the addition of a window, changes in the filler gas (e.g. nitrogen versus helium), and other alterations in the configuration of the chamber are being evaluated by the algorithm which calculates the S/N ratio.

DEPTH PROFILE STUDIES

One of the goals of the IR-PAS technique is its anticipated use in depth profile studies. To this end, preliminary mirror velocity studies were undertaken on a sample of poly(ethylene terephathalate) (PET) film, 2 mils thick (51 m). KBr powder was placed below the film in order to scramble any reflections from below. Spectra were taken for mirror velocities ranging from 0.176 cm/s to 0.640 cm/s. Figures 4,5 and 6 illustrate spectra obtained from this sample at three different mirror velocities, i.e. 0.220 cm/s (Figure 4), 0.470 cm/s (Figure 5) and 0.503 cm/s (Figure 6). Note the "negative" carbonyl band (~1730 cm^{-1}) in Figure 5.

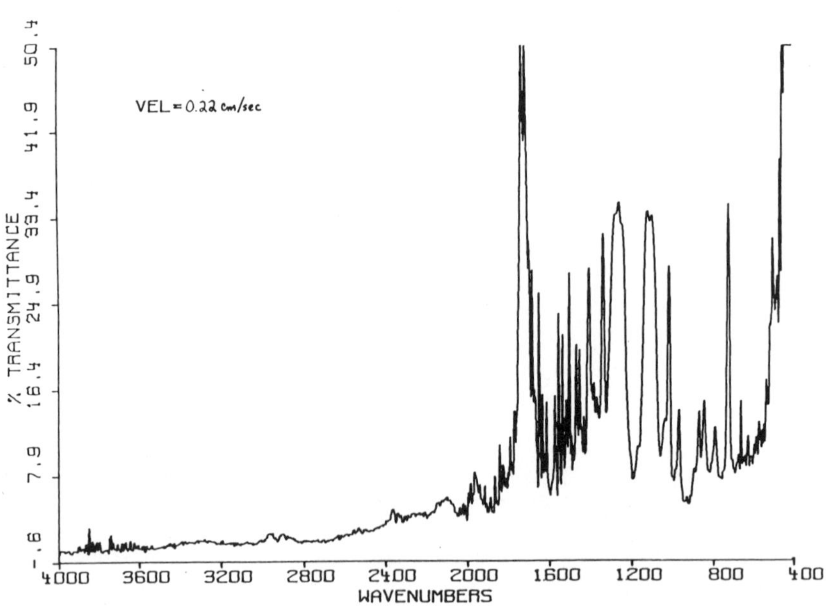

Figure 4. IR-PAS spectrum of PET at a mirror velocity of 0.220 cm/s.

A trend in the intensity of the carbonyl band as a function of mirror velocity may be visualized with the aid of Figure 7. Noteworthy is the suggestion that a minimum in percent transmittance occurs at about 0.47 cm/s for this sample.

We have noted that a "negative" intensity has been observed for the carbonyl band at a mirror velocity of 0.47 cm/s. It may possibly occur at a different mirror velocity when there is a film coating over the polyester substrate. This phenomenon, if present, could be due to a phase delay between a given interferometer modulation frequency and the corresponding acoustical wave. If, indeed, there is a phase delay, then an appropriate phase delay function may allow one to treat the interferogram mathematically before the Fourier transform, and thus actually determine the phase delay for a given band. Preliminary studies are currently in progress on thin coatings on polymeric films.

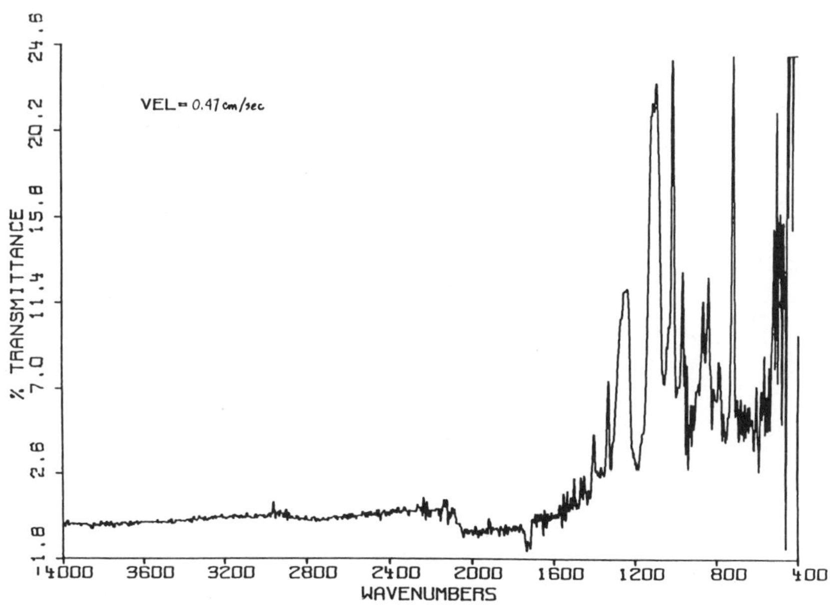

Figure 5. IR-PAS spectrum of PET at a mirror velocity of 0.470 cm/s.

A program is being written based on the following mathematical interpretation of the phase delayed interferogram. With A_f the amplitude of the largest component at a given frequency, θ the phase delay from the largest component, ϕ the phase delay from the mirror generated signal, K_f the exponential decay rate of the amplitude, and f the frequency as the experimental variables, a given frequency contributes the following value to the interferogram at time t:

$$I(f, t) = \sum_{\theta=0}^{\infty} A_f e^{-k_f \theta} \cos (2 \pi f \ (t + \theta + \phi_f)) \ .$$

The entire interferogram is then

$$I (t) = \sum_{f_{min}}^{f_{max}} I \ (f, t).$$

Thus the thickness of a thin coating on the surface of a sample might be determinable by the IR-PAS technique, based on the phase delay phenomenon. A three-dimensional plot of peak intensity vs. wavelength vs. phase delay (depth) could be useful for depth profile studies of polymeric samples.

CONCLUSION

Attenuated total reflectance (ATR) spectroscopy using either a germanium or KRS-5 crystal is also available for surface analysis. Why then study PAS if ATR might be superior? One important reason is the sampling limitations of ATR; one needs a smooth surface for ATR studies, whereas this is not a requirement for PAS. A second reason is that information regarding the layers in a sample is in the interferogram taken at different mirror velocities, and thus should be mathematically extractable for depth profile studies.

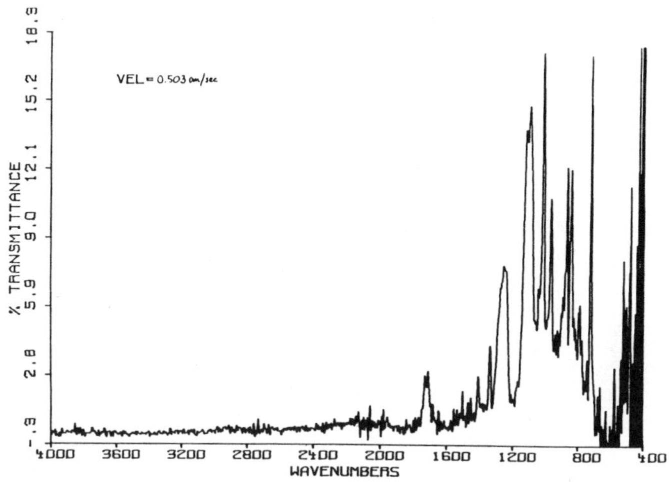

Figure 6. IR-PAS spectrum of PET at a mirror velocity of 0.503 cm/s.

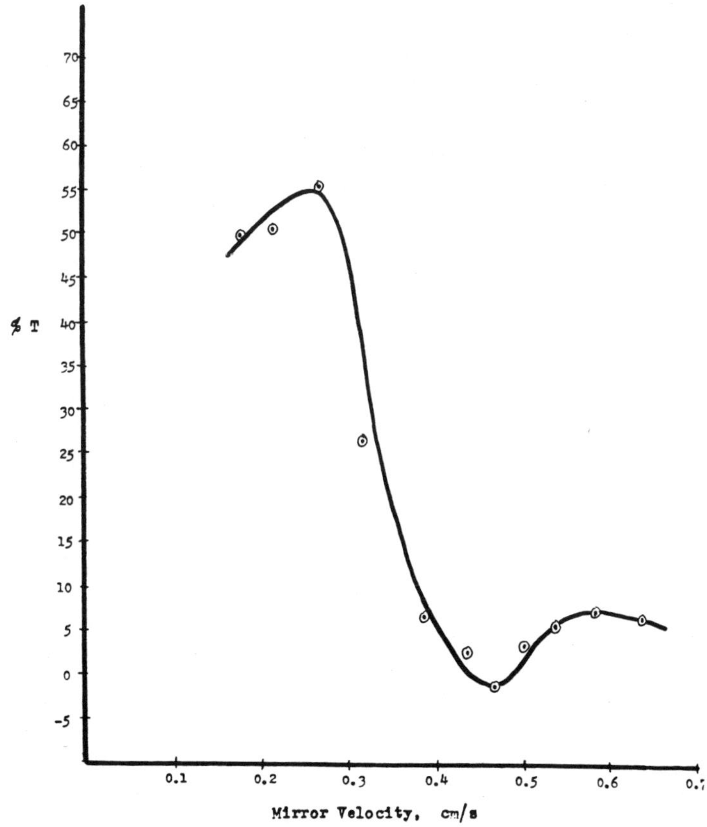

Figure 7 Percent Transmittance for the C=O peak of a PET Film as a Function of Mirror Velocity.

ACKNOWLEDGEMENT

The authors would like to thank R.J.Wann, W.L.Stebbings W.A.Peters, W.E.Breneman, D.Huppler, and P.Griffiths for assistance in this effort.

REFERENCES

1. J.R.Ferraro and L.J.Basile, Ed.,"Fourier Transform Infrared Spectroscopy-Techniques using Fourier Transform Interferometry", Vol.3, Academic, New York (1982).

2. A.Rosencwaig,"Photoacoustics and Photoacoustic Spectroscopy", John Wiley & Sons, New York (1980).

3. J.F.McClelland, Anal. Chem., 55, 89A (1983).

4. A.Rosencwaig and A.Gersho, J. Appl. Phys., 47, 64 (1976).

5. F.A.McDonald, J. Opt. Soc. Amer., 70, 555 (1980).

6. J.F.McClelland and R.N.Kniseley, Appl. Phys. Lett., 28, 467 (1976).

7. J.B.Kinney and R.H.Staley, Ann. Rev. Mater. Sci., 12, 285 (1982).

8. P.Mahmoodi, R.W.Duerst and R.A.Meiklejohn, Appl. Spectrosc., 38, 437 (1984).

RECENT ADVANCES IN RHEO-OPTICAL FOURIER-TRANSFORM INFRARED SPECTROSCOPY OF POLYMERS

H.W. Siesler

Bayer AG, Werk Dormagen
Research & Development
Postfach 100140
D 4047 Dormagen
Federal Republic of Germany

ABSTRACT

In the late seventies rheo-optical Fourier transform Infrared (FT-IR) spectroscopy has emerged as an extremely valuable tool to study deformation and relaxation phenomena in polymeric solids. With the wide-spread availability of FT-IR systems this technique will in the near future become a standard method to monitor structural changes on-line to the mechanical treatment of a polymer. The present contribution is intended to review the state-of-the-art of rheo-optical FT-IR spectroscopy with reference to recent experimental results obtained by this technique.

INTRODUCTION

The increased need for more detailed experimental data and a better understanding of the molecular mechanisms involved in polymer deformation and relaxation has led to the search for new experimental techniques for the characterization of molecular structure during mechanical processes. Primarily the rapid-scanning capability of the FT-IR technique offered the possibility to combine vibrational spectroscopy with mechanical measurements in analogy to other rheo-optical methods [1-6]. Basically, in rheo-optics a mechanical test is combined with one of various types of optical measurements (for example birefringence, light scattering, x-ray diffraction or infrared absorption) and the relation between stress, strain and an optical quantity measured simultaneously with stress and strain as a function of time is established [7,8]. Thus, spectroscopic parameters of the conformation, crystallization and orientation of the polymer under examination can be monitored by rapid-scanning FT-IR spectroscopy simultaneously to the deformation and relaxation processes in very small strain or time intervals relative to the total elongation or time scale of the experiment.

The potential of the technique will be demonstrated with reference to recent studies of the structural changes induced by the mechanical treatment of various thermoplastic and elastomeric polymer systems.

EXPERIMENTAL AND SOFTWARE

Rheo-optical spectra were obtained on a Nicolet 7199 FT-IR spectrometer equipped with a Nicolet 1280 computer. The electromechanical apparatus and the polarizer unit constructed for the simultaneous measurement of FT-IR polarization spectra and stress-strain diagrams during elongation and stress-relaxation of polymer films at variable temperatures have been desribed in detail elsewhere [5,9,10]. Basically, the stretching machine is mounted on an x-y-stage in the sample compartment of the FT-IR spectrometer and the specimen to be tested is held between two clamps which are attached to force and displacement transducers, respectively. For orientation measurements, the polarization direction of the incident radiation can be alternately adjusted parallel and perpendicular to the stretching direction by a pneumatically rotatable polarizer unit which is controlled by the computer of the FT-IR system. The stretching apparatus is also equipped with a variable temperature cell and deformation and stress-relaxation phenomena can be studied under variable temperature conditions.

For the routine analysis of the spectra series acquired during the rheo-optical experiment, the automated information processing capability of the dedicated computer in the FT-IR system is exploited with the aid of specifically developed BASIC software [11]. For the evaluation of the individual spectra taken with light alternately polarized parallel and perpendicular to the direction of elongation, programs are applied which are based on the integral or peak-maximum intensity (with or without automatic peak search) and which automatically calculate the following spectroscopic parameters:

1. The structural absorbance A_o [6,12,13]:

$$A_o = \frac{A_{\parallel} + 2A_{\perp}}{3} \qquad (1)$$

2. The dichroic ratio R [6,12-16]:

$$R = \frac{A_{\parallel}}{A_{\perp}} \qquad (2)$$

3. The dichroic function DF [17]:

$$DF = \frac{R - 1}{R + 2} \qquad (3)$$

4. In the case of well-defined transition moment directions the frequently used orientation function f [6,10,12-17]:

$$f = \frac{(R - 1)\ (R_o + 2)}{(R + 2)\ (R_o - 1)} \qquad (4a)$$

where R_o is the dichroic ratio for perfect uniaxial order [6,1,12-15]. For an absorption band having its transition moment parallel or perpendicular to the chain axis f reads:

$$f_{\parallel} = \frac{R - 1}{R + 2} \qquad (4b)$$

and

$$F_{\perp} = -2\,\frac{R - 1}{R + 2} \qquad (4c)$$

For a detailed discussion of the theory of orientational measurements in polymers using infrared dichroism the reader is referred to the pertinent literature [6,12-17]. The values of the above-mentioned parameters for specified absorption bands of the individual spectra were determined by appropriately correlating the successively measured absorbance values $A_{\parallel}$ and $A_{\perp}$. The structural absorbance has been chosen as intensity parameter because it eliminates the influence of changing orientation on the actual intensity of an absorption band [6,12,13]. Changes in sample thickness during elongation were compensated by comparing against a suitable reference band. The representation of the orientation data in terms of the dichroic function was chosen because of its proportionality to the frequently used orientation function despite the lack of knowledge of the exact transition moment directions of the investigated absorption bands. Upon data processing the individual parameters can be plotted as a function of strain or time in an operator-selected format with a separate software routine [11].

RESULTS

A. Polyethylene

Numerous publications are available in the literature in which infrared spectroscopy has been applied to monitor the orientation of the crystal axes of polyethylene as a function of strain [18-23]. However, only few data have been published where dynamic methods have been employed for this purpose [24-26]. In fact, the rheo-optical FT-IR technique has so far only been applied by two research groups [6,10,27,28] to characterize the crystal-axes orientation and the crystal-phase transformation in high-density polyethylene during elongation up to several hundred percents of strain. Here, the comparative study of high- and low-density polyethylene by this method is reported with reference to the correlation of the macroscopic mechanical properties and the microstructural changes of crystal-axes orientation and phase transformation from the orthorhombic to the monoclinic phase.

On the basis of the well-established band assignment of the infrared spectrum of polyethylene the band doublet at 730/720 cm^{-1} has been utilized to monitor the orientation of the crystallographic a-, b-, and c-axes of the orthorhombic phase relative to the stretching direction by the corresponding orientation functions f_a, f_b, and f_c on-line to the elongation procedure. Thus, it has been shown [19,24] that f_a and f_b are related to the dichroic ratios of the 730 cm^{-1} (B3u) and 720 cm^{-1} (B_{2u}) absorption bands which are polarized along the a- and b-axes, respectively, by

$$f_a = \frac{R_{730} - 1}{R_{730} + 2} \qquad (5a)$$

and

$$f_b = \frac{R_{720} - 1}{R_{720} + 2} \qquad (5b)$$

The orientation function for the crystal c-axis can then be evaluated from the relation [19]:

$$f_a + f_b + f_c = 0 \qquad (5c)$$

The investigated high- and low-density polyethylene films were prepared by blown extrusion and had densities of 0.958gcm^{-3} and 0.921 gcm^{-3}, respectively. From the polarization spectra of the original specimens it could be derived that the b-axis is preferentially aligned parallel to the transverse direction of the film geometry while the a- and c-axes are distributed about the b-axis [29,30]. In different experiments film specimens with gauge dimensions 8x4 mm and a thickness of 0.025 mm (high-density) and 0.055 mm (low-density) were stretched in the machine direction at constant rate of elongation (95% strain/min) at 300°K and 10-scan spectra (resolution 4 cm^{-1}) were taken in about 8%-strain intervals with radiation polarized alternately parallel and perpendicular to the direction of stretch. The stress-strain diagrams recorded at 300°K for the high and low-density polyethylene films are shown in Figure 1. The large differences observable in the stress-strain diagrams are also reflected upon visual inspection of the samples during the mechanical treatment. While high-density polyethylene elongates via formation of a neck which propagates over the entire sample area no significant inhomogeneities could be detected during deformation of the low-density polyethylene specimen. Thus, high-density polyethylene exhibits a substantial decrease of stress beyond the yield point at about 15% strain and subsequently runs through a minimum stress value at

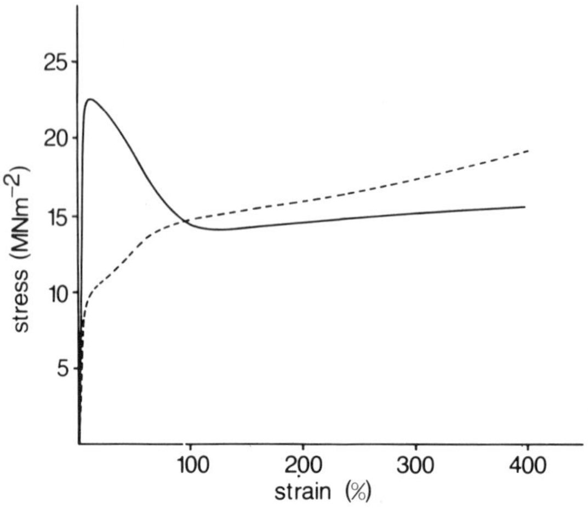

Figure 1 Stress-strain diagrams of high-density (——) and low-density (---) polyethylene measured at 300°K.

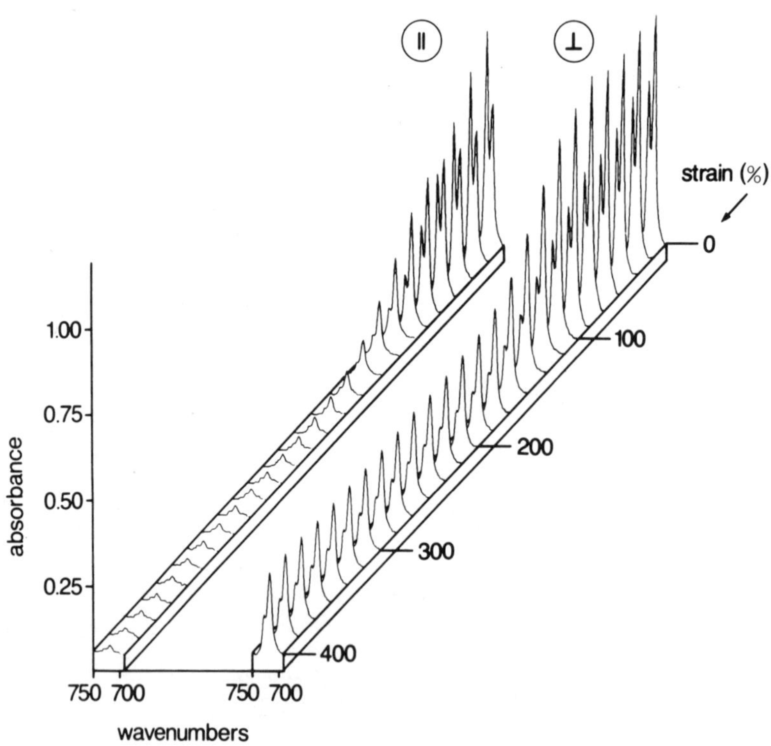

Figure 2 FTIR polarization spectra taken at 300°K during elongation of high-density polyethylene film up to 400% strain.

120% strain whereas low-density polyethylene shows only two diffuse yield points at about 15% and 70% strain. The FT-IR spectra recorded during elongation of a high-density polyethylene specimen up to 400% strain in the 750-690 cm^{-1} region are shown separately for the parallel and perpendicular polarization directions in Figure 2.

To correlate the macroscopic properties and the structural changes occurring during deformation of the two types of polyethylene the orientation functions evaluated from Eqs. 5a-c have been plotted as a function of strain in Figure 3. Onogi and Asada [24] have discussed in detail the expected changes of infrared dichroism and orientation functions due to the different molecular processes of lamellar structure which is stretched in the direction parallel to the b-axis [31,32]. These authors have shown that for an original lamellar geometry with the b-axis oriented predominantly perpendicular to the subsequent direction of stretch orientational changes of the lamellar units as a whole induce an increase of f_b and a decrease of f_a whereas for the unfolding mechanism a decrease of f_a and f_b can be predicted. Despite the differences in the stress-strain diagrams the orientation function-strain curves for high- and low-density polyethylene are very similar with the only exception that the crystal c-axes in high-density polyethylene are finally more perfectly aligned in the stretching direction than those of low-density polyethylene. The orientation function-strain curves for both polymers can be roughly separated into four different regions. Thus, no significant orientation function changes are initially observed in the elastic deformation region. In the subsequent strain interval f_b further remains constant whereas f_a and f_c show a substantial decrease and increase, respectively. In the case of high-density polyethylene the onset of these changes on the strain scale corresponds to the propagation of the neck past the sampling area. In the third region f_b then also decreases and eventually the curves level off when the polymer reaches the final state of fibrillar orientation in which the c-axis is preferentially aligned parallel and the a- and the c-axes are oriented perpendicular to the direction of stretch. While the second region is primarily representative of a rotational motion of the c-axis parallel and the a-axis perpendicular to the direction of stretch the third interval characterizes the predominance of chain-unfolding. Finally, the increase of stress at higher strains may be attributed to a further improvement of the orientation of unfolded chains.

The aforementioned structural changes occurring during uniaxial elongation are additionally superimposed by a crystal phase transformation. Several authors have shown, that under conditions of stress the orthorhombic phase of crystalline polyethylene is partially transformed to a monoclinic structure [6,10,27,28,33,

34]. The progress and extent of this transformation in dependence of strain can be studied with the aid of the CH_2-rocking vibration. Unlike the 730/720 cm^{-1} band doublet of the orthorhombic phase the monoclinic structure is characterized by a single band near 717 cm^{-1} since the correlation splitting is expected to be small. Due to the overlap of the CH_2-rocking vibrations of the orthorhombic and monoclinic phases we have applied absorbance subtraction and band fitting for the separation of the individual contributions of these phases in previous publications [6,10,28]. Here, Fourier self-deconvolution was employed to study the formation of the monoclinic phase during mechanical treatment. Basically, this new method provides a means of computationally resolving overlapped absorption bands that cannot be instrumentally resolved due to their intrinsic line width [35,36].

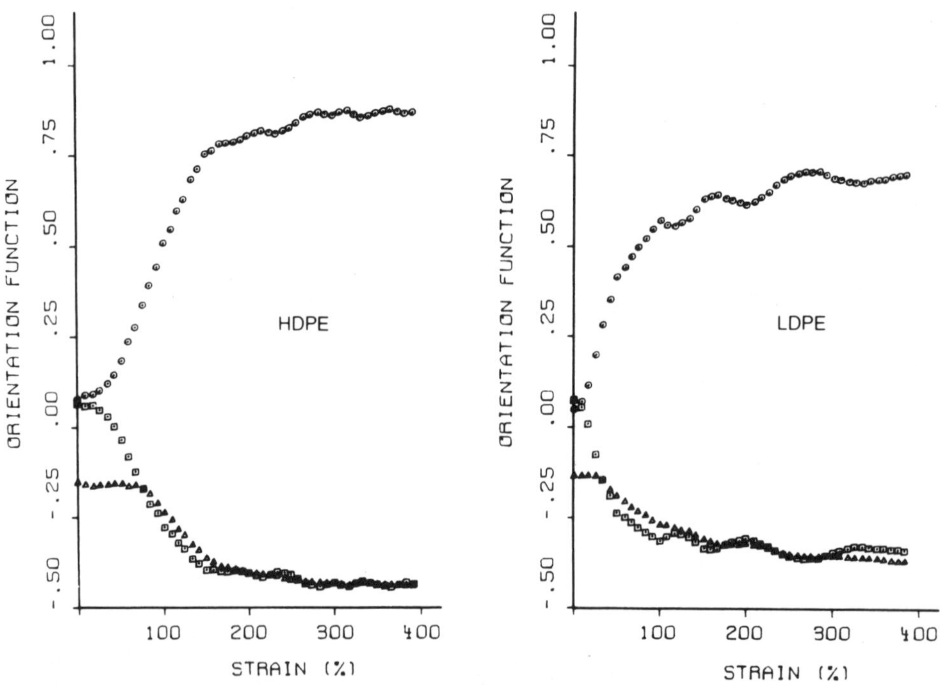

Figure 3. Orientation function/ strain-plots of high-density (HDPE) and low-density (LDPE) polyethylene (f_a: -⊡-, f_b: -△-, f_c: -⊙-).

Since, for a condensed phase sample, the observed spectrum of an infrared vibration usually has a line shape which is close to

Lorentzian, the spectral resolution obtainable is limited by the width of these Lorentzian lines and not by the instrumental resolution. In mathematical terms, the infrared band contour is the result of a convolution of a sharp line with a Lorentzian line shape, and the removal of this line shape to enhance the apparent resolution is called deconvolution. There are two ways of performing the correct deconvolution procedure, working either in frequency space on raw spectral data or in Fourier space on an interferogram [36]. The approach previously mentioned is more efficient, where convolution functions become simple multiplication, in comparison to the frequency space, where a complex function is involved [35,36]. One important feature of deconvolution is that the integrated band area is not altered by the process, which means that quantitative information can be taken from the spectra with enhanced resolution, often leading to improved quantitative data.

For the application of this mathematical procedure to the present problem, rheo-optical FT-IR spectra had to be recorded with the higher resolution of 2 cm^{-1}. In Figure 4a part of the FT-IR spectra taken with unpolarized radiation in 16.6% strain intervals during elongation of high-density polyethylene at 95% strain/min up to 400% are shown alongside their deconvolved analogues in the 750-690 cm^{-1} region. For the automatic deconvolution of a spectra series the deconvolution program has been included in a macro-routine. The program takes an operator-selected region of the absorbance spectrum containing less than 512 data points, creates an interferogram of this region, carries out the deconvolution by multiplying this interferogram by a function and then transforms the data back to give the deconvolved spectrum. The quality of the deconvolution procedure is controlled by two variables, one of which is the half-band width of the Lorentzian line being used for deconvolution and the second is the resolution enhancement achieved [35,36]. The accentuation of the 717 cm^{-1} absorption band in the deconvolved spectra of high-density polyethylene up to 150% strain is demonstrated in Figure 4b. The deconvolution procedure has been applied to both spectra series of high- and low-density polyethylene acquired during elongation to 400% strain and for the characterization of the onset and progress of the formation of monoclinic phase the 717/720 cm^{-1} peak-maximum intensity ratio of the deconvolved spectra has been plotted versus strain in Figure 5. As the most significant difference between the two types of polyethylene, this parameter runs through a maximum in the case of high-density polyethylene at about the strain interval where the neck has propagated across the sampling area. This result is in reasonable agreement with previous data derived from FT-IR studies by the band separation technique on the formation of monoclinic phase during uniaxial elongation of high-density polyethylene [10,28].

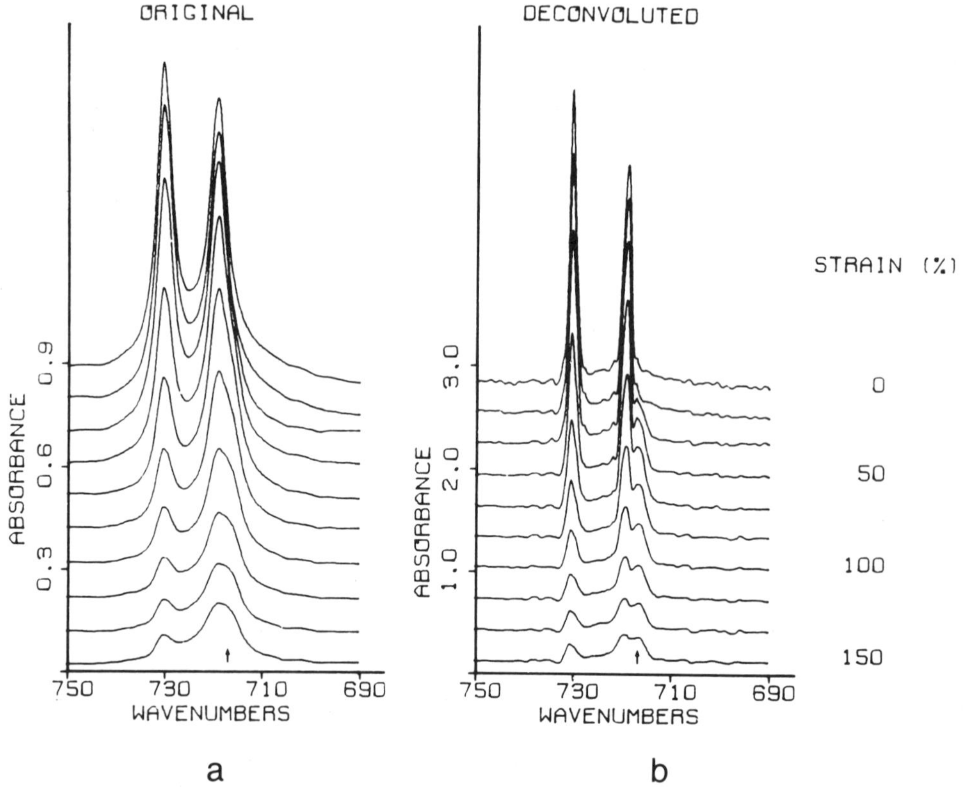

Figure 4 Accentuation of the 717 cm^{-1} CH_2-rocking absorption band (↑) of monoclinic polyethylene by deconvolution. (a) FTIR spectra taken with unpolarized radiation in the 0 - 150% strain region during elongation of high-density polyethylene up to 400% strain. (b) deconvolved spectra (VFO: half bandwith of Lorentzian line being used for deconvolution = 3.2, VF1: resolution enhancement = 2.8).

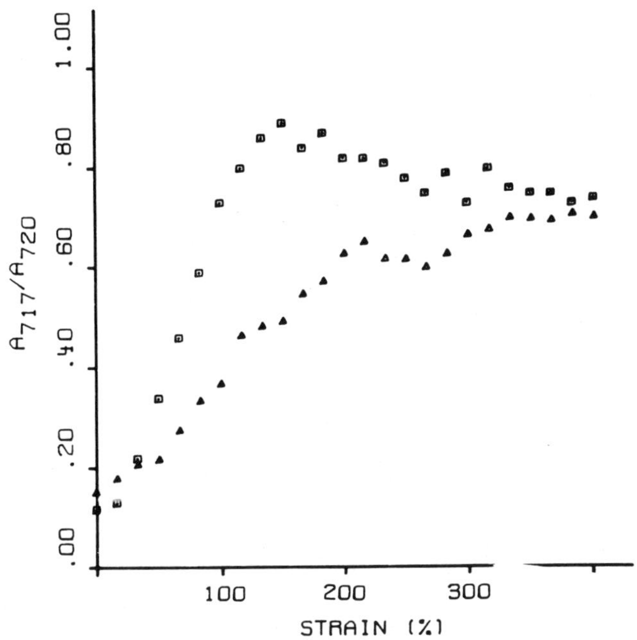

Figure 5a and b 717 cm^{-1}/720 cm^{-1} peak-maximum intensity ratio of deconvolved spectra of Fig. 4b as a function of strain (a: high-density polyethylene, b: low-density polyethylene).

In contrast, the 717/720 cm^{-1} intensity ratio shows an almost gradual increase with strain for the low-density polyethylene specimen. From peak area measurements of the deconvolved spectra a value of about 20% monoclinic phase could be assigned to the 717/720 cm^{-1} intensity ratios of high- and low-density polyethylene at the maximum elongation of 400% strain.

B. Poly(ethylene terephthalate)

Heat-setting of oriented polymeric fibers and films is an important technological operation. There has been an increasing interest understanding the structural changes that take place during drawing or annealing of poly(ethylene terephthalate) (PET) and various techniques including infrared spectroscopy [12,37-64] have been employed. Earlier studies showed [54] that there were considerable differences in the stress-strain behaviour of oriented, crystalline PET yarns which had been prepared by heat-setting the commercial drawn yarn around 473°K in the taut and free conditions. The present rheo-optical FT-IR investigations made on PET film samples prepared via a similar route as the yarn samples were primarily aimed at understanding the transient structural changes that occur in such samples during the stress-strain test [64].

The starting material was a Mylar® PET film obtained from E. I. du Pont Company. The film was semi-crystalline and had biaxial orientation. This PET film was stretched in an Instron tensile tester at 413°K and 25% strain/min to a draw ratio of $\lambda=2.5$. The stretched film was heat-set in a silicone oil bath maintained at 493°K for 15 min under two conditions, viz. while free to relax and when held taut at constant length. The density of the film samples was determined in a gradient column with a carbontetrachloride/xylene mixture. From the measured values of 1.399 gcm^{-3} for the free-annealed and 1.400 gcm^{-3} for the taut-annealed sample weight-percentage crystallinities of about 54% and 55%, respectively, were derived. The film strips used for the rheo-optical FT-IR measurements had gauge dimensions of 20x4 mm and a thickness of about 0.015 mm. The samples were stretched at 300°K with an elongation rate of 33%/min and simultaneously to the recording of the stress-strain curve 6-scan spectra (resolution 4 cm^{-1}) were taken in 2.12-second intervals with light polarized alternately parallel and perpendicular to the drawing direction.

The stress-strain curves obtained for a taut-annealed (TA) and a free-annealed (FA) PET film are represented in Figure 6. In Figure 7, the FT-IR spectra taken during elongation of a free-annealed sample up to fracture are shown separately for the two polarization directions.

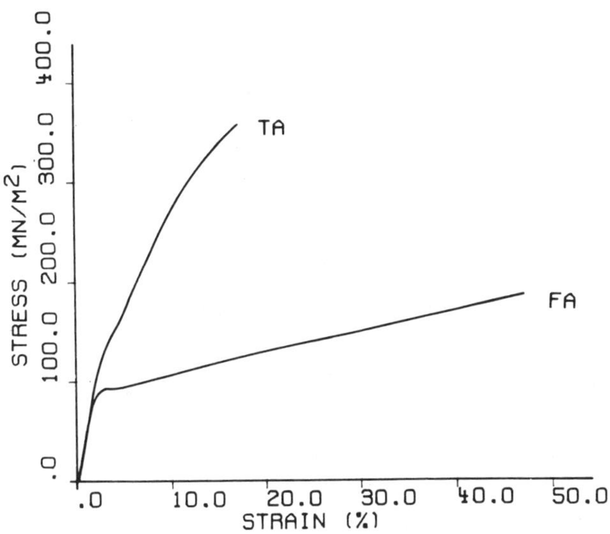

Figure 6. Stress-strain curves of a free-annealed (FA) and a taut-annealed (TA) PET film.

From wide-angle x-ray measurements and the FT-IR data to be discussed below it was found that while crystallinity and crystallite orientation are quite high in both samples, amorphous orientation is initially much lower in the free-annealed sample. PET yarn samples also show similar features [43,44]. Moreover, the arrangement of the crystalline and amorphous regions has been proposed to be different in the two cases [54]. Thus, the stress-strain behavior would be expected to be related to these structural and morphological features. Some comments on the molecular origins of the stress-strain behavior will be made after the data on structural changes occurring during elongation have been presented.

From the polarization spectra series acquired during elongation the following data were derived:

1. Changes in the trans/gauche conformations of the ethylene glycol segments were monitored by the ratio of the structural absorbances of the 848 cm^{-1} (trans) and 898 cm^{-1} (gauche) $\gamma_r(CH_2)$ absorption bands [59,60].

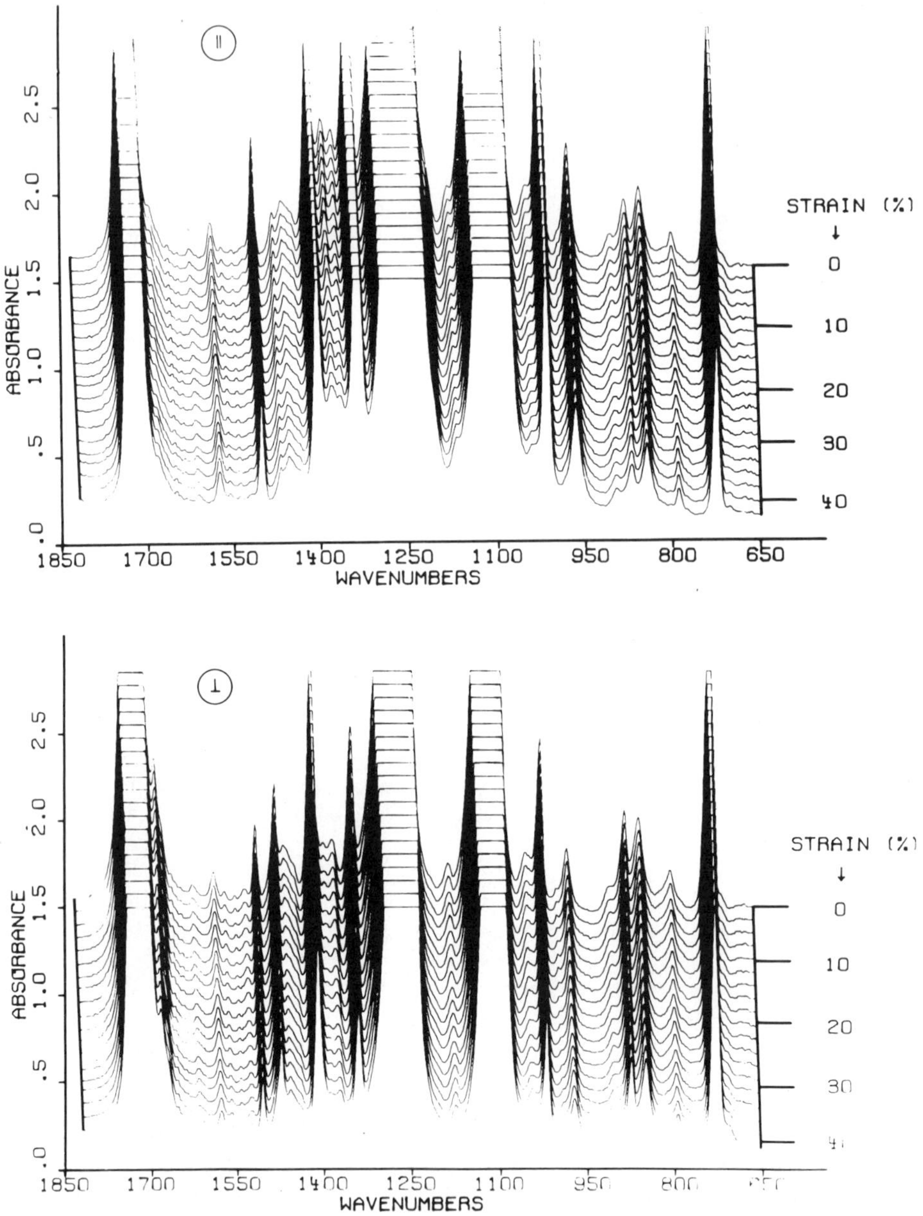

Figure 7 FTIR polarization spectra taken alternatively with light polarized parallel and perpendicular to the stretching direction during elongation of a free-annealed PET film.

2. The transient changes in crystalline and amorphous orientation were derived from the dichroic functions of the 972 cm^{-1} ν(C-O-C) and 1578 cm^{-1} $\nu_{8A}(A_1)$ [12,37,38] absorption bands, respectively.

3. Chain unfolding was detected using the structural absorbance of the absorption band at 988 cm^{-1} [62,63]. The decreasing sample thickness was compensated by the reference band at 1506 cm^{-1} (completely analogous results were also obtained with the frequently used 793 cm^{-1} reference band).

The data on the structural absorbances arising from the changes in trans (848 cm^{-1}) and gauche (898 cm^{-1}) conformations are presented in Figure 8. In both samples an increase in the trans/gauche ratio can be observed in the vicinity of the yield point. While this increase extends up to fracture in the taut-annealed film, elongation beyond 20% strain does not further alter the trans/gauche proportion in the free-annealed specimen. The transient variations in crystalline and amorphous orientation are illustrated in terms of the dichroic functions of the 972 cm^{-1} and 1578 cm^{-1} absorption bands, respectively (Figure 9). Here, the comparatively lower initial orientation of the amorphous regions of the free-annealed relative to the taut-annealed sample becomes obvious. In the free-annealed sample a slight improvement in crystallite orientation takes place below 5% strain while a gradual increase up to fracture is observed for the amorphous regions. The taut-annealed sample, on the other hand, exhibits slight increases in amorphous as well as crystallite orientation over the entire strain range. The data on the structural absorbance of the 988 cm^{-1} band arising from transient changes in the degree of chain folding are shown in Figure 10. It is observed that during elongation this band decreases in intensity in both samples. However, in the taut-annealed sample the decrease takes place at low elongations while in the free-annealed sample it does not occur before 20% strain. It should be noted, however, that due to the extremely low intensity of the 988 cm^{-1} band and its interaction with the adjacent dichroic band at 972 cm^{-1}, the value of the 988/1506 cm^{-1} intensity ratio is only of limited value for a comparative estimation of the degree of chain folding in different samples.

On the basis of the data presented, it appears that in the free-annealed sample there is a substantial increase in the orientation of molecular chains in the amorphous regions throughout the test. Superimposed on this is a slight improvement of crystallite orientation in the initial stages and chain unfolding beyond 20% strain. In the taut-annealed sample, on the other hand, chain unfolding occurs very early during its deformation and

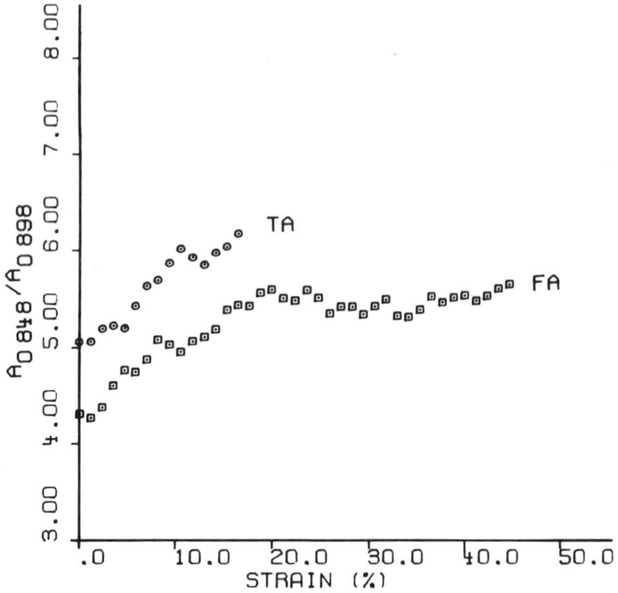

Figure 8 Changes in the structural absorbance ratio of the 848 cm^{-1}/898 cm^{-1} (trans/gauche) absorption bands during the stress-strain test (FA: -⊡-, TA: -⊙-).

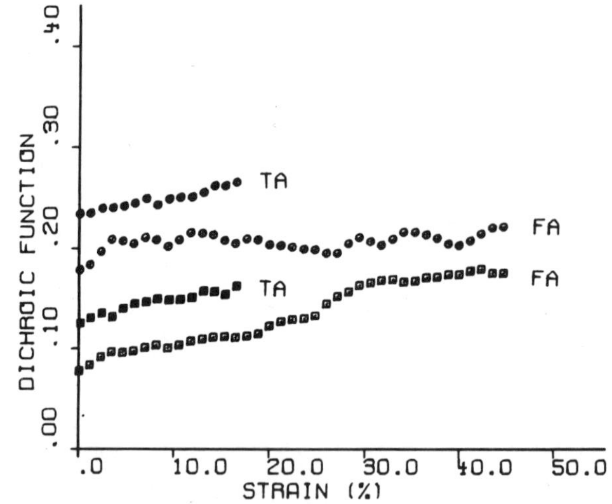

Figure 9 Variations in the orientation of the crystalline and amorphous phases monitored by the dichroic functions of the 972 cm^{-1} (-⊙-) and 1578 cm^{-1} (-⊡-) absorption bands, respectively, during the stress-strain test (PA: open symbols, TA: closed symbols).

a gradual increase in amorphous and crystallite orientation can be observed during elongation.

Detailed studies on PET yarns prepared by similar techniques had indicated [54] that while in the free-annealed samples the crystalline and amorphous phases are quite distinct and are stacked in series, in taut-annealed samples there are no sharp boundaries and the crystallites are distributed within the amorphous matrix. Added to these morphological features is the comparatively low amorphous orientation of the free-annealed sample. Based on these models and the present results, the following picture emerges about the molecular origin of stress-strain curves.

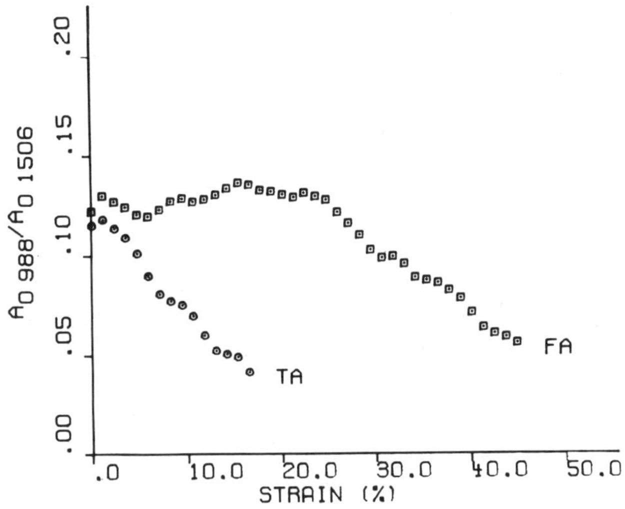

Figure 10. Characterization of chain unfolding during elongation by the 988 cm^{-1}/1506 cm^{-1} structural absorbance ratio (FA: -⊡-, TA: -⊙-).

In the free-annealed sample onsiderable chain uncoiling occurs in the amorphous regions up to the breaking point. Since there is predominantly a series coupling between the crystalline and amorphous regions, the initial improvement in amorphous orientation is also accompanied by a slight improvement in crystallite orientation. By the time the sample has been elongated 20% or more, the molecules in the amorphous regions would have oriented to a considerable extent and longitudinal slip processes can occur, resulting in chain unfolding. Further studies will be required to specify the longitudinal slip process.

In the taut-annealed sample, the initial amorphous orientation is higher and, moreover, since the amorphous and crystalline regions are coupled in series and in parallel, the load-sharing is better. Here, the longitudinal slip processes occur early in the deformation history and result in a concomitant increase of trans conformations. Additionally, a further slight, more uniform, improvement of the initially high amorphous and crystalline orientation is observed.

C. Poly(vinylidene fluoride)

Poly(vinylidene fluoride) (PVF_2) is a polymer with interesting technological and scientific properties. Depending upon the crystallization conditions the polymer can exist in three different phases denoted I(β), II(α) and III(γ). The crystal structures of these three forms, the conditions of their formation and mutual transformation by tensile force, heat treatment or high pressure have been reported in detail by several authors [65-80].

The crystal system of phase I is orthorhombic and its molecular conformation is of the TT-type. Unoriented samples can be prepared by film casting from hexamethylphosphoroamide whereas oriented film specimens of phase I are obtained by stretching melt-crystallized specimens of phase II at room temperature [76,85,89]. Form II can be produced by melt crystallization at atmospheric pressure or by film casting from acetone [76,85,89]. Its crystal system is monoclinic with a $TGT\bar{G}$ molecular conformation. Phase III can be obtained by film casting from dimethylsulfoxide or dimethylacetamide and has a monoclinic crystal system with a $T_3GT_3\bar{G}$ molecular conformation.

Although a number of infrared and Raman studies have been done on the various phases of PVF_2 [69,72,75,81-92] no data on dynamic investigations of the crystal transformation by vibrational spectroscopy are so far available. In view of the well-established band assignments [84,85,87] and the significant differences in the IR spectra of the II(α) and I(β) phases (see also Figure 14), the rheo-optical FT-IR technique seemed particularly attractive to study the II(α)-I(β) crystal transformation under tensile stress as a function of temperature. In this respect, previous rheo-optical wide-angle x-ray measurements [72] had revealed that this II(α)-I(β) crystal transformation occurred below 413°K with the sample deforming by formation of a neck while above this temperature no changes in crystal structure were observed upon application of stress. For a more detailed correlation between the deformation mechanism and the crystal transformation PVF_2 film specimens in the unoriented phase II(α) were subjected to uniaxial elongations up to 400% strain with an elongation rate of 95%/min at 348, 373, 398 and 423°K and 10-scan FT-IR polarization spectra were taken with light alternately polarized parallel and perpendicular to the direction of stretch in 5-second intervals at a resolution of 4 cm^{-1}. The films had been prepared from PVF_2 granulate (kindly supplied by Kureha Chem. Ind., Japan)

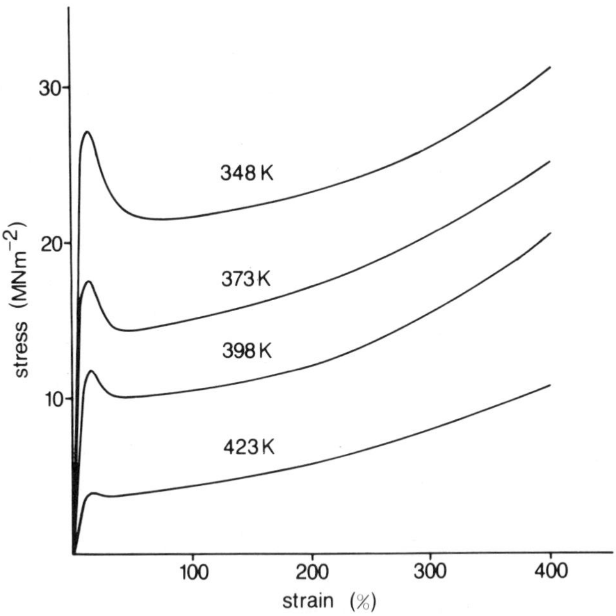

Figure 11. Stress-strain diagrams of unoriented PVF_2 films in the II(α) modification at different temperatures.

by hot-pressing at 483°K and melt crystallization at room temperature. From these original specimens, film strips with a thickness between 0.030-0.035 mm and gauge dimensions of 8x4 mm were prepared for the rheo-optical FT-IR investigations. Wide-angle x-ray diagrams were taken at room temperature of the original film specimen and of the samples drawn at different temperatures up to 400% strain in the stress-relaxed state. From the intensities of the combined (100)/(020) and (110) reflections of phase II(α) at about 2 =18° and 20°, respectively, and the combined (200)/(110) reflections of phase I(β) at about 2Θ=21° conclusions in terms of the effect of the mechanical treatment on the relative amounts of the two crystal forms could be drawn [72,78,80].

The stress-strain diagrams of the individual experiments at different temperatures are shown in Figure 11. Apart from the drastic reduction of the stress level with increasing temperature the most obvious feature is the almost complete disappearance of the stress-decreasing region beyond the yield point. The corresponding wide-angle x-ray diagrams are presented in Figure 12

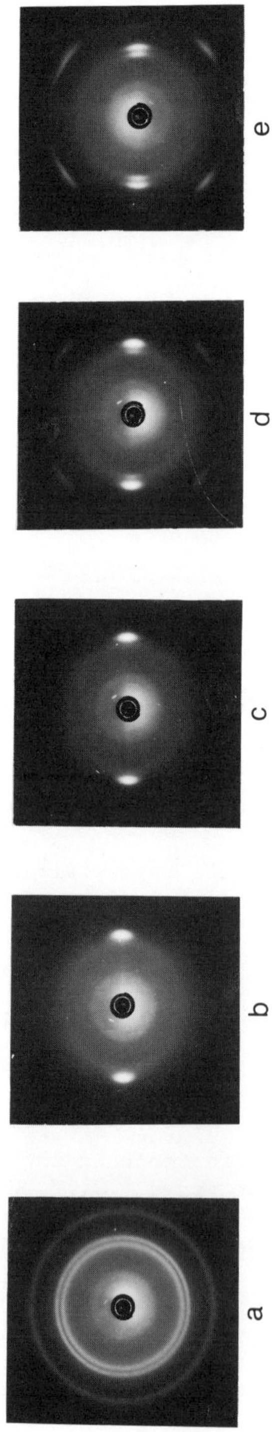

Figure 12 Wide-angle x-ray diagrams of the original PVDF film in the unoriented II (α) modification (a) and of the stress-relaxed samples elongated to 400% strain at 348°K (b), 373°K (c), 398°K (d) and 423°K (e).

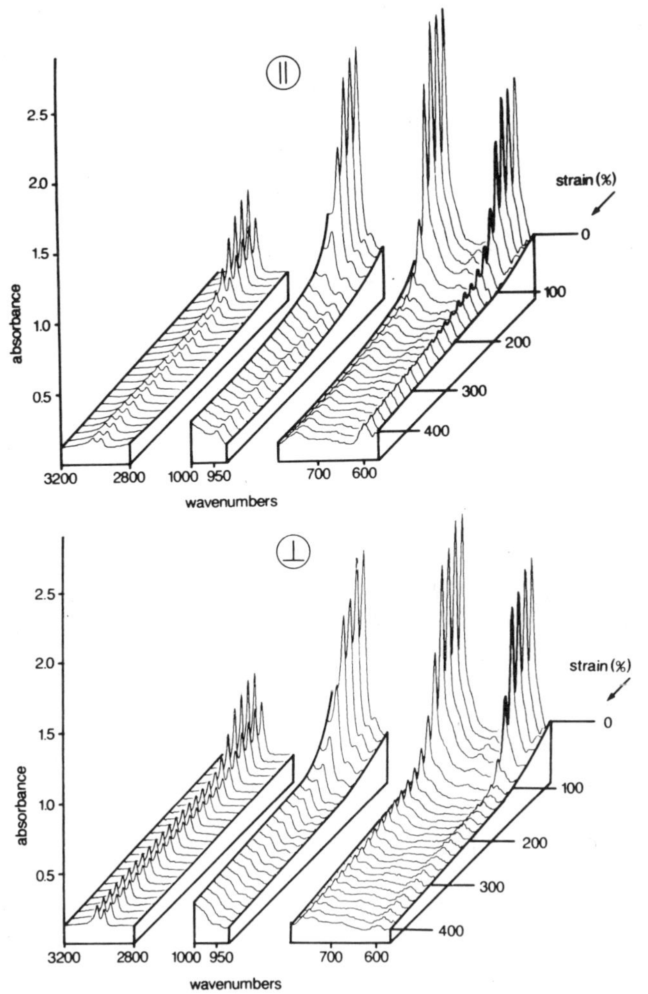

Figure 13 FTIR polarization spectra taken during elongation at 348°K of a PVDF film with an initial II (α) crystal structure.

and reflect the decreasing tendency of crystal II(α)-I(β) transformation with increasing temperature. The wavenumber regions of specific interest for the investigated crystal transformation of the polarization spectra taken at 348°K as a function of strain are shown in Figure 13. In Figure 14 the structural absorbance spectra corresponding to 0% and 400% elongation at 348°K, respectively, clearly indicate the complete disappearance of phase II(α) as a consequence of uniaxial elongation at this temperature. This result is also supported by the absence of the (100)/(020) and (110) reflections of the II(α) modification in the wide-angle x-ray diagram of Figure 12b. The approximate normal mode assignment of the absorption bands which are characteristic of the phase II(α) is given in Table 1.

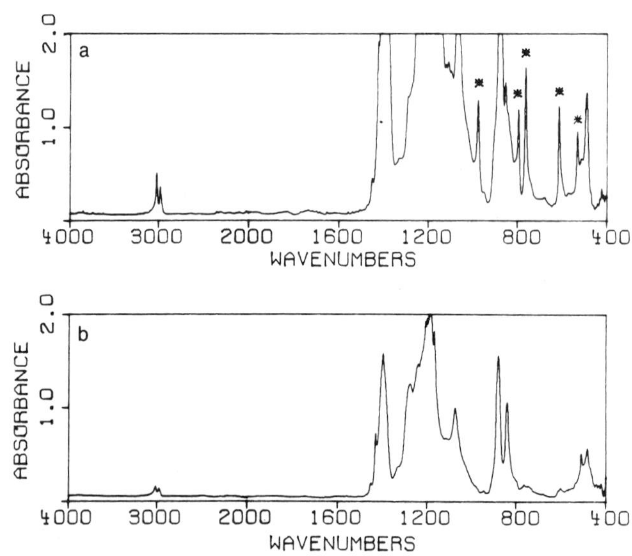

Figure 14. Structural absorbance spectra of PVF_2 elongated at 348°K (a) 0% strain, modification II(α), (b) 400% strain, modification I(β).

For the detailed representation of the structural changes during elongation at different temperatures the structural absorbance of the 975 cm^{-1} absorption band of the phase II(α) has been ratioed against the structural absorbance of the $\nu_a(CH_2)$ thickness reference band at 3022 cm^{-1} and plotted as a function of strain in Figure 15a-d. While regions with a drastic drop in the intensity ratio are observable at 348°K and 373°K, a more or less linear decrease occurs at 398 K and no significant changes take place at 423°K. Thus, according to the FT-IR measurements at 348°K, phase II has been completely transformed to phase I upon elonga-

TABLE I

Approximate normal mode assignment of some characteristic absorption bands of form II (α) of PVDF [85,87]

Frequency (cm^{-1})	Polarization	Approximate normal mode assignment
3022	$\perp$	$\nu_a(CH_2)_i$
2984	$\perp$	$\nu_s(CH_2)_i$
975	$\perp$	$\tau(CH_2)_i + \nu_s(CF_2)_i$
796	$\perp$	$r(CH_2)_i$
763	$\perp$	$\nu_s(CF_2)_o + w(CF_2)_o$
614	$\parallel$	$w(CF_2)_i + \delta(CF_2)_i + \delta(CCC)$
532	$\perp$	$\delta(CF_2)_o$

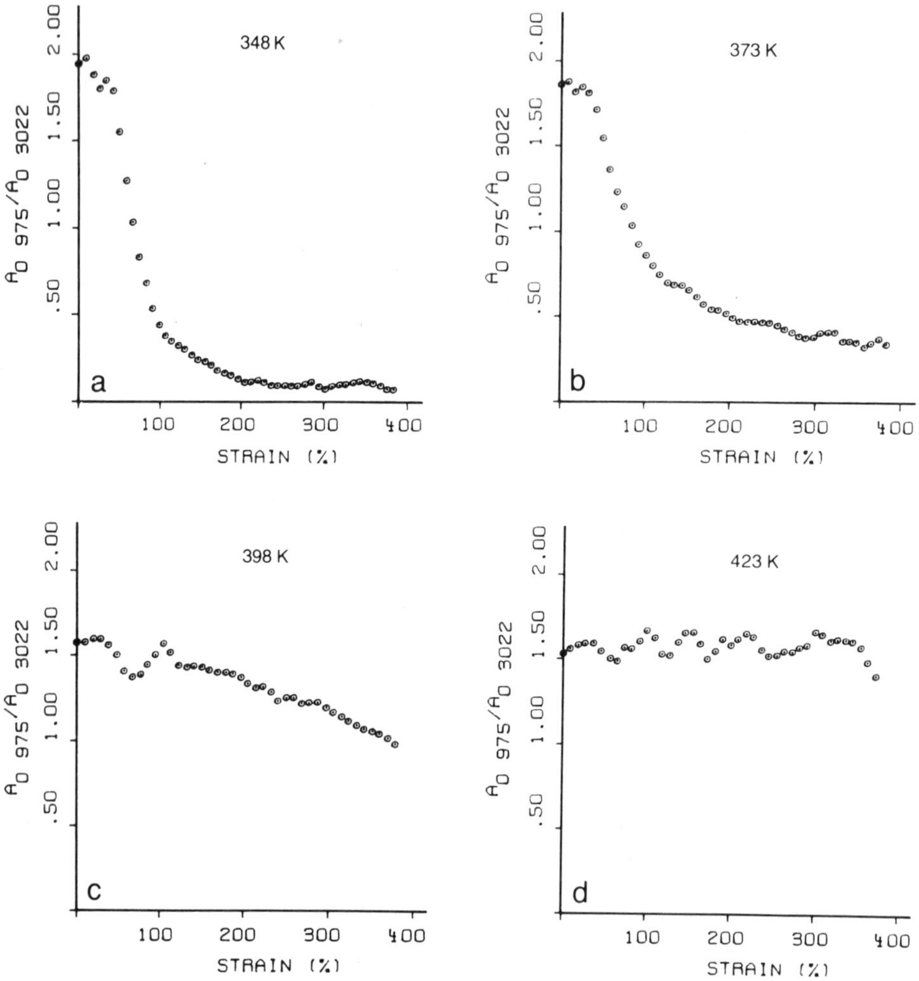

Figure 15 Variations in the structural absorbance ratio A_{o975}/A_{o3022} as a function of strain at different temperatures (a) 348°K, (b) 373°K, (c) 398°K and (d) 423°K.

tion to 400% strain, whereas small amounts of phase II seem to have been retained at 373°K. On the other hand, an appreciable proportion of phase II has resisted the mechanical treatment at 398°K and deformation at 423°K only leads to an orientation of phase II with no concomitant transformation into phase I (see also Figure 12a-e). From a separate rheo-optical experiment at 348°K up to 100% strain only, it could be shown that the steep decrease of the intensity ratio observed between 50 and 100% strain at 348°K and 373°K is a consequence of the propagation of the neck past the sampling area in the IR beam. Thus, at 100% strain, the sample (Figure 16a) consists of a neck region (B) which according to wide-angle x-ray diffraction (Figure 16b) and IR spectroscopy (Figure 16c) has already been transformed to a large degree into the phase I(β) and a thicker, unoriented region (A) which is almost completely retained in the phase II(α). The initiation of the crystal transformation is therefore based on a heterogeneous stress distribution during neck formation whose decreasing influence with increasing temperature is reflected in an enhancement of the proportion of retained phase II(α). Further research with regard to a more quantitative evaluation of the spectroscopic data in combination with wide-angle x-ray measurements is in progress.

D. Natural Rubber

Apart from the degree of crosslinking the mechanical properties of a polymer having network structure are strongly influenced by strain-induced crystallization [93-104]. This phenomenon is of great practical importance both during processing [99] and with regard to the tehnological properties of the product such as tear strength and maximum extensibility [100]. A schematic representation of strain-induced crystallization within a polymer network which has been elongated by a force in the specified direction is shown in Figure 17. Crystallites thus formed act as crosslinks of high functionality and since they are nondeformable at the stress levels involved, diminish the amount of elastomeric material able to respond to the imposed stress [93,100]. Additionally, the crystallites act as filler particles which generally increase the modulus of a rubber-like material [100,101]. In the stress-strain diagram, strain-induced crystallization affects a drastic increase of elastic force with strain as a consequence of a significant self-reinforcement of the elastomer during elongation [93,95,97,99-103].

The stress-strain data are customarily represented using the semi-empirical equation of Mooney and Rivlin [105,106]:

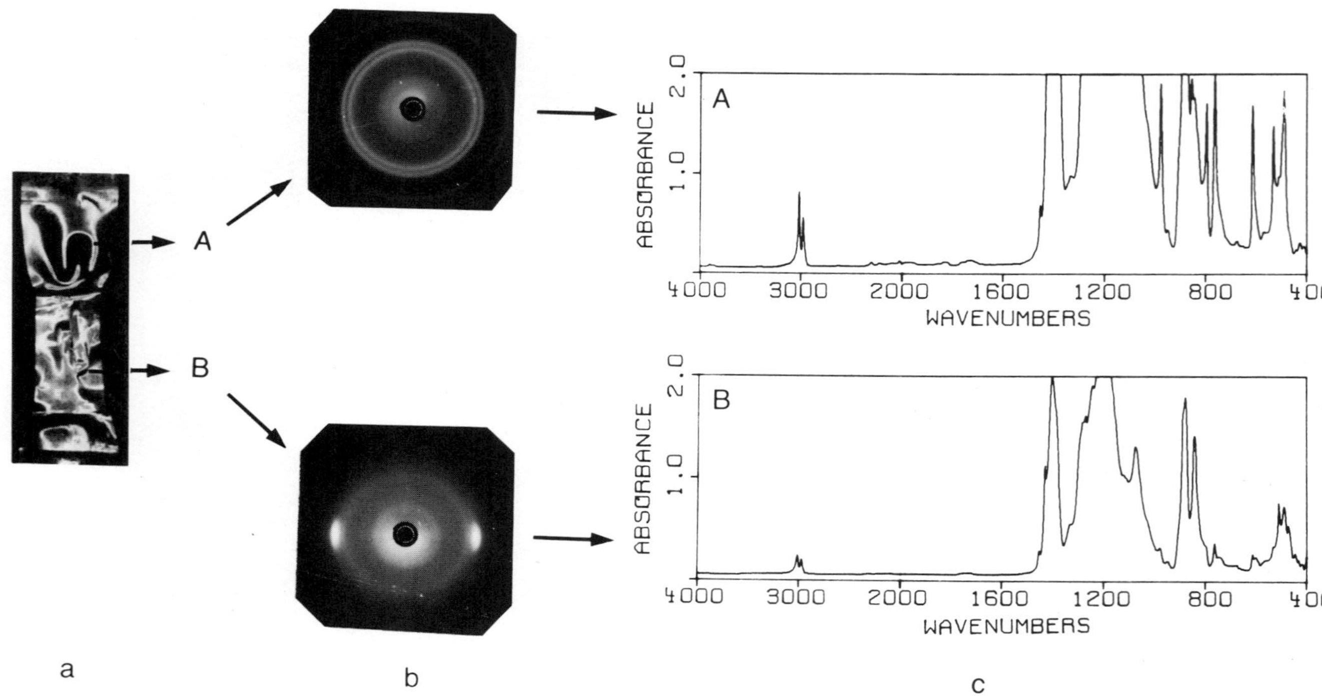

Figure 16 Heterogeneous stress distribution during elongation of a PVDF film in the II(α) form at 348°K

(a) stress-relaxed sample after elongation to 100% strain

(b) wide-angle x-ray diagrams of the two regions A and B

(c) structural absorbance spectra of region A and B

$$[f^*] = 2 \cdot C_1 + 2 \cdot C_2 \cdot \lambda^{-1} \tag{6}$$

in which C_1 and C_2 are constants independent of the elongation ratio λ and $[f^*]$ is the reduced stress or modulus [107-109]. A linear relationship between $[f^*]$ and λ^{-1}, however, holds only at small elongations and an upturn in the reduced stress occurs at small reciprocal elongations less than 0.4. The deviation from the linearity is controversially interpreted on the one hand by the limited chain extensibility [95,110-112] and on the other hand by strain-induced crystallization [97,98,100].

In this context, rheo-optical FT-IR spectroscopy has proved a valuable technique to study strain-induced crystallization on-line to the deformation process of the elastomer under examination. Thus, the onset and progress of strain-induced crystallization during elongation and its disappearance upon recovery can be readily monitored simultaneously to the stress-strain measurements by the intensity of an absorption band which is characteristic of the crystalline phase. Here, the results obtained by the rheo-optical FT-IR technique with sulfur-crosslinked (1.8% S) natural rubber (100% 1,4-cis-polyisoprene) in cyclic elongations will be discussed in some detail.

In the mechanical treatment the film sample with gauge dimensions 10x4 mm and a thickness of 0.0950 mm was subjected to three successive loading-unloading cycles up to 545%, 545% and 650% strain, respectively, with an elongation and recovery rate of 85% strain per minute and 15-scan spectra were taken with light alternately polarized parallel and perpendicular to the direction of stretch at 300°K in about 8-second intervals with a resolution of 4 cm^{-1}. The total number of FT-IR spectra acquired simultaneously to the three loading-unloading cycles was 300.

The stress-strain diagram corresponding to this mechanical treatment is shown in Figure 18. In the first cycle the wide-angle x-ray diagrams of the stress-relaxed samples taken at 200% and 500% strain, respectively, have been included to demonstrate the change in the state of order during elongation. The polarization spectra monitored during the elongation-recovery cycles are shown separately for the two polarization directions in Figure 19. Based upon previous absorbance subtraction procedures [113], the 1126 cm^{-1} crystallinity-sensitive absorption band which has been assigned to a $C\text{-}CH_3$ in-plane deformation vibration [114-116] was used in conjunction with the $\nu(C{=}C)$ absorption band at 1662 cm^{-1} as thickness reference band to monitor the onset, extent and disappearance of strain-induced crystallization during the mechanical

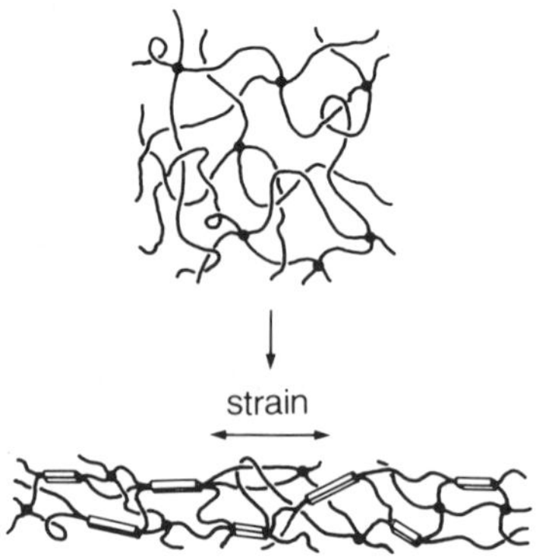

Figure 17 Schematic of strain-induced crystallization in a cross-Stitched elastomer.

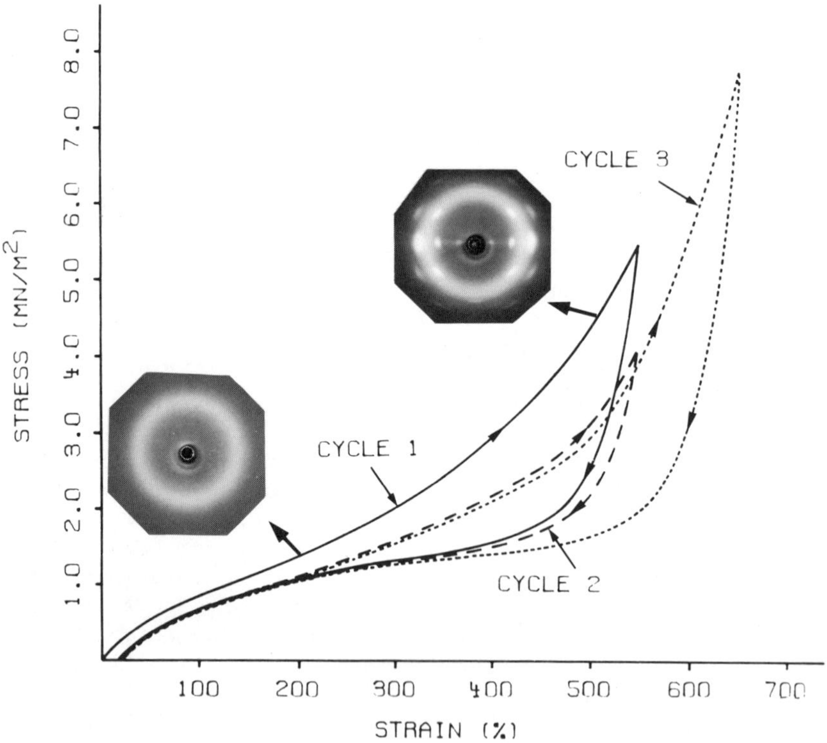

Figure 18 Stress-strain diagram of three successive loading-unloading cycles of sulfur-crosslinked natural rubber at 300°K.

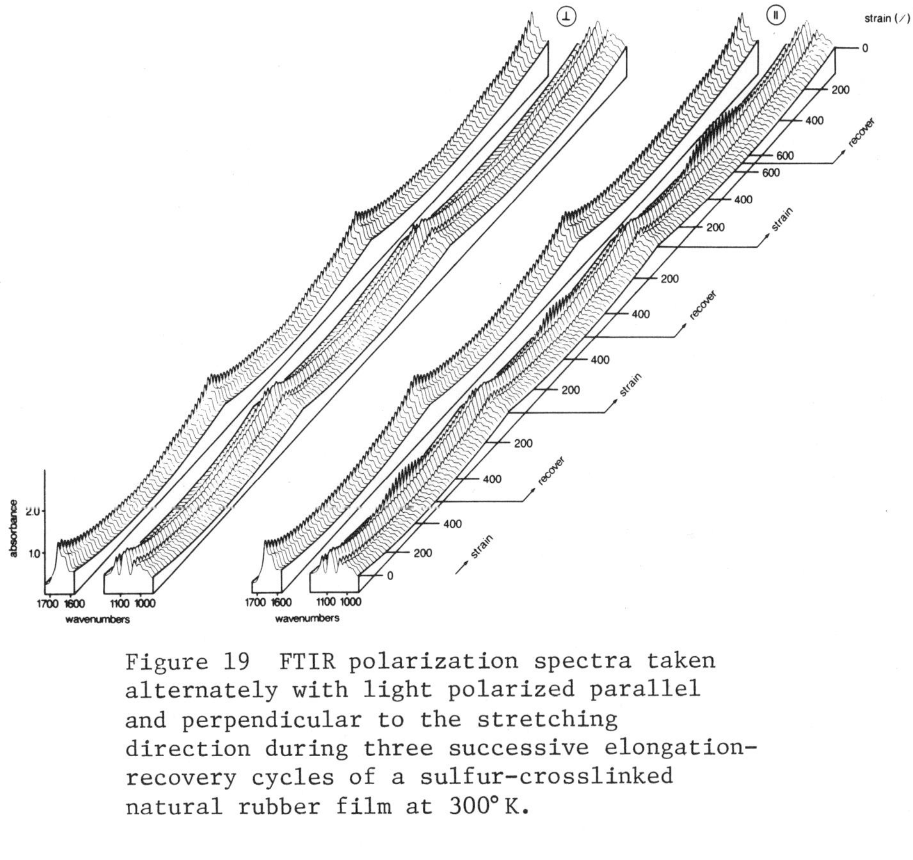

Figure 19 FTIR polarization spectra taken alternately with light polarized parallel and perpendicular to the stretching direction during three successive elongation-recovery cycles of a sulfur-crosslinked natural rubber film at 300° K.

treatment. It should be mentioned, that in actual fact the data derived from the 1126 cm^{-1} band refer to the conformational regularity of the individual polymer chains, which is, however, a necessary prerequisite for three-dimensional order. Thus, the ratio of the structural absorbances of the 1126 cm^{-1} and 1662 cm^{-1} bands was evaluated with the above mentioned software routine for the spectra taken during the loading-unloading cycles and plotted as a function of strain in Figure 20. The data clearly demonstrate the reversible nature of strain-induced crystallization and allowed a strain value of about 230% to be assigned for the onset of crystallization in the first cycle [113]. Additionally, a significant retention of the strain-induced crystallinity relative to the loading half-cycle is observable during recovery. A further increase of the maximum degree of crystallization relative to the preceding cycle is observed only when the maximum elongation of a loading-unloading cycle exceeds the corresponding maximum elongation of the preceding cycle (see Figures 18 and 20, cycle 3).

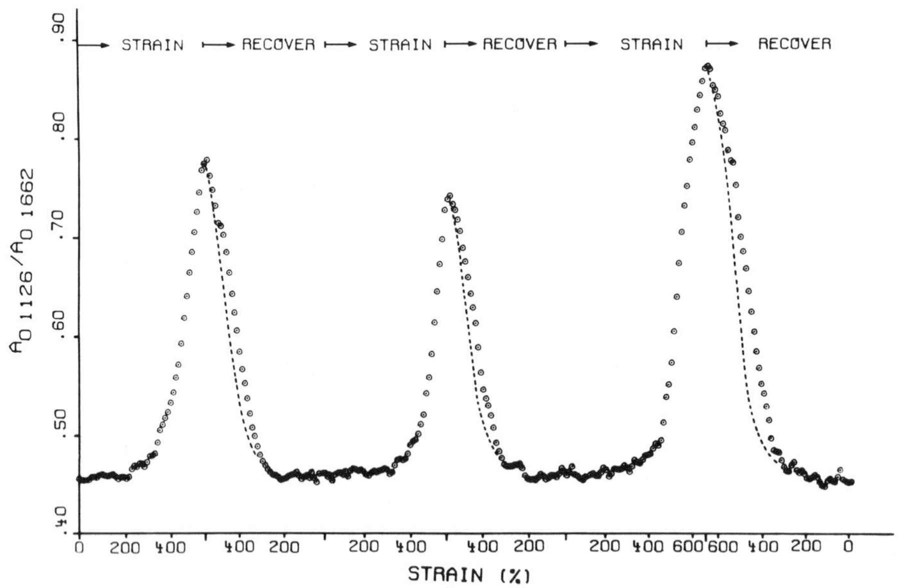

Figure 20. Plot of the structural absorbance ratio $A_{o\,1126}/A_{o\,1662}$ as a function of strain for three successive loading-unloading cycles of a natural rubber film at 300°K.

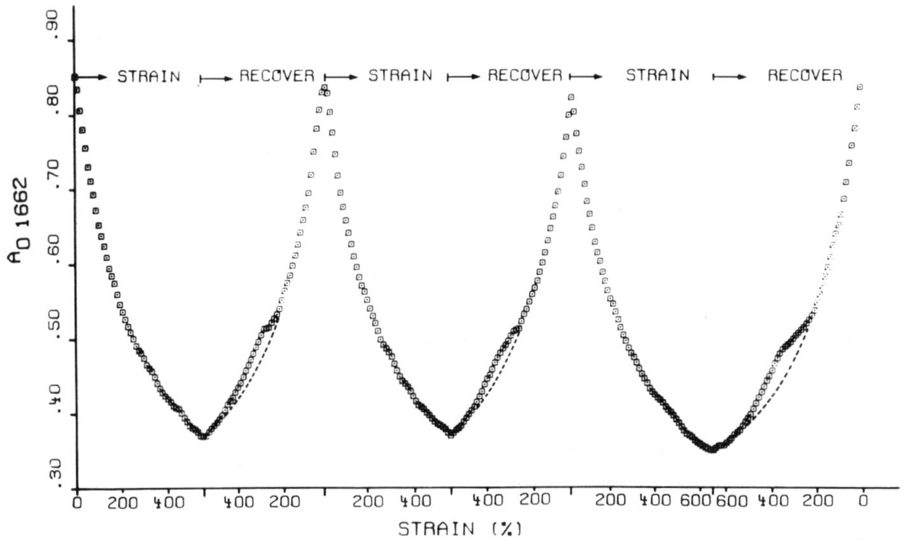

Figure 21 Structural absorbance/strain-plot of the 1662 cm^{-1} thickness reference band of sulfur-crosslinked natural rubber monitored at 300°K in three successive elongation-recovery cycles.

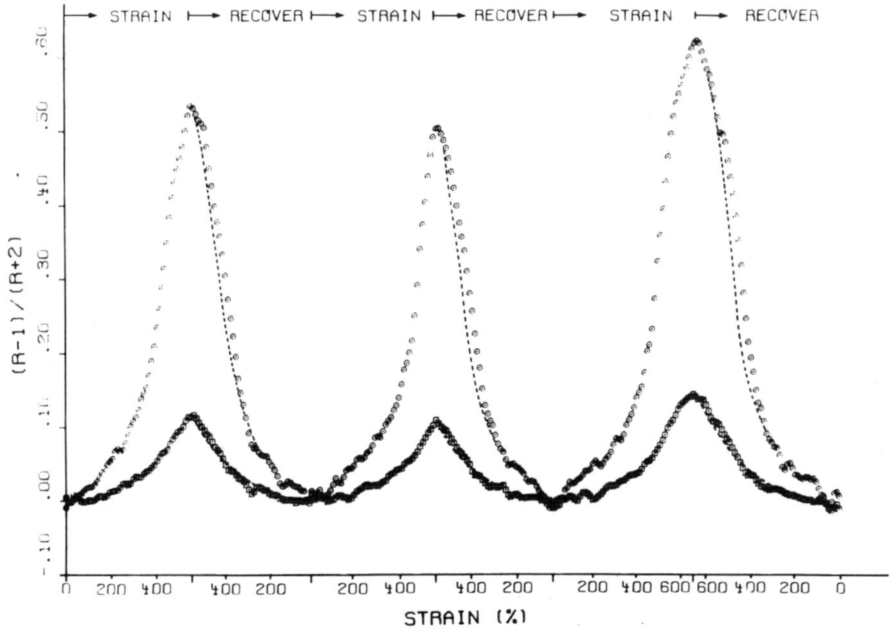

Figure 22 Dichroic function/strain-plot of the 1126 cm^{-1} (⊙) and 1662 cm^{-1} (▣) absorption bands for three successive loading-unloading cycles of a natural rubber film at 300°K.

An interesting effect becomes evident when the structural absorbance of the ν(C=C) thickness reference band is plotted as a function of strain (Figure 21) for the loading-unloading cycles. Thus, a distinct asymmetry in the recovery of the original sample geometry could be detected whenever the polymer under examination exhibited extensive strain-induced crystallization. The phenomenon may be correlated with the composite nature of such a strain-crystallized system and can be interpreted in terms of a preferential recovery of the thickness relative to the transverse dimension in the stress-strain plateau region of the unloading half-cycle between about 400% and 200% strain.

To monitor the orientation of the crystalline phase and the average polymer in the elastomer under investigation, the dichroic functions of the 1126 cm^{-1} and 1662 cm^{-1} bands, have been plotted versus strain in Figure 22. The representation of the spectroscopic data in this form has been chosen bacause no values for the transition moment directions of these absorption bands were available from the literature. Nevertheless, the onset of crystallization and its retention during unloading as well as the enhancement of chain-alignment in the strain-crystallizing relative to the amorphous regions are readily reflected from Figure 22. In a previous detailed orientation analysis [113], it has been shown that, based upon the assumption of a zero transition moment angle for the 1126 cm^{-1} band (the dichroic function is then equivalent to the orientation function of this absorption band), average inclination angles of approximately 40° and 30° can be calculated for the average polymer and the polymer chains in the conformationally regular phase, respectively, and the direction of stretch in the 650 % drawn sample.

Rheo-optical FT-IR spectroscopy, therefore, not only provides a means to monitor strain-induced crystallization on-line to the mechanical treatment but also yields detailed information in terms of the orientation of the polymer chains in the amorphous relative to those in the strain-crystallizing domains.

ACKNOWLEDGEMENTS

The author gratefully acknowledges helpful discussions with Dr. U.Eisele, Prof. Dr. V.B.Gupta and Dr. G.Spilgies and thanks Bayer AG for the permission to publish the experimental data.

REFERENCES

1. H.W.Siesler, J. Polym. Sci. Poly. Lett. Ed., 17, 453 (1979).

2. H.W.Siesler, in "Proceedings of the 5th European Symposium on Polymer Spectroscopy", D.O.Hummel, Ed., Verlag Chemie, Weinheim (1979) p.137.

3. H.W.Siesler, J. Mol. Struct., 59, 15 (1980).

4. W.Stach and K. Holland-Moritz, J. Mol. Struct., 60, 49 (1980).

5. H.W.Siesler, Polym. Prepr. Amer. Chem. Soc., Div. Polym. Chem., 21, 163 (1980).

6. H.W.Siesler and K.Holland-Moritz,"Infrared and Raman Spectroscopy of Polymers," Marcel Dekker, New York (1980).

7. D.G.Le Grand and P.F.Erhardt, Trans. Soc. Rheol., 6, 301 (1962).

8. R.S.Stein, J. Polym. Sci., C15, 185 (1966).

9. H.W.Siesler, Polym. Bull., 9, 382 (1983).

10. Ibid., Adv. Polym. Sci., 65, 1 (1984).

11. H.W.Siesler and H.P.Schlemmer, unpublished results.

12. J.Dechant,"Ultrarotspektroskopische Untersuchungen an Polymeren", Akademie Verlag, Berlin (1972).

13. A.Peterlin, Ed.,"Plastic Deformation of Polymers", Marcel Dekker, New York (1971).

14. R.Zbinden,"Infrared Spectroscopy of High Polymers", Academic, New York (1964).

15. B.Jasse and J.L.Koenig, J. Macromol. Sci. Rev. Macromol. Chem., C17, 61 (1979).

16. R.J.Samuels,"Structured Polymer Properties", Wiley-Interscience, New York (1974).

17. Ibid., Makromol. Chem. Suppl., 4, 241 (1981).

18. R.Kaiser, Kolloid-Z. Z. Polymere, 149, 84 (1956).

19. B.E.Read and R.S.Stein, Macromolecules, 1, 116 (1968).

20. W.Glenz and A.Peterlin, J. Polym. Sci., A2/9, 1191 (1971).

21. M.A.McRae and W.F.Maddams, J. Appl. Polym. Sci., 22, 2761 (1978).

22. J.L.Koenig, S.W.Cornell and D.E.Witenhafer, J. Polym. Sci., A2/5, 301 (1967).

23. A.R.Wedgewood and J.C.Seferis, Pure Appl. Chem., 55, 873 (1983).

24. S.Onogi and T.Asada, in "Progress in Polymer Science, Japan", M.Imoto and S.Onogi, Eds., 2, 261 (1971).

25. I.Noda, A.E.Dowrey and C.Marcott, J. Polym. Sci. Polym. Lett. Ed., 21, 99 (1983).

26. Y.Uemura and R.S.Stein, J. Polym. Sci. Polym. Phys. Ed., 10, 1691 (1972).

27. K.Holland-Moritz, I.Holland-Moritz and K.van Werden, Colloid Polym. Sci., 259, 156 (1981).

28. H.W.Siesler, Infrared Phys., 24, 239 (1984).

29. W.F.Maddams and J.E.Preedy, J. Appl. Polym. Sci., 22, 2721 (1978).

30. Ibid., 22, 2739 (1978).

31. K.Sasaguri, S.Hoshino and R.S.Stein, J. Appl. Phys., 35, 47 (1964).

32. T.Oda, H.Nomura and H.Kawai, J. Polym. Sci., A3, 1993 (1965).

33. Y.Kikuchi and S.Krimm, J. Macromol. Sci. Phys., B4, 461 (1970).

34. P.C.Painter, J.Havens, W.W.Hart and J.L.Koenig, J. Polym. Sci. Polym. Phys. Ed., 15, 1237 (1977).

35. J.K.Kauppinen, D.J.Moffat, H.H.Mantsch and D.G.Cameron, Appl. Spectrosc., 35, 271 (1981).

36. D.Compton, FT-IR Spectral Lines, Nicolet Instr. Corp., 5, 4 (1983).

37. A.Miyake, J. Polym. Sci., 38, 479 (1959).

38. S.Krimm, Adv. Polym. Sci., 2, 51 (1960).

39. C.J.Heffelfinger and P.G.Schmidt, J. Appl. Polym. Sci., 9, 2661 (1965).

40. A.unningham, I.M.Ward, H.A.Willis and V.Zichy, Polymer, 15, 749 (1974).

41. L.D'Esposito and J.L.Koenig, J. Polym. Sci. Polym. Phys. Ed., 14, 1731 (1976).

42. H.G.Zachmann, Polym. Eng. Sci., 19, 966 (1979).

43. V.B.Gupta and S.Kumar, J. Polym. Sci. Polym. Phys. Ed., 17, 1307 (1979).

44. Ibid., J. Appl. Polym. Sci., 26, 1865 (1981).

45. R.S.Moore, J.K.O'Loane and J.Shearer, Polym. Eng. Sci., 21, 903 (1981).

46. R.Qian, J.Shen and L.Zhu, Makromol. Chem. Rapid Commun., 2, 499 (1981).

47. M.Matsuo, M.Tamada, T.Terada, C.Sawatari and M.Niwa, Macromolecules, 15, 988 (1982).

48. S.B.Lin and J.L.Koenig, J. Appl. Polym. Sci. Polym. Phys. Ed., 20, 2277 (1982).

49. a) Ibid., 21, 1539,(1983).

b) Ibid., 21, 2076 (1983).

50. M.Ito, J.R.C.Pereira, S.L.Hsu and R.S.Porter, J. Polym. Sci. Polym. Phys. Ed., 21, 389 (1983).

51. A.H.Khan and G.M.Venkatesh, J. Polym. Sci. Polym. Phys. Ed., 19, 589 (1981).

52. S.R.Padjibo and I.M.Ward, Polymer, 24, 1103 (1983).

53. S.S.Sikka and H.H.Kausch, Colloid Polym. Sci., 257, 1060 (1979).

54. a) V.B.Gupta and S.Kumar, J. Appl. Polym. Sci., 26, 1885 (1981).

b) Ibid., 26, 1897 (1981).

55. E.W.Fischer and S.Fakirov, J. Mater. Sci., 11, 1041 (1976).

56. H.J.Biangardi, Makromol. Chem., 179, 2051 (1978).

57. S.M.Aharoni, R.K.Sharma, J.S.Szobota and D.A.Vernick, J. Appl. Polym. Sci., 28, 2177 (1983).

58. J.Stokr, B.Schneider, D.Doskocilova, J.Lovy and P.Sedlacek, Polymer, 23, 714 (1982).

59. B.von Falkai, W.Gie ler, F.Schultze-Gebhardt and G.Spilgies, Angew. Makromol. Chem., 108, 9 (1982).

60. S.A.Baranova, P.N.Pahomow, P.N.Kljusnik, S.A.Gribanov, V.E.Geller and M.V.Sablygin, Vysokomol. Soed., Ser. B23, 104 (1981).

61. F.Rietsch and B.Jasse, Polym. Bull., 11, 287 (1984).

62. I.M.Ward, J. Polym. Sci. Polym. Symp., 58, 1 (1977).

63. J.L.Koenig and M.J.Hannon, J. Macromol. Sci. Phys., 1, 119 (1967).

64. V.B.Gupta, C.Ramesh and H.W.Siesler, J. Polym. Sci. Polym. Phys. Ed., 23, 405 (1985).

65. E.L.Gal'Perin, L.V.Strogalin and M.P.Mlenik, Vysokomol. Soed., 7, 933 (1965).

66. J.B.Lando, H.G.Olf and A.Peterlin, J. Polym. Sci., A1/4, 941 (1966).

67. a) J.B.Lando and W.W.Doll, J. Macromol. Sci. Phys., B2, 205 (1968).

b) Ibid., B2, 219 (1968).

68. a) W.W.Doll and J.B.Lando, J. Macromol. Sci. Phys., B4, 309 (1970).

b) Ibid., B4, 889 (1970).

c) Ibid., B4, 897 (1970).

69. a) R.Hasegawa, M.Kobayashi and H.Tadokoro, Polymer J., 3, 591 (1972).

b) Ibid., 3, 600 (1972).

70. Y.Takahashi and H.Tadokoro, Macromolecules, 13, 1318 (1980).

71. A.J.Lovinger, Polymer, 21, 1317 (1980).

72. K.Matsushige, K.Nagata, S.Imada and T.Takemura, Polymer, 21, 1391 (1980).

73. A.J.Lovinger, Macromolecules, 15, 40 (1982).

74. S.Weinhold, M.A.Backmann, B.H.Litt and J.B.Lando, Macromolecules, 15, 1631 (1982).

75. K.Tashiro, K.Takano, M.Kobayashi, Y.Chatani and H.Tadokoro, Polymer, 24, 199 (1983).

76. K.Matsushige and T.Takemura, J. Polym. Sci. Polym. Phys. Ed., 16, 921 (1978).

77. B.Servet, D.Broussox and F.Micheron, J. Appl. Phys., 52, 5926 (1981).

78. a) U.Kofer, R.Hirte and P.Weigel, Acta Polymerica, 33, 486 (1982).

b) Ibid., 33, 534 (1982).

79. B.P.Kosmynin, E.L.Gal'Perin and D.J.Tsvankin, Vysokomol. Soed., A12, 1254 (1970).

80. U.Kofer, R.Hirte and Ch.Ruscher, Acta Polymerica, 34, 352 (1983).

81. a) G.Cortili and G.Zerbi, Spectrochim. Acta, 23A, 285 (1967).

b) Ibid., 23A, 2216 (1967).

82. S.Enomoto, Y.Kawai and M.Sugita, J. Polym. Sci., A2/9, 1517 (1968).

83. N.I.Makarevic and V.N.Nikitin, Vysokomol. Soed., 7, 1673 (1965).

84. F.J.Boerio and J.L.Koenig, J. Polym. Sci., A2/9, 1517 (1971).

85. M.Kobayashi, K.Tashiro and H.Tadokoro, Macromolecules, 8, 158 (1975).

86. G.L.Cessak and J.G.Curro, J. Polym. Sci. Polym. Phys. Ed., 12, 695 (1974).

87. M.A.Bachmann and J.L.Koenig, J. Chem. Phys., 74, 5896 (1981).

88. C.Leonard, J.L.Halary, L.Monnerie, D.Brossoux, B.Servet and F.Micheron, Polym. Commun., 24, 110 (1983).

89. K.Tashiro, M.Kobayashi and H.Tadokoro, Macromolecules, 14, 1757 (1981).

90. M.Latour, A.Montaner, M.Galtier and G.Geneves, J. Polym. Sci. Polym. Phys. Ed., 19, 1121 (1981).

91. S.Osaki and Y.Ishida, J. Polym. Sci. Polym. Phys. Ed., 13, 1071 (1975).

92. R.Danz, Acta Polymerica, 33, 1 (1982).

93. P.J.Flory "Principles of Polymer Chemistry", Cornell University Press, Ithaca (1953).

94. E.H.Andrews and A.N.Gent, in "The Chemistry and Physics of Rubberlike Substances", L.Bateman Ed., Wiley, New York (1973).

95. L.R.G.Treloar "The Physics of Rubber Elasticity", 3rd Ed., Clarendon, Oxford (1975).

96. C.K.L.Davies, S.V.Wolfe, I.R.Gelling and A.G.Thomas, Polymer, 24, 107 (1983).

97. J.E.Mark, M.Kato and J.H.Ko, J. Polym. Sci., C54, 217 (1976).

98. D.S.Chiu, T.K.Su and J.E.Mark, Macromolecules, 10, 1110 (1977).

99. U.Eisele, Progr. Colloid Polym. Sci., 66, 59 (1979).

100. a) J.E.Mark, Polym. Eng. Sci., 19, 254 (1979).

b) Ibid., 19, 409 (1979).

101. T.L.Smith, Polym. Eng. Sci., 17, 129 (1977).

102. P.J.Flory, Chem. Rev., 35, 51 (1944).

103. G.Gee, J. Polym. Sci., 2, 451 (1947).

104. Y.Shimomura, J.L.White and J.E.Spruiell, J. Appl. Polym. Sci., 27, 3553 (1982).

105. M.Mooney, J. Appl. Phys., 19, 434 (1948).

106. R.S.Rivlin, Phil. Trans. Royal. Soc. (London), A241, 379 (1948).

107. A.Ciferri and P.J.Flory, J. Appl. Phys., 30, 1498 (1959).

108. J.E.Mark and P.J.Flory, J. Appl. Phys., 37, 4635 (1966).

109. J.E.Mark, Rubber Chem. Technol., 48, 495 (1975).

110. W.O.S.Doherty, K.L.Lee and L.R.G.Treloar, British Polym. J., 3, 19 (1980).

111. J.Furukawa, Y.Onouchi, S.Inagaki and H.Okamoto, Polym. Bull., 6, 381 (1981).

112. L.R.G.Treloar and G.Riding, Proc. Roy. Soc., A369, 281 (1979).

113. H.W.Siesler, Colloid Polym. Sci., 262, 223 (1984).

114. G.B.B.M.Sutherland and H.V.Jones, Trans. Farad. Soc., 9, 281 (1950).

115. J.L.Binder, J. Polym. Sci., A1, 37 (1963).

116. Ibid., Appl. Spectrosc., 23, 17 (1969).

FT-IR SPECTROSCOPIC STUDIES ON THE DEFORMATION OF POLYMERS BY MEANS OF COMPUTERIZED INSTRUMENTATION

Kurt Holland-Moritz

Institute for Physical Chemistry
University of Cologne
and
Heyden Datasystems
Cologne
West Germany

INTRODUCTION

In recent years the range of applicability of IR spectroscopy to chemical and physical problems has enromously expanded by the revival of Fourier transform infrared (FT-IR) spectroscopy. The most frequently used basic optical component of FT-IR instruments, the Michelson interferometer, has been known for almost a century, It was A.A.Michelson [1] who could postulate in 1891 from the interference fringes generated by his interferometer, that the red Balmer line in the hydrogen spectrum was in reality a doublet. As early as 1892, Lord Rayleigh [2] recognized that the interferogram (intensity as function of optical path difference) could be related to the frequency of the radiation passing through the interferometer (intensity as funtion of frequency) by a mathematical operation called Fourier transformation. However, it was beyond the mathematical and technical possibilities of that time to verify this assumption. Because of these difficulties interferometric measurements were not applied to a large extent. It was nearly half a century later in 1949 when P.Fellgett [3] used a Michelson interferometer in his astronomical observations to analyze weak radiation from outer space. He actually performed a numerical Fourier transformation to calculate the frequency distribution of the incident radiation. The applicability of the interferometric technique has been enormously increased after the introduction of the Cooley-Tukey algorithm for fast Fourier transformation [4] in 1966. The rapid development of digital computers greatly facilitated the mathematical procedures necessary for the collection and correction of the experimental data (interferogram) and enormously decreased the actual measure time and the time for the Fourier transformation. This progress in the field of digital eletronics together with the well-known advantages of interferometric measurements has opened new horizons in infrared spectroscopy.

EXPERIMENTAL

Today modern fast scanning FT-IR spectrometers allow one to obtain the data necessary for a complete infrared spectrum within seconds. In spite of the energy throughput (Jacquinot) advantage, multiplex (Fellgett) advantage, and wavelength accuracy, this speed proves to be the most powerful argument to use an FT-IR spectrometer, since the measure time for a complete infrared spectrum on a modern computer equipped or microprocessor controlled dispersive instrument is still three orders of magnitude larger, so far as spectra of comparable resolution are considered. However, we should keep in mind that the resolution in commercially available dispersive instruments is wavenumber dependent and may vary by more than 100 percent. For routine measurements the optimum resolution on grating instruments is obtained near 1000 cm^{-1}, whereas FT-IR spectrometers exhibit constant resolution over the whole spectral region. All the data, necessary to perform the Fourier transformation, are available after less than a second. For lower resolution this time decreases further. Accumulation of interferograms only increases the signal-to-noise ratio which proves to be necessary for strong absorbing samples like coals or samples in very low concentrations where sometimes hundreds or thousands scans have to be taken to obtain reasonable results. However, in the daily routine usually a measure time of 30 to 60 seconds is applied for standard preparations. To scan a spectrum comparable in quality on a modern dispersive instrument takes about 10 to 20 minutes, which still is a factor of 20 or more slower than on a FT-IR machine. However, further reduction of the measure time on dispersive instruments is only possible at the expense of resolution and therewith has to be paid with loss of information.

In addition to the above discussed "time advantage" which is common to all small, medium and large size FT-IR instruments, we have to consider the "computer advantage" of the large scale instruments. Here, the optical part of the spectrometer can be completely controlled by the operator via a more or less powerful data system. The basic units of the data system are a computer, a monitor, one or more moving head disk drives, a plotter and input/output devices (teletype and/or data terminal) for the communication with the computer. All accesories together allow easy data handling and storing. In addition to the software provided by the manufacturers of FT-IR spectrometers, individual software programs can be prepared for special applications and evaluation problems. These programs can be written in machine language, in BASIC, FORTRAN, PASCAL or any other high level language (if the respective computer is available) and stored together with the other software and experimental data on the disk cartridge. Building up a personal software library is of extreme importance for quantitative and qualitative investigations which require a series of subsequent spectra to be recorded and evaluated.

Before outlining the experimental and evaluation possibilities of such a system, let us briefly discuss the storage facilities. We will restrict ourselves here to the Nicolet System model 7199 which is installed in the Department of Physical Chemistry at the University of Cologne. According to the sampling theorem a spectrum from 8000 to 0 cm^{-1} requires approximately 4000, 8000 or 16000 data points (intensities) to be stored for a resolution of 4, 2 or 1 cm^{-1}, respectively. Each intensity value is stored in a 20-bit word, where 352 words form one block or sector. Since the wavenumber difference between the data points within one spectrum is constant and depends on the frequency of the laser used to digitize the signal of the main interferometer, only the number of data points has to be stored in the file status block together with the other parameters necessary to reproduce the spectrum. The file status block (352 words) and the intensity values (necessary number of words=16384/resolution (cm^{-1}) form one file). However, in most cases only a selected wavenumber region (mainly 4000 to 400 cm^{-1}) is of interest and the data points representing the other parts of the spectrum can be deleted. Thus, for the conventional MIR region from 4000 to 400 cm^{-1} the following amounts of 20-bit-words or blocks have to be stored:

1 cm^{-1} resolution:	25 blocks=8800 words
2 cm^{-1} resolution:	13 blocks=4576 words
4 cm^{-1} resolution:	7 blocks=2464 words

The binary information of a file representing the experimental data can be stored on a disk cartridge. Each cartridge has a maximum capacity of 6496 blocks. This amounts to 2286592 words of information or 433 spectra with a resolution of 2 cm^{-1} between 4000 and 400 cm^{-1}. However, since there are generally 2 cartridges on line, one fixed and one removable, the total capacity is doubled. More sophisticated equipment uses two disk-drives and thus can store about 1700 complete spectra. The storage capacity for spectra is only insignificantly reduced by the software programs necessary for data evaluation or manipulation. This capacity is not only helpful for building up a library of complete spectra on the disk cartridges, it is absolutely necessary for kinetic studies. Thus, inspection of stress-strain phenomena with the stretching device mentioned later, requires 100 to 2500 spectra to be scanned during only one experiment.

Besides these advantages of the computer assisted IR spectrometers in the field of data storage and retrieval, the computer equipment can be used to control additional accessories when performing vibrational spectroscopic studies of short-time phenomena in polymers during elongation, stress relaxation, crystallization, temperature changes, and other similar experiments.

FT-IR SPECTROSCOPY OF DEFORMATION PHENOMENA AND TEMPERATURE DEPENDENT EFFECTS

The modern fast scanning FT-IR spectrometer allows easy monitoring of short-time phenomena which are not accessible with the conventional dispersive infrared spectrometers. Dispersive instruments record only small wavenumber regions or intensity variation at constant wavenumber during such a short time. On FT-IR instruments a series of complete infrared spectra (6000-400 cm^{-1}) can be recorded during the respective process. The possibilities of such FT-IR instrument for the investigation of the changes in the state of order in polymer films during streching, relaxation, and changes of temperature are discussed in this section. To monitor the spectral changes occuring during elongation, relaxation and temperature change measurements, special stretching and heating devices were developed.

The experimental arrangement of the FT-IR spectrometer (Nicolet model 7199) with two Nicolet 1180 computers in combination with other data processing accessories and the special devices for stress-strain and temperature dependent experiments is outlined in Figure 1 [5-8]. This equipment, which is installed in our laboratory, allows complete computer control of time-dependent experiments and is successfully applied to the study of short-time phenomena in polymers during elongation, stress relaxation, crystallization, temperature changes, chemical reactions, and other similar experiments.

Two stress-strain machines developed in our workshop allow simultaneously recording a stress-strain diagram and infrared spectra with polarized light within preprogrammed time intervals. One machine (Figure 2) is designed for stretching velocities between $0<v<0.5$ mm/sec. With the other apparatus (Figure 3) polymer films can be expanded by means of a pneumatic mechanism within tens of second up to 3000 %. With this apparatus relaxation and crystallization phenomena after very rapid expansion are studied. The velocity of the elongation can be controlled via air pressure between 10 mm/sec and 500 mm/sec. The direction of the transmitted light of the polarizer can be changed automatically after an operator-selected number of scans. The corresponding

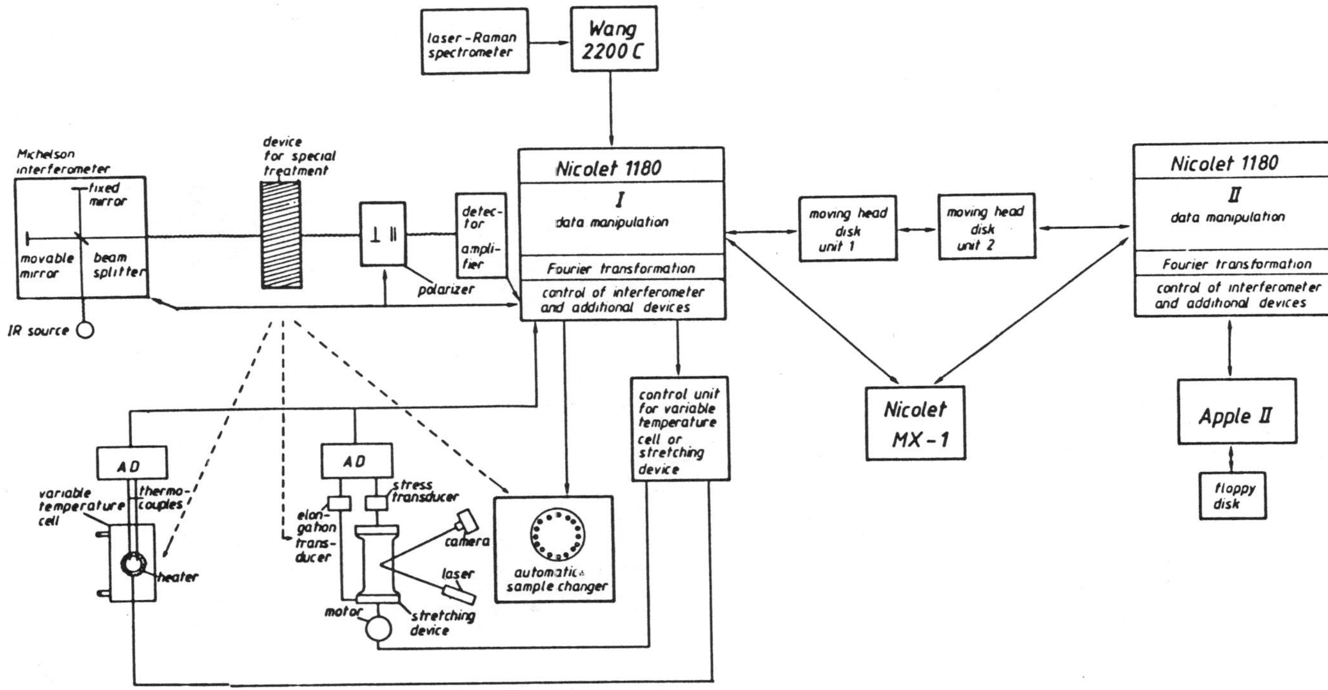

Figure 1 Schematic representation of experimental and computational possibilities of rapid scanning FT-IR spectrometers for studies on time-dependent phenomena in polymers.

stress-elongation values are stored in the file status block of each spectrum. In this way it is possible to assign each spectrum to a specified point of the stress-strain diagram. Stretching and relaxation in the machine (a) is controlled via the computer. A computer-controlled camera takes pictures from that region of the polymer film which is irradiated by the infrared beam. The exact position is determined by means of a small power helium-neon laser. In this way, misinterpretation of IR spectra due to macroscopic defects in the polymer film and asymmetric neck formation can be avoided. Furthermore, for every spectrum the corresponding sample geometry is known. This is of extreme importance when performing measurements in the necking region of the deformed polymer.

The various connections to the computer and its control units allow the elongation and relaxation experiments to be exactly preprogrammed. This was possible by slight modification of the original software delivered by the manufacturer and additional programs. Thus at the end of such an experiment the following information and data are available:

1. A stress-strain diagram.

2. A series of infrared polarization spectra for the complete stress-strain process.

3. A photograph of the observed sample area for each spectrum.

4. A film thickness-elongation diagram.

This diagram can be obtained since most polymer films exhibit interference fringes which generally are not desired. However, here they help to monitor the changes in the film thickness during stretching or relaxation. After the determination of the respective film thickness the interference fringes are removed from the spectra by means of a special computer program, which generates from the observable fringes an averaged sine curve and subtracts these artificial data points from the original spectral data. Although this procedure has its limitation becuase of the influence of the refractive index, it helps reduce errors in quantitative evaluations.

The large number of infrard spectra scanned during a temperature change program or during a stress-strain experiment cannot be evaluated in the conventional manner. Fortunately, the FT-IR instrument is equipped with a powerful computer system which

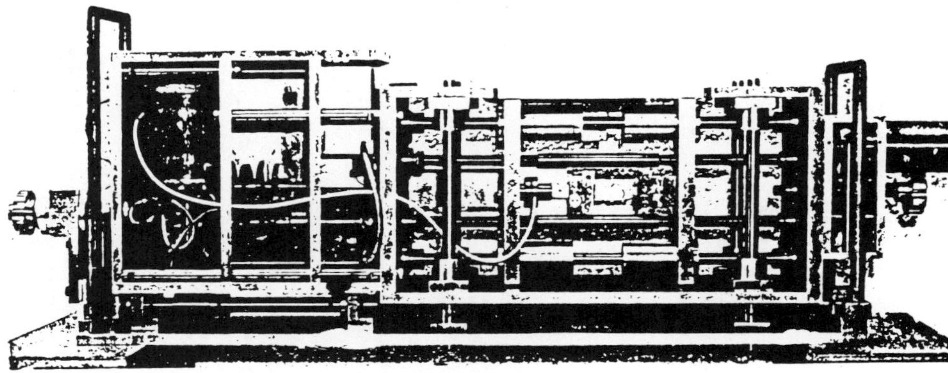

Figure 3

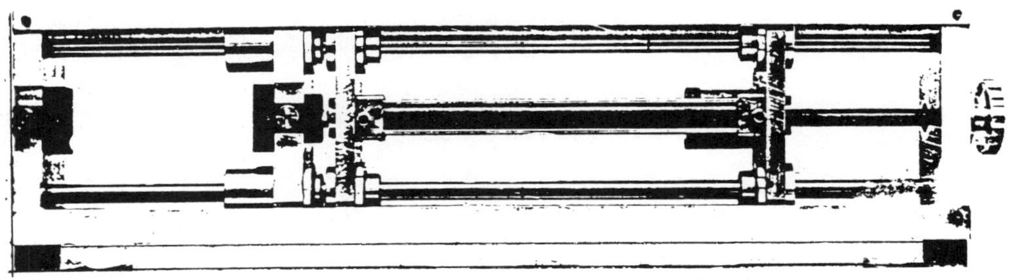

Figure 2

Stretching device for elongation velocities between 0 to 0.5 mm/s (Figure 2) and 10 to 500 mm/s (Figure3).

allows a computer supported spectra evaluation by application of personally developed computer programs. Basically, the following automatical pre-evaluation procedure can be performed:

1. The integral intensities (optionally beseline corrected) of selectable spectral regions of all spectra are plotted versus the number of the corresponding spectrum. Thus, changes in the spectra can easily be recognized and only the necessary spectra have to be plotted on the recorder.

2. To avoid misinterpretation, a second program controls the first one, because simultaneous increase and decrease of the intensities of two neighboring bands (which sometimes occurs during changes in the modification) cannot be detected by the first program. The control program is based on a comparison of wavenumber and intensity of each band of a spectrum with the respective values of the preceding spectrum. This program cannot be applied alone, since it does not consider changes in the half-band-width.

3. The dichroic ratios of operator selected bands can be automatically plotted together with the respective stress values versus a strain or time scale. Thus, to produce Figures 4 and 5 from the base-line corrected band intensities of the B_{3u}-mode at 730 cm^{-1} and B_{2u}-mode at 720 cm^{-1} of 152 polyethylene spectra, the dichroic ratios were automatically plotted together with the respective stress values via a specially designed computer program which finally finishes its output by drawing the frame including the used scaling.

For experiments of temperature change, the stretching device simply has to be interchanged with the heating and cooling cell. Instead of the measured quantities of the stress-strain transducer, the electromotive force (EMF) of the thermocouples is fed into the computer. Via special software programs the temperature of the cell can be controlled. Thus, experiments can be performed in this cell with cooling or heating rates equivalent to those applied in differential thermal analysis (DTA) or differential scanning calorimetry (DSC) experiments.

APPLICATIONS TO SELECTED PROBLEMS

A. Stress-Strain Experiments

Polyethylene

The above outlined experimental equipment was used to study the deformation behaviour of varous polymers. The mechanism of deformation of polymers during the stretching process has been the subject of numerous papers. Peterlin [9] pointed out a three stage process on cold drawing of crystalline polymers: The plastic deformation of the original spherulitic structure ahead of the neck, the discontinuous transformation of the spherulitic into the fiber structure in the neck, and the plastic deformation of the fiber structure. The first process involved shear, slip, and rotation of the stacked lamellae, phase transformation and twinning chain tilt and slip inside every single lamellae. During the whole first step the individuality of spherulites, stacks of lamellae and of single lamellae remains preserved.

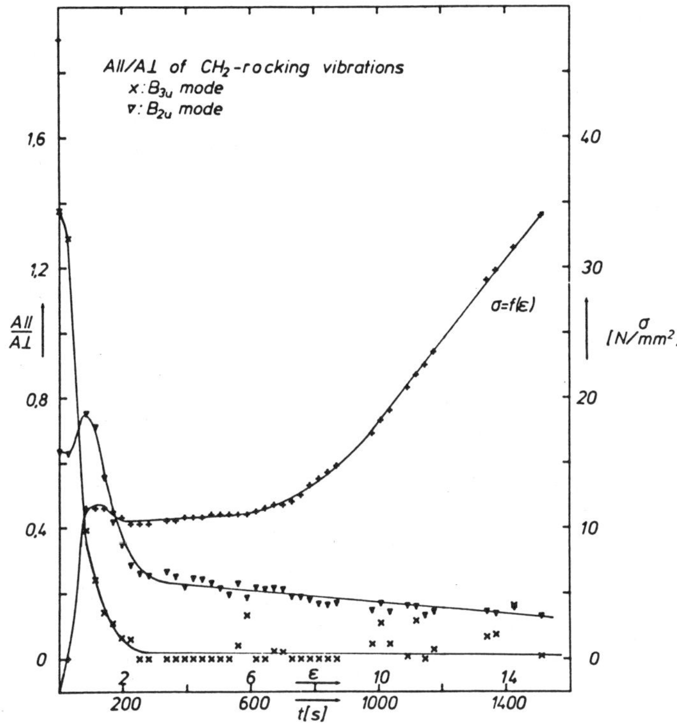

Figure 4. Dichroic ratios ($A_{\parallel}/A_{\perp}$) and stress (σ) as a function of strain (ε) (stretching velocity: 0.1 mm/s). x: CH_2-rocking band (B_{3u}) between 736 and 726 cm^{-1}, : CH_2-rocking band (B_{2u}) between 726 and 710 cm^{-1}, and +: stress-strain diagram.

The second step proceeds by transformation of the lamellae primarily by chain slip in the boundary layers between adjacent mosaic blocks into a bundle of microfibrils. The plastic deformation of the fiber structure, finally, occurs by sliding motion of the microfibrils past each other. Kobayashi [10] explained the second process by complete unfolding of the lamellae in the necking zone. In the necked region, the molecules are expected to occur in extended chains parallel to the stretching direction. According to their electron microscopic studies on thin films of polyethylene, Petermann [56] suggests a deformation process that is somewhere between the processes proposed by Peterlin and Kobayashi.

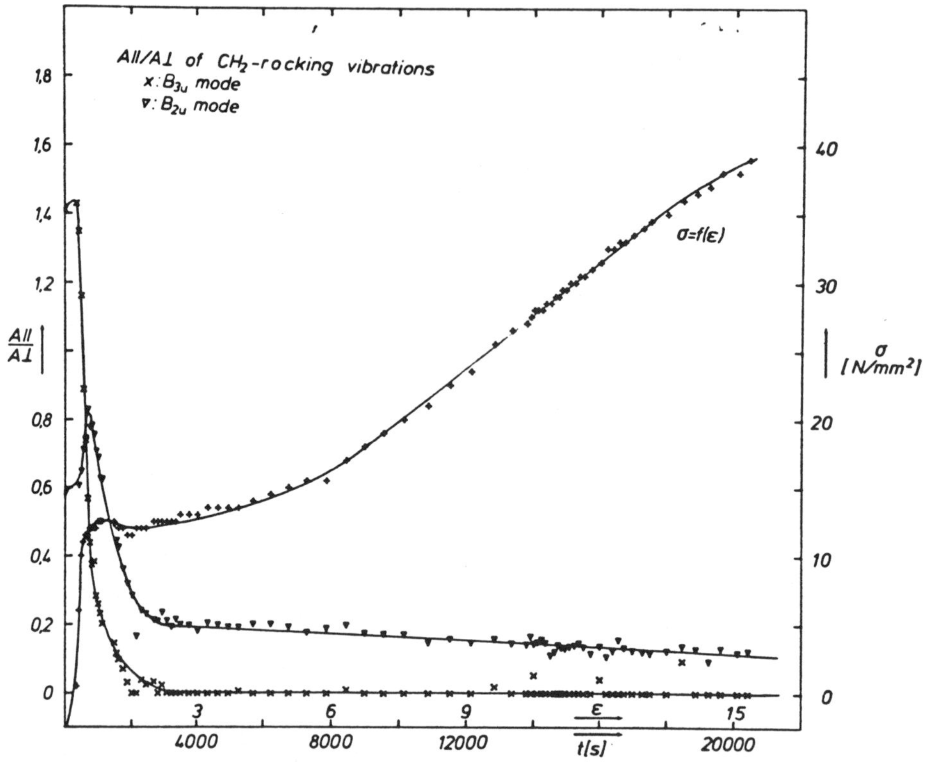

Figure 5. Dichroic ratios ($A_{/\!/}/A_{\perp}$) and stress (σ) as a function of strain (ε) (stretching velocity: 0.008 mm/s), x: CH_2-rocking band (B_{3u}) between 736 and 726 cm^{-1}, : CH_2-rocking band (B_{2u}) between 726 and 710 cm^{-1}, and +: stress-strain–diagram.

Recently, Kilian [11,12] discussed the deformation mechanism on the basis of a "cluster" model. Here, the orientation in partially crystalline polymers is described by a molecular network. Pope and Keller [13] interpreted their results on low-density polyethylene up to strains of 0.5 % of preoriented and annealed samples by three reversible processes: Lamellar slip, chain slip, and lamellar separation. For high-density polyethylene a low degree of reversibility was found.

To elucidate the mechanism of the deformation of polyethylene infrared spectroscopic studies were performed with the above described experimental equipment on blown polyethylene films. These films (Vestolen A 3512, A 6016, and A 6142) used for stretching experiments were kindly provided by Chemische Werke Huls (Marl, West-Germany). Density, average moleular weight, CH_3-content, Young's modulus and some other physical constants are given in Table 1. X-ray and IR studies established that the polyethylene unit-cell c and a-axes possess a preferred orientation perpendicular to the film plane and parallel to the machine direction (which coincides with the later stretching direction), respectively.

Let us first focus our attention on the spectral changes occuring during the stretching process. For these studies the middle of the PE-film was always kept in the IR-beam (beam diameter: 1.4 mm). The symmetrical formation of the neck was proved via the optical observation system and only those experiments which fulfilled this requirement were completed. Figure 6 illustrates, as introduction into, the discussion of the occuring phenomena in the infrared spectra of a polyethylene film (Vestolen A. 0.04 mm) in the region of the CH_2 bending modes (1500-1400 cm^{-1}) and CH_2 rocking modes (800-700 cm^{-1}). The infrared spectra can be correlated to the corresponding stress-strain diagram as indicated in Figure 6. The last shown spectrum (upper spectrum) was recorded 160 s (250 % strain) after the start of the stretching experiment.

Before discussing the changes in the spectra during elongation let us shortly summarize the known assignment of the CH_2 bending and CH_2 rocking vibrations [14]. The band doublets arise from in-phase and out-of-phase vibrations of the ethylene groups of neighbouring chains in a unit cell. The respective movement of the hydrogen atoms is outlined in Figure 6. The transition moments of the B_{2u} and B_{3u}-modes are perpendicular and parallel to the a-axis, respectively.

The 730 cm^{-1} band of the spectrum recorded before starting the stretching process exhibits parallel dichroism (//), thus, indicating a preferred orientation of the a-axis along the stretching direction. For perpendicular polarization (σ-dichroism) the 720 cm^{-1} band gradually disappears for parallel polarization. Also for perpendicular polarization the intensity of this band decreases, however, it is still present behind the yield point as a well pronounced shoulder. Now, both bands of the rocking vibration show apparently σ-dichroism. The shoulder at 730 cm^{-1} compeltely disappears on further stretching. Figures 7a and 7b show, as example, the spectral changes occuring during the complete stretching process of two series of selected IR spectra for parallel (//) and perpendicular ($\perp$) polarization in the region

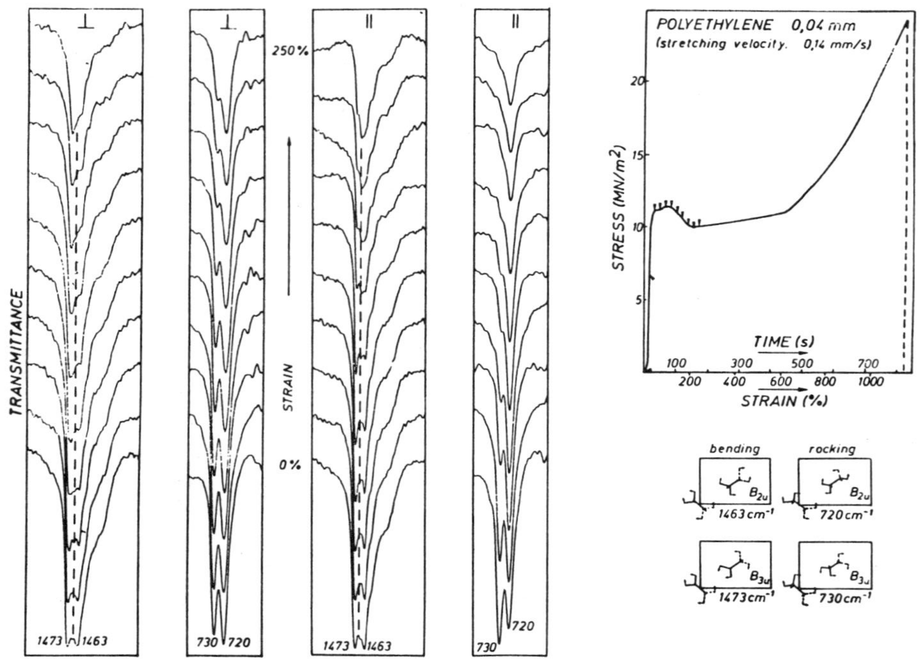

Figure 6 Stress-strain diagram and infrared spectra of polyethylene film in the region of the CH_2-rocking and CH_2-bending vibrations.

of the CH_2 rocking vibrations. The stretching velocities were 0.1 mm/sec (Figure 7a) and 0.25 mm/sec (Figure 7b). The upper spectra of the unstressed samples show the mentioned preferred orientation of the a and b-axes of the PE unit cells parallel and perpendicular to the later stretching direction.

For small strains (<0.5) this orientation increases slightly. With the beginning of the neck formation characteristic changes in the dichroic ratio (B_{2u}: 0.6--0.8, B_{2u}: 1.4--0.4) indicate a small rotation of the b-axis into the stretching direction and an opposite movement of the a-axis. The dichroic ratio of the B_{2u}-mode always reaches its maximum for 1. In the interval $1<\varepsilon<3$ both dichroic ratios (B_{2u} and B_{3u}) decrease in a comparable manner, indicating a rotation of a- and b-axes in directions perpendicular to the stretching direction. Simultaneously, the intensity of the B_{2u}-modes decreases in comparison to that of the B_{2u}-mode. This behavior seems to indicate the beginning of the distortion of the orthorhombic structure of the unit cells by unfolding the lamellae. On further stretching the B_{3u}-mode completely disappears for parallel polarization and the respective dichroic ratios become zero. The respective values for the B_{2u}-modes approach 0.1. Thus, application of the respective formulae for ideal uniaxial orientation results in an angle smaller than 13 degrees between c-axis and stretching direction in the necked region. This description holds for all stretching velocities between 0.008 mm/s and 0.3 mm/s. The spectra differ only for large ε values. Higher stretching velocities favour the disappearance of the B_{3u}-mode in the spectra scanned immediately before rupture. Obviously, relaxation processes can occur on slower drawing which restore a certain orthorhombic structure. This process can be experimentally monitored by relaxation of the stressed samples. Within approximatelly 2 h of relaxation of the IR spectra resemble those spectra scanned during the stretching process at $\varepsilon=2$.

In another series of experiments the spectra were scanned from areas of the polymer film in the necking region during the stretching process as indicated in Figure 8. The spectra from the boundary between the zone ahead of the neck and the necking zone resemble very closely those spectra scanned for $0.5<\varepsilon<1$ (Figure 7). In the necking zone we find spectra characteristic for the strain interval $1<\varepsilon<2$. In the boundary between necking zone and necked zone the B_{3u} and B_{2u} modes show strong perpendicular polarization, and the splitting is still well pronounced, indicating an alignment of the c-axis parallel to the stretching direction and a non-destroyed orthorhombic structure. The spectra scanned in the necked region (not shown in Figure 8), similar to those of Figure 7 for $\varepsilon<4$, finally reflect the distortion of the orthorhombic arrangement of the methylene sequences.

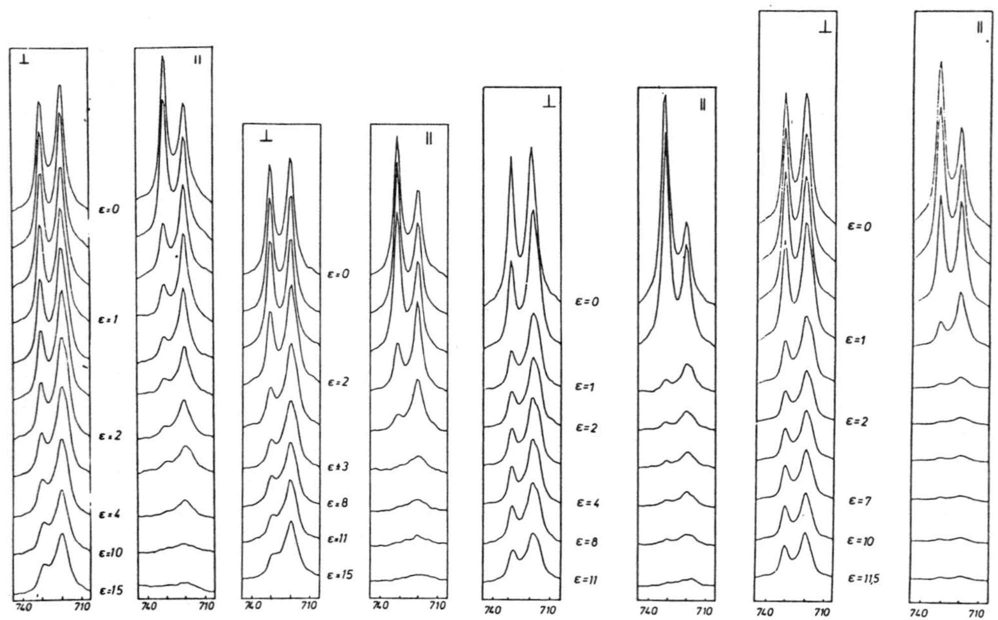

Figure 7 Selected infrared spectra in the region of the CH_2- rocking vibrations for perpendicular and parallel polarization in dependence of the applied strain

a) stretching velocity: 0.1 mm/s Vestolen A 3512
b) stretching velocity: 0.25 mm/s Vestolen A 3512
c) stretching velocity: 0.1 mm/s Vestolen A 6042
d) stretching velocity: 0.1 mm/s Vestolen A 6016.

The higher density polyethylene (Vestolen A 6016 and A 6042) exhibit nearly the same qualitative spectral behavior with respect to the initial orientation of the a- and b-axes and the subsequent rotation of c-axis into the stretching direction. However, there are two important differences (Figure 7c and 7d):

1. The correlation splitting of the CH_2 rocking (730/720 cm^{-1}) and CH_2 bending vibration (1473/1463 cm^{-1}) does not significantly decrease for higher elongation rates.

2. During the stretching process the significant shoulder appears at 716 cm^{-1}.

This shoulder is more pronounced for faster stretching velocities. Figure 9 shows the stress-strain diagram of Vestolen A 6042 film together with experimental and sythesized spectra in the region of the CH_2 rocking vibration. The selected experimental spectra (circled number) of the stress-strain experiment. This wavenumber region was resolved by application of Lorentzian shaped bands (row III). The synthesized spectra are reproduced in row II.

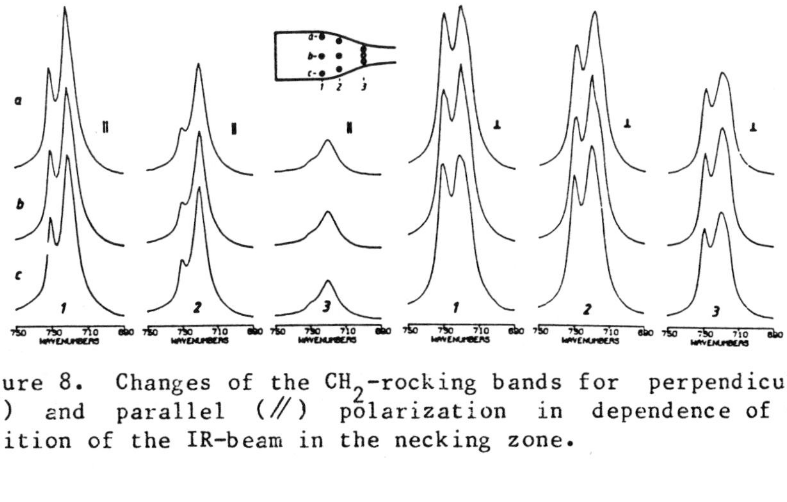

Figure 8. Changes of the CH_2-rocking bands for perpendicular (⊥) and parallel (//) polarization in dependence of the position of the IR-beam in the necking zone.

Several authors have shown that under conditions of stress the orthorhombic modification of crystalline polyethylene is partially conformed into a monoclinic structure [15,16]. The progress and extent of this morphological change in dependence of strain can now be studied by FT-IR spectroscopy. Unlike the 730/720 cm^{-1} doublet of the orthorhombic modification the monoclinic curve resolution (Figure 9) of the CH_2 rocking regions it can be derived that the integrated absorbance of the 716 cm^{-1} band amounts to approximately 20 % of the total integrated absorbance of all three rocking bands. Instead of the sophisticated curve resolving procedure simple digital subtraction of absorbance spectra will enhance the 716 cm^{-1} band, too [5].

Poly(tetramethylene terephthalate)

The methylene sequence in poly(tetramethylene terephthalate) (PTMT) can form two different conformations depending on the thermal or mechanical pretreatment. Reversible transitions between these conformations occur by stretching and relaxation of the polymer. Applied strain forces the approximately gauche-trans-gauche conformation (α-form) of PTMT into an approximately all-trans conformation (β-form) which exists only under stress. During complete relaxation or after fracture of the polymer film the interactions of the terephthalate residues drive the chains back to the gauche-trans-gaushe (GTG) conformation of the α-form.

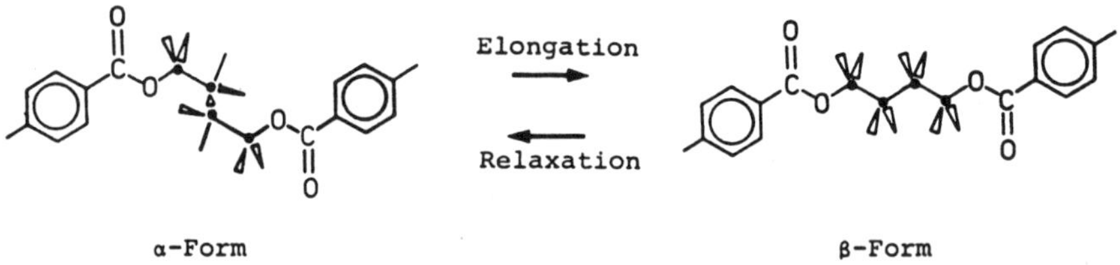

Several authors studied these conformations by x-ray diffraction [17-25] and infrared [25-35] and Raman [27,31,33] spectroscopy. Yokouchi [18] performed x-ray investigation on the reversible transformations from one conformation into another depending on the sample pretreatment. The applied treatment is reproduced in Figure 10. The deviation of the fiber identity period (1.16 nm) of the unstressed, fully relaxed α-conformation from the length of the completely extended repeat unit (1.325 nm) has been primarily attributed to an approximately GTG conformation of the aliphatic chain segments. Under stress the relaxed form transforms to an extended β-conformation with a fiber identity

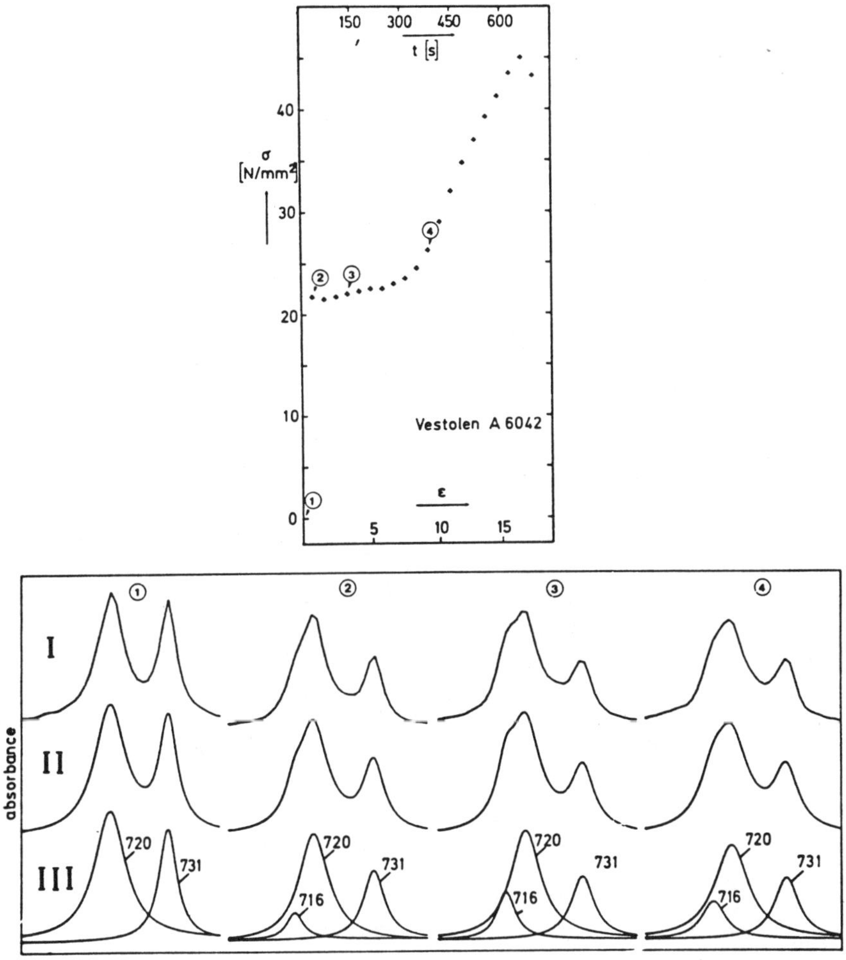

Figure 9 a) Stress-strain-diagram of a Vestolen 6042 film stretching velocity: 0, 26 mm/s

b) Experimental (row I), synthesized (row II), and resolved (row III) bands of the CH_2-rocking bands. The experimental spectra were scanned at the indicated positions (circled numbers) of the stress-strain-diagram (a).

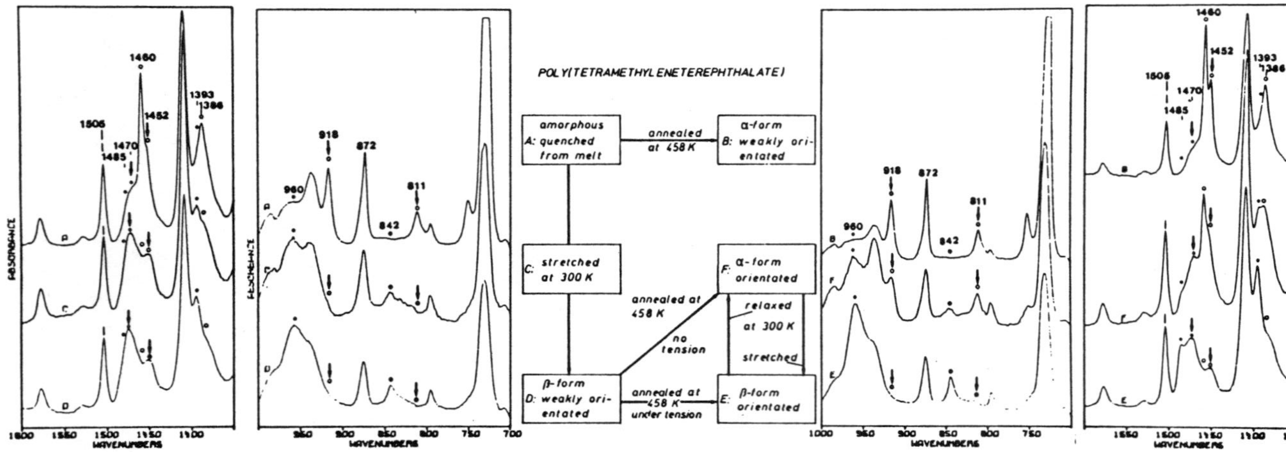

Figure 10 Infrared spectra from 1600 to 1350 cm^{-1} and 1000 to 700 cm^{-1} of PTMT films prepared according to Yokouchi (64).

period of 1.3 nm in which the methylene groups form an essentially planar all-trans conformation. α- and β-conformations can reversibly be transformed into each other by stretching, relaxation and temperature changes. Figure 10 shows the characteristic infrared spectra together with the respective sample preparation which was also used [18] for x-ray studies. The spectral data [31,32] and the normal coordinate calculations [33] indicate that the most characteristic spectral changes upon elongation of PTMT occur in the methylene bending, wagging and rocking region because of intramolecular interactions within the methylene sequence. The spectra of PTMT (Ultradur B 4500) of Figure 11 were scanned during the stretching process and illustrate the spectral changes occuring during elongation. The lower (0 s) and upper spectra (225 s) of Figure 11 represent the most characteristic bands of the α- and β-form, respectively.

The reversibility of the transition between the two modifications was proved by successive stretching and relaxation of the polymer film at room temperature. According to the theory we expect two IR active CH_2-bending, two wagging and two rocking modes in a tetramethylene sequence. In the twisted α-form we find the two bending and wagging modes at 1460/1452 cm^{-1} (Fig.11a) and at 1386/1173 cm^{-1} (Fig.11b). The bands at higher wavenumbers have to be assigned to vibrations of the CH_2-groups (α-CH_2) adjacent to the oxygen atoms, the lower wavenumber bands to methylene groups having methylene groups as next neighbours (β-CH_2). This assignment is based on the potential energy distribution (PED) and the corresponding cartesian displacements coordinates of the atoms [33]. The bands at 1720 cm^{-1} and 1505/1408 cm^{-1} arise from the carbonyl stretching vibration and coupled ring vibrations, respectively. In the CH_2-rocking region of the α-form (Fig.11c) the bands at 918/811 cm^{-1} show a complicated potential energy distribution because of the strong coupling effects with the carbon skeleton. These bands can be assigned to highly coupled skeleton vibrations of the GTG-conformation. The bands at 1018 cm^{-1} and 872 cm^{-1}, whose intensity does not change on mechanical treatment are assigned to in-plane deformation and CH out-of-plane vibrations of the ring, respectively.

Upon elongation the GTG conformation of the α-form is transformed into TTT conformation of the β-form. Thus, the bands of the CH_2 bending (1460/1452 cm^{-1}), CH_2 wagging (1386/1173 cm^{-1}) and coupled rocking-skeleton modes (918/811 cm^{-1}) of the α-form decrease in intensity whilst the bands of the β-form appear at higher wavenumbers: 1485/1470 cm^{-1}, 1393/1208 cm^{-1}, and 960/842 cm^{-1}.

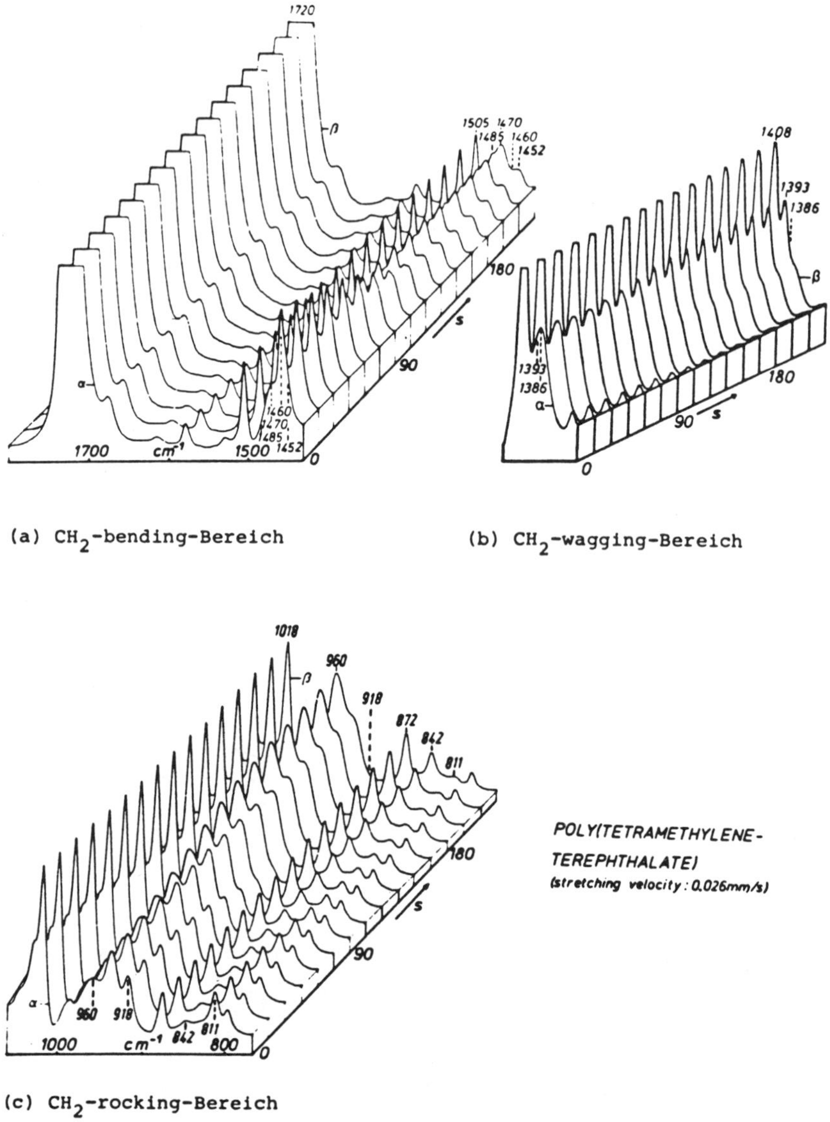

Figure 11 FT-IR spectra of PTMT scanned during the elongation process
a) CH_2-bending region
b) CH_2-wagging region
c) CH_2-rocking region.

Inspection of the frequencies of the α- and β-form shows that the bands of the wagging and coupled skeleton modes shift more upon applied stress than the bending modes. This behavior finds its explanation in a smaller interaction of the bending vibrations with the carbon skeleton. The potential energy distribution (PED) of both conformations is similar. The assignment of the calculated frequencies in terms of force constants does not seriously differ and only changes in the numerical values occur. As in the α-form the higher frequency bending and wagging modes of the β-form are caused by vibrations of the α-CH_2 groups and the lower frequency modes by the β-CH_2 groups. The conformational transition within the methylene chain does not affect the bands at 1720 and 1505/1408/1018/872 cm^{-1} caused by carbonyl and ring vibrations, respectively. The small change in intensity of these bands is due to the alteration of the sample thickness upon elongation.

Comparison of the spectra of Figure 11 and simultaneously recorded stress-strain diagram shows that the most drastic changes in the band intensity of the conformationally sensitive bands occur for $l/S-L < 0.44$ (L; length at time t, S; initial length). On further stretching no significant differences can be observed.

Notwithstanding of an optimum sample treatment the spectra of the α - or β -form always contain weak bands of the corresponding other form. Thus, the bands of the β-form at 1470 and 960 cm^{-1} appear as small shoulders in the spectra of the α-form and vice versa. We observe bands of the α-form at 1452 and 1386 cm^{-1} which do not completely disappear upon elongation to rupture (Figure 12). This can be explained by the fact that macromolecules seldom can be oriented perfectly because of different chain length, entanglement, branches and other defects within the polymer chain. However, appropriate subtraction of the spectra E and F by means of the FT-IR computer software can help to produce spectra of the pure α - and β-form. Figure 13 shows the result of the subtraction, spectrum E - C x spectrum F and spectrum F - C′ x spectrum E (Figure 10). This mathematical procedure results in a much better pronounced difference between the spectra of the two conformations, which was helpful in the assignment of the calculated frequencies. Figure 12 demonstrates the rapid relaxation after rupture of PTMT film. The upper spectrum was scanned just before rupture at the indicated time. The above discussed characteristic bands show their maximum intensity in the first spectrum. By removal of the tension the interactions of the polar groups of the terephthalate residue drive the methylene sequence back from a TTT to a GTG conformation as can clearly be seen from the disappearance of the β-bands.

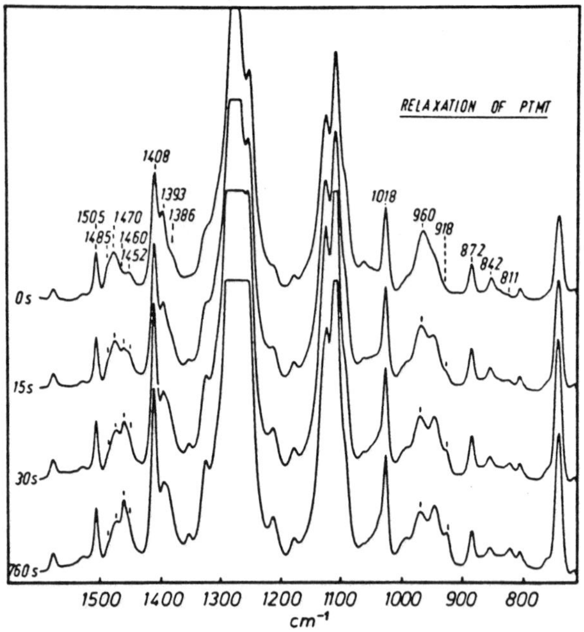

Figure 12 FT-IR spectra scanned after rupture of the PTMT-film.

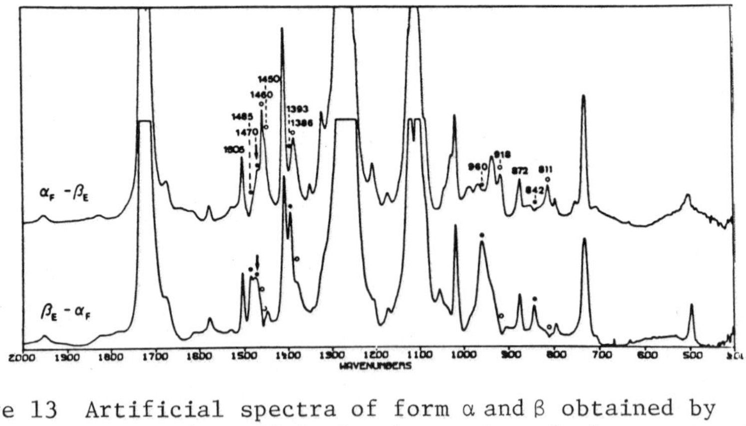

Figure 13 Artificial spectra of form α and β obtained by appropriate digital subtraction of the spectra E and F (Figure 21).

A detailed picture [32] of the response of amorphous PTMT film to the applied stress and any reversibility of the structural changes during uniaxial elongation and recovery in cyclic loading-unloading processes can be derived from simultaneous FT-IR and stress-strain measurements (Figure 14).

The isotropic, amorphous film samples were prepared from PTMT chips characterized by an intrinsic viscosity of $|\eta|=51\ cm^3 g^{-1}$ determined on solutions in phenol/1,1,2,2,-tetrachloroethane (60/40) at 298 K. the polymer was hot pressed at 24.5 MPa and rapidly quenched in ice water.

For the representation of the spectroscopic effects the most sensitive 1500-1400 cm^{-1} $\delta(CH_2)$ region has been selected. As a further illustration of the orientation effects the wide-angle x-ray diagrams of the original sample and the 600 % elongated sample have been included in Figure 14. The stress-strain curve of the applied mechanical treatment is shown in Figure 14a. The initial steep rise of stress with strain is followed by a plateau region between 50 to about 350 % strain. In this region the neck whose formation starts at a threshold value of about 50 MNm^{-2} propagates through the entire film specimen with a concomitant reduction in sample thickness from 0.04 mm to about 0.015 mm. The spectroscopic changes corresponding to this necking region are illustrated in the spectra 3 to 5 (Figure 14b) where the neck has been monitored with a laser reference system to move through the sampling area. Apart from a general intensity decrease as a consequence of the reduction in sample thickness the intensities of the 1460/1455 cm^{-1} bands (not resolved here) of the crystalline, crumpled α-form drastically decrease while the intensity of the 1475/1470 cm^{-1} band complex exhibits a relative increase. This result can be attributed to the predominant formation of a fibrilar structure with the above mentioned imperfect all-trans conformation of the aliphatic chain segments. Once the neck has completely moved across the specimen (at about 350 % strain) stress increases linearly with strain. To study the reversibility of the deformation the sample has then been subjected to an unloading-loading cycle. The spectra taken during this cycle have been labelled 18-23 and the small intensity changes of the 1460/1455 cm^{-1} bands relative to the 1475/1470 cm^{-1} bands and the ν(C-C) aromatic ring reference band at 1505 cm^{-1} clearly reflect that there is only an extremely small portion of polymer chains reversibly recovering to the crumpled conformation upon unloading. This is also demonstrated in the wide-angle x-ray diagram of such an unloaded sample where the (104) reflex of the elongated crystal β-modification is still observable alongside the corresponding reflex of the residual α-form.

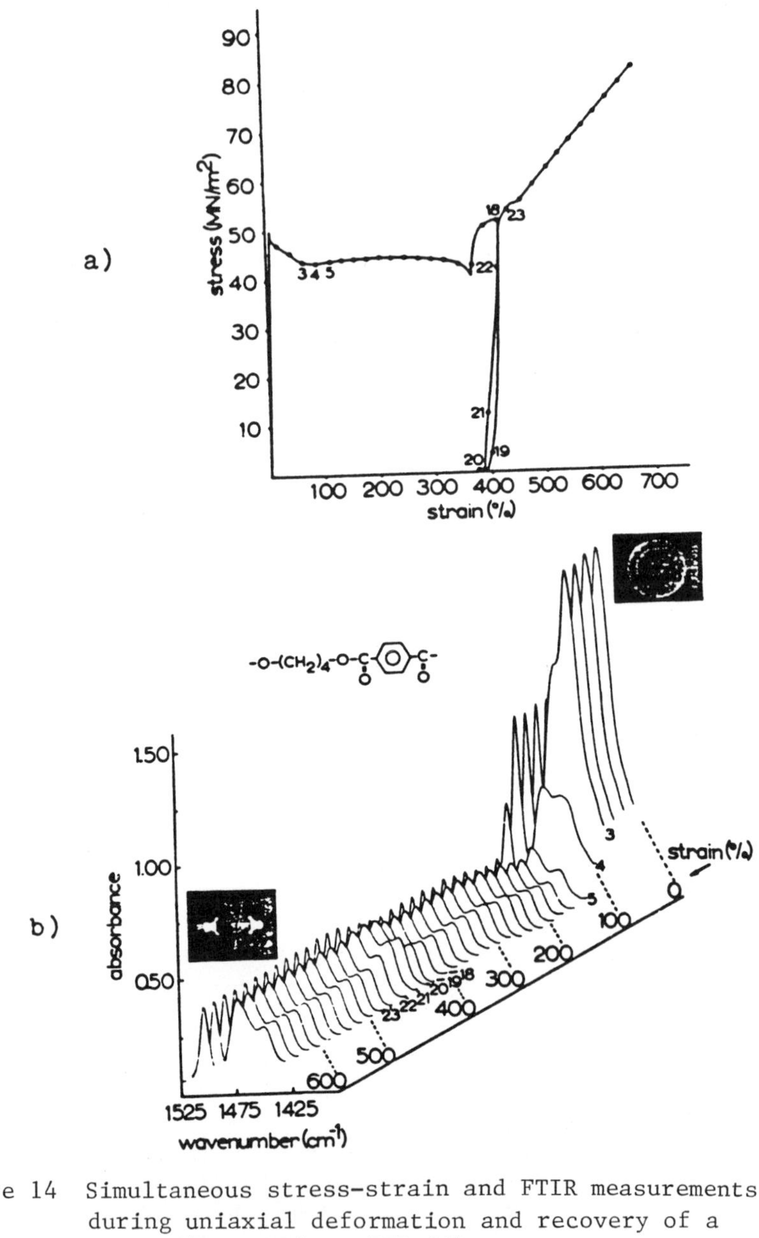

Figure 14 Simultaneous stress-strain and FTIR measurements during uniaxial deformation and recovery of a primarily amorphous PBT film:

a) stress-strain curve of the mechanical treatment (elongation rate: 1.66 %/s).

b) FTIR spectra in the 1500-1400 cm^{-1} $\delta(CH_2)$ region alongside the wide-angle x-ray diagrams of the original and a 600 % sample.

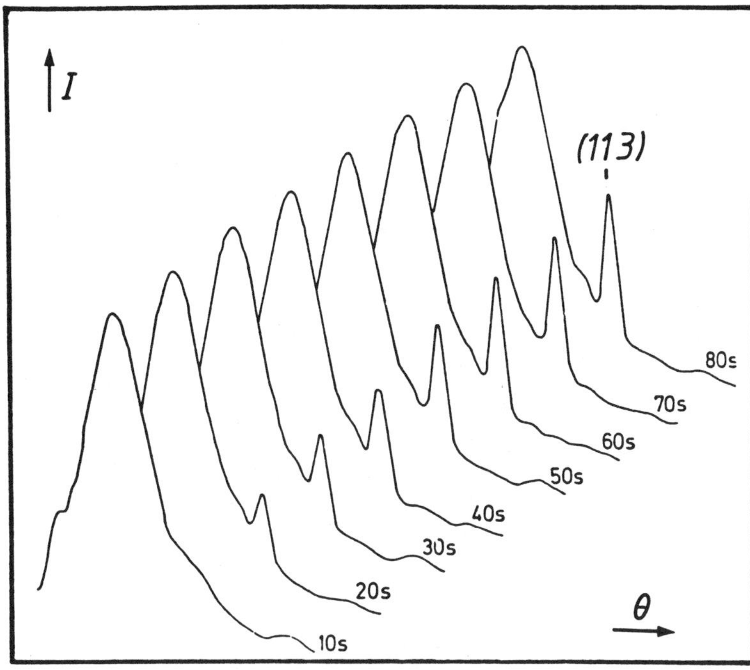

Figure 15. Schematic representation of experimental and computational possibilities of rapid scanning FT-IR spectrometers for studies on time-dependent phenomena in polymers.

B. Temperature Dependent Experiments

For the characterization of polymer structure the study of vibrational spectra recorded at various temperatures has become of increasing importance. Any changes observed in absorption intensity, wavenumber position, and band shape directly reflect the temperature dependence of the vibrational behavior of polymers as a consequence of changes in the inter- and intramolecular interactions and the state of order [39,40]. The vibrational spectra of polymers scanned at certain temperatures are of special interest for the studies of melting and recrystallization processes, thermal degradation, hydrogen bonding, and polymorphism. They facilitate the assignment of conformational regularity and crystallinity bands. Furthermore, when the experiments are performed on modern FT-IR spectrometers (Figure 1) with appropriate control electronics the sample treatment can be applied as in DTA or DSC experiments. Thus, for every DTA or DSC curve a series of

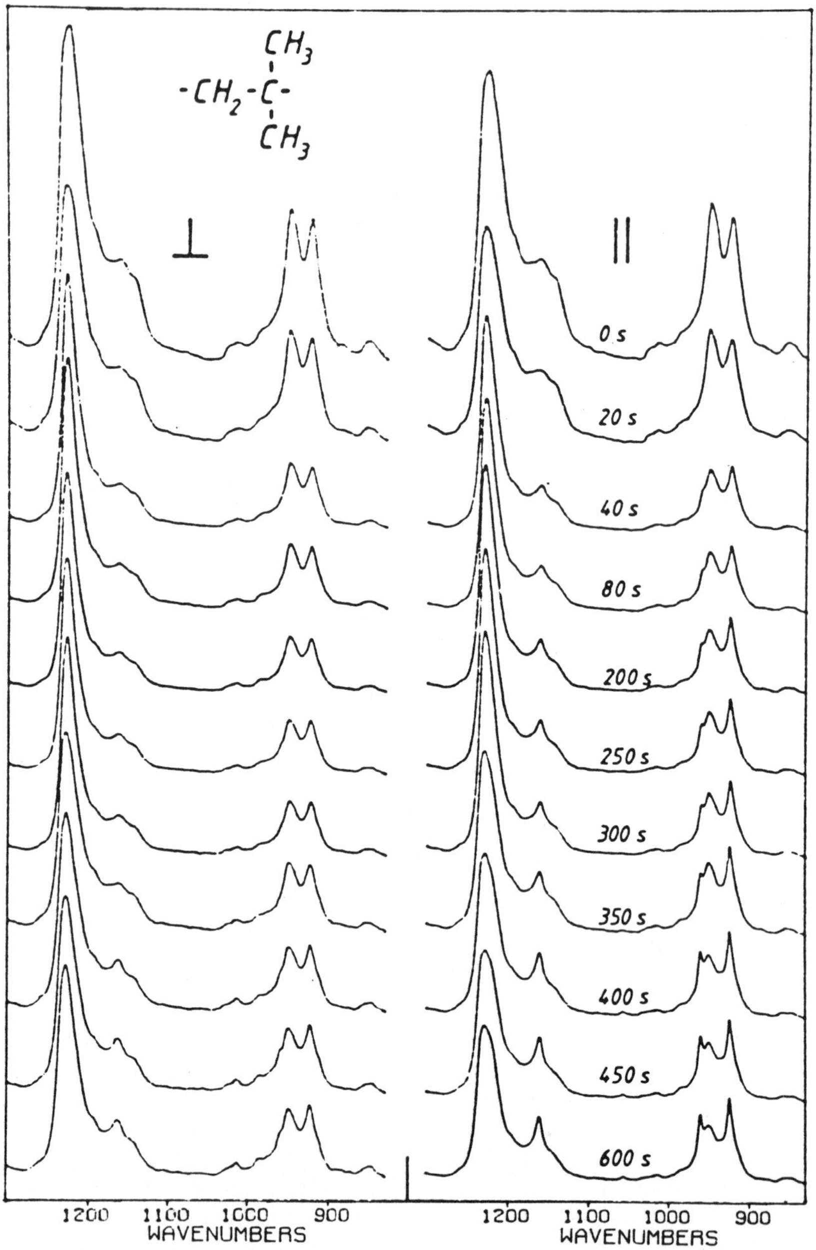

Figure 16 FTIR spectra of a poly (1,1-dimethylethylene) film for || and ⊥ polarization scanned after fast stretching (o.5 s)

infrared spectra are accessible which support the interpretation of the peaks observed in the thermal analysis experiment and thus help to give a detailed picture on the related changes of the microstructure.

The FT-IR spectra of Figure 17 were scanned in time intervals of 8 s during the recrystallization process of an 1,8-octane-diol-adipic acid polyester film on a KBr disk. The experiment was started, after the KBr disk with the molten polyester film was fixed on the sample holder in the FT-IR spectrometer at room temperature. The two lower spectra are due to the spectra of the melt. In the following spectra the intensities of the typical conformational regularity and crystallinity bands gradually increase. Conformational regularity characterizes a physically caused geometrical arrangement of a polymer chain, when every chemical repeat unit or sequence of repeat units with a well defined conformation can be transformed into the following unit (s) by means of a screw axis operation. Conformational regularity bands appear if a well defined order is possible within the polymer. Thus, the typical band contour for liquid or molten esters at 1165/1245 cm^{-1} disappears on crystallization and new bands at 1180 and 1265 cm^{-1} appear.

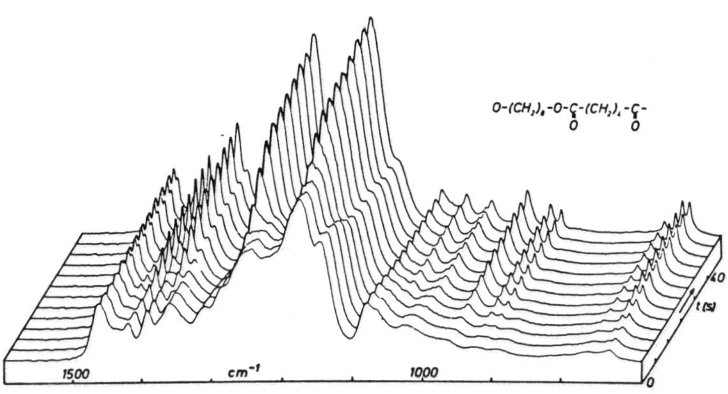

Figure 17. FT-IR spectra of a 1,8-octanediol-adipic acid polyester scanned during recrystallization.

These bands and additional ones at 1415, 1400, 1372, and 1292 cm^{-1} characterize the planar zigzag conformation of the methylene sequences. Furthermore, we observe two groups of crystallinity bands at 730/720 cm^{-1} and 1470/1460 cm^{-1}. These doublets are given in [38].

These data lead to the conclusion that under the experimental conditions (elongation rates varied between 0.26 and 3.33 %/s) the reversibility of the conformational transitions occuring during elongation is strongly hampered. For the polymer chains in the amorphous domains this can be understood in terms of the lack of a driving force such as the improved molecular packing efficiency of the α-crystal form [23]. The loss of the reversibility of the conformational change for the small crystalline regions distributed in the amorphous domains may be readily explained by the entanglement of the polymer chains in the amorphous matrix during the elongation process.

Poly(1,1-dimethylethylene)

Poly(1,1-dimethylethylene) (PIB) is an amorphous rubber-like polymer which only crystallizes when stress is applied.

The crystallization process strongly depends on the stretching velocity, stretching rate, and applied tension [36]. Until today this stress-induced crystallization in poly(1,1-dimethylethylene) was less studied by infrared and x-rays probably because of the time dependence of the effects involved. Recently, x-ray measurements on rapidly stretched PIB-cables [32,38] were performed by means of the Synchrotron-radiation (DESY, Hamburg). The use of this radiation allows one to obtain good quality x-ray diagram within seconds after stretching the cable in a pneumatic stretching device by 800 % in 50 ms. The x-ray diagrams (Figure 15) [37] show that these experimental conditions lead to a stress induced crystallization (indicated by the (113) reflex) within the first 20 s after the elongation. The intensity of the (113) reflex reaches its maximum in about 250 s after elongation had initiated.

The FT-IR spectra between 1300 and 800 cm^{-1} of a PIB film stretched in the pneumatic cell of Figure 16b within 0.5 s by 1000 % are shown in Figure 16 for parallel and perpendicular polarization. The most prominent spectral changes occur for parallel polarization at 1161 and 962 cm^{-1}. The intensity of the two new bands increases in time indicating the formation of conformational regular chains within the film. The bands can be assigned to C-C stretching vibrations which strongly couple with the CH_2 and CH_3 groups, respectively. Contrary to the above described x-ray results the start of the stress induced crystallization can be detected by infrared spectroscopy (under these applied experimental conditions) after approximately 40 s. In addition, the 1161 cm^{-1} band seems to increase faster in intensity than the 962 cm^{-1} band. A more detailed discussion of the phenomena occuring in dependence of the experimental parameters like stretching velocity, stretching rate and applied tension will be

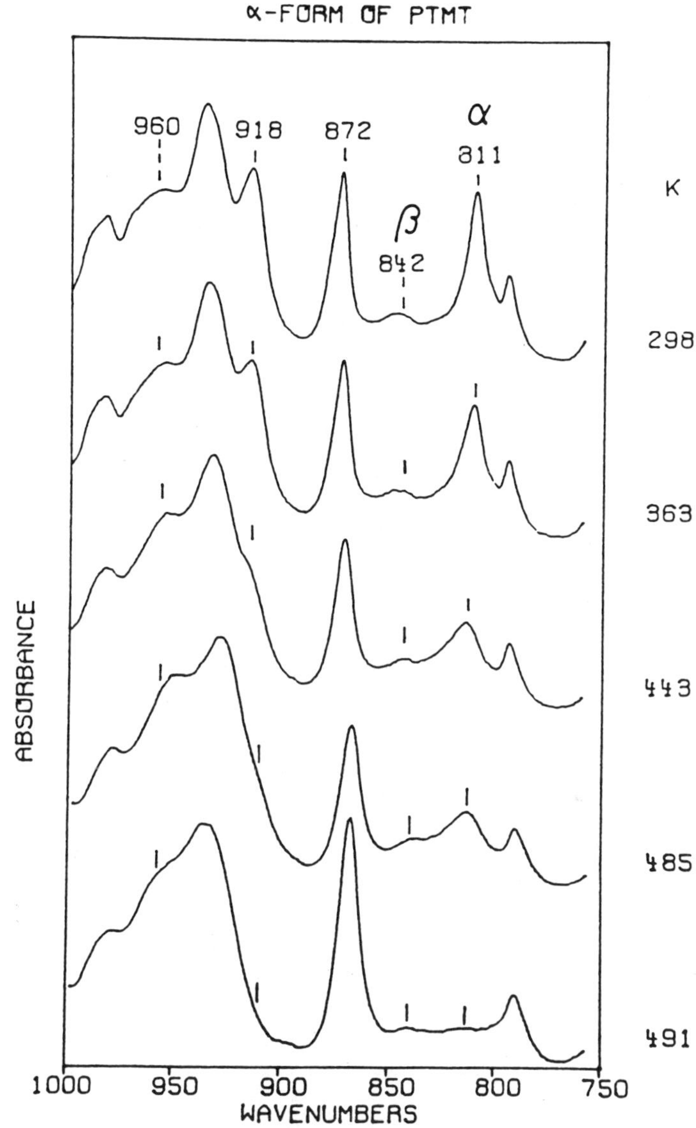

Figure 18a FT-IR spectra between 1000 and 750 cm^{-1} of the α- and β-form of poly(tetramethyleneterephtalate) measured during heating the sample from room-temperature to melting temperature.

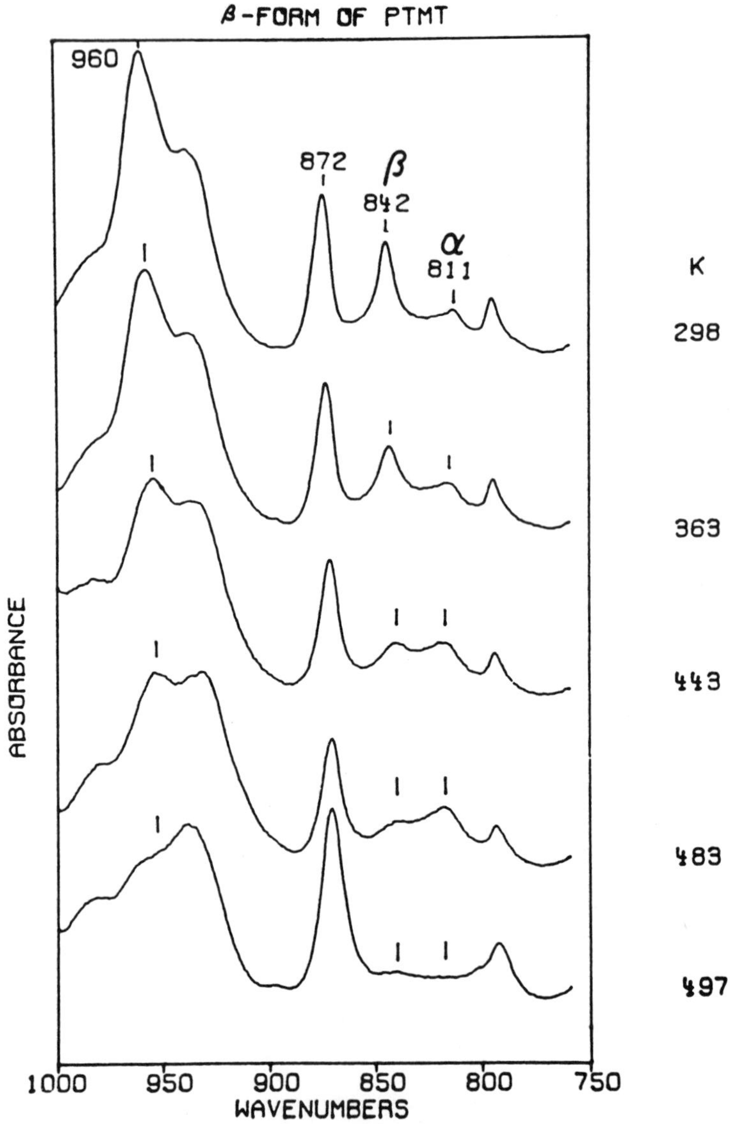

Figure 18b FT-IR spectra between 1000 and 750 cm^{-1} of the α- and β-form of poly(tetramethyleneterephthalate) measured during heating the sample from room-temperature to melting temperature.

caused by in-plane and out-of-plane CH_2 rocking and bending vibrations of two neighboring methylene sequences running through a crystallographic unit cell.

As discussed earlier, poly(tetramethylene terephthalate) can form two different conformations of its methylene sequences in dependence of the sample treatment (Figure 11). The FT-IR spectra of Figure 18 were scanned during heating the α- and β-form with a heating rate of 0.025 K/s. The most obvious spectral changes can be observed in the range of the CH_2 bending (1500-1400 cm^{-1}) (not shown) and the coupled skeletal CH_2 rocking vibrations (1000-750 cm^{-1}) while the bands of the ring vibrations do not alter their position and intensity. A more detailed inspection of the infrared spectra [33] shows that the bands of the skeletal vibrations at 842/811 cm^{-1} (TTT/GTG) disappear on melting the sample. The disappearance justifies their assignment to conformational regularity bands. The melting points at 491/497 K as determined by infrared spectroscopy coincide with those values derived from DSC measurements performed with the same heating rate. α- and β-form contain at room temperature always some amounts of the other form which can be seen by ocurrence of the weak bands at 811 and 842 cm^{-1}. When heating up the β-form (Figure 18b), the intensity of the weak 811 cm^{-1} band remains constant whereas that of the 842 cm^{-1} decreases. From 443 to 483°K the GTG band at 811 cm^{-1} increases, indicating the formation of more GTG sequences. However, on melting this band and the rest of the 842 cm^{-1} vanishes completely. The conformationally sensitive bands of the CH_2 bending vibrations show an analogous behavior.

REFERENCES

1. A.A.Michelson, Phil. Mag., 31 (1891).

2. Lord Rayleigh, Phil. Mag. Ser. 34, 409 (1892).

3. P.B.Fellget, in "Aspen International Conferene on Fourier Spetrosopy 1970", V.A.Vanasse Stair, A.T., Baker, D.J., Eds., AFCRL-71-0019 (1970) p.139.

4. J.W.Cooley and J.W.Tukey, Math. Comput., 19, 297 (1965).

5. H.W.Siesler and K.Holland-Moritz,"Infrared and Raman Spectroscopy of Polymers", Marcel Dekker, New York (1980).

6. K.Holland-Moritz, W.Stach and I.Holland-Moritz, J. Mol. Struc., 60, 1 (1980).

7. K. Holland-Moritz, I. Holland-Moritz and K. van Werden, Colloid Polym. Sci., 259, 156 (1981).

8. K.Holland-Moritz, W.Stach and I.Holland-Moritz, Progr. Colloid Polym. Sci., 67, 161 (1980).

9. A.Peterlin, J. Mat. Sci., 6, 490 (1971).

10. K.Kobayashi, cited in P.H.Geil,"Polymer Single Crystals", Wiley, New York (1963) p.473.

11. B.Heise, H.-G.Kilian and W.Wulf, Progr. Colloid Polym. Sci., 67, 143 (1980).

12. H.G.Kilian, Makromolekulares Kolloquium, Freiburg (1980).

13. D.P.Pope and A.Keller, J. Polym. Sci. Polym. Phys. Ed., 13, 533 (1975).

14. S.Krimm, Adv. Polym. Sci., 2, 51 (1960).

15. P.C.Painter, J.Havens, N.W.Hart and J.L.Koenig, J. Polym. Sci. Polym. Phys. Ed., 15, 1257 (1977).

16. Y.Kikuchi and S.Krimm, J. Macromol. Sci. -Phys., 64, 461 (1970).

17. Z.Mencik, J. Polym. Sci. Polym. Phys. Ed., 13, 2173 (1975).

18. M.Yokouchi, K.Sakakibara, Y.Chatani, M.Tadokoro, T.Tanaka and K.Yoda, Macromolecules, 9, 266 (1976).

19. R.Jakeways, T.Smith, I.M.Ward and M.A.Wilding, J. Polym. Sci. Polym. Lett. Ed., 14, 41 (1976).

20. J.H.Hall and M.G.Pass, Polymer, 17, 807 (1976).

21. I.J.Desborough and J.H.Hall, Polymer, 18, 825 (1977).

22. M.G.Brereton, G.R.Davis, R.Jakeways, T.Smith and I.M.Ward, Polymer, 19, 17 (1978).

23. U.Alter and R.Bonart, Colloid Polym. Sci., 258, 332 (1980).

24. B.D.Stambaugh, J.L.Koenig and J.B.Lando, J. Polym. Sci. Polym. Phys. Ed., 17, 1053 (1979).

25. K.Tashiro, Y.Nakai, M.Kobayashi and H.Tadokoro, Macromolecules, 13, 137 (1980).

26. B.D.Stammbaugh, J.L.Koenig and J.B.Lando, J. Polym. Sci. Polym. Lett. Ed., 15, 299 (1977).

27. I.M.Ward and M.A.Wilding, Polymer, 18, 327 (1977).

28. B.D.Stammbaugh, J.B.Lando and J.L.Koenig, J. Polym. Sci. Polym. Phys. Ed., 17, 1063 (1979).

29. H.W.Siesler, J. Polym. Sci. Polym. Lett. Ed., 17, 453 (1979).

30. K.Holland-Moritz, W.Stach and I.Holland-Moritz, Progr. Colloid Polym. Sci., 67, 161 (1980).

31. W.Stach and K.Holland-Moritz, J. Mol. Structure, 60, 49 (1980).

32. K.Holland-Moritz and H.W.Siesler, Polym. Bull., 4, 165 (1981).

33. W.Stach, Ph.D. Thesis, University of Cologne, Cologne (1982).

34. H.W.Siesler, Makromol. Chem., 180, 2261 (1980).

35. W.Stach and K.Holland-Moritz, J. Mol. Structure, (in press).

36. K.Holland-Moritz (in preparation).

37. T.Tanaka, Y.Chatani and H.Tadokoro, J. Polym. Sci., Polym. Phys. Ed., 12, 515 (1974).

38. K.Holland-Moritz and W.Stach (in preparation).

39. K.Holland-Moritz,"Proceedings of the 5th European Symposium on Polymer Spectroscopy", Verlag Chemie, Weinheim (1979) p.93.

40. K.Holland-Moritz and H.W.Siesler, Appl. Spectrosc. Rev., 11, 1 (1976).

FT-IR AND THERMAL-MECHANICAL CURE CHARACTERIZATION OF BLOCKED ISOCYANATE CONTAINING COATINGS

G.M.Carlson, C.M.Neag, C.Kuo and T.Provder

The Glidden Co
16651 Sprague Rd
Strongsville, Ohio 44136

INTRODUCTION

Isocyanates have been used for many years as crosslinkers in coatings, foams, and elastomers. The desirability of one pack coatings systems for many end use applications has led to the development of "blocked" isocyanate crosslinkers. Cure in these materials follows a two step mechanism [1,2] consisting of thermal deblocking of the blocked isocyanate to give free isocyanate followed by reaction of free isocyanate with a reactive functionality, often hydroxyl, to provide the actual chemical crosslinks.

$$R'\text{-NH-COR} \xrightarrow{k_1} R'\text{-NCO} + R\text{-H}$$

$$R'\text{-NCO} + P\text{-OH} \xrightarrow{k_2} R'\text{-NH-CO-OP}$$

Cure response can be improved by the addition of various catalysts including tin compounds and tertiary amines. While these materials are often reported to be deblocking catalysts their actual mode of action is often poorly understood.

It is now possible, taking advantage of the rapid scanning ability of the FT-IR, to monitor the chemical changes which occur when a sample is cured in a heated cell [3-6]. Conversion curves for each species involved in the cure process as a function of time and temperature can be generated from the infrared absorbance.

Methodologies have been developed, previously, to determine

cure kinetics parameters from conversion curves for a single dynamic temperature scan obtained using differential scanning calorimetry, DSC [6,7], and dynamic mechanical analysis, DMA [6,8]. These kinetics parameters can then be used to predict the cure behavior under a variety of cure temperature-time profiles. These methodologies have proven useful as a tool for the development of improved coatings as well as quality control. The determination of kinetics parameters from a single dynamic temperature scan [9] is based on the general n-th order rate expression in which the conversion, F, as a function of time, t, and temperature, T, is given by

$$dF(t,T)/dt = k[1-F(t,T)]^n$$

where n is the reaction order. Assuming that the temperature dependence of the rate constant is described by the Arrhenius equation

$$k = A\ e^{-E/RT}$$

the conversion curve can be used to determine the reaction kinetics parameters, E, n, and ln A.

This paper describes the use of FT-IR to monitor the deblocking of conventional blocked isocyanates. The effect of catalysts on the deblocking parameters and reaction products will be discussed. Evidence from gel permeation chromatographic analysis of the blocked isocyanates before and after deblocking will be shown. Preliminary investigations concerning the cure of blocked isocyanate containing coatings using DMA and FT-IR will be described.

EXPERIMENTAL

Materials

Blocked isocyanates were prepared from trimerized IPDI (Veba-Chemie T-1890), which was used as received to synthesize blocked derivatives. In a typical blocking reaction, 100g of T-1890, 0.15 moles was dissolved in 100 g of dry ethyl acetate and 0.1 g of dibutyltin dilaurate was added. Methylethylketoxime, 48.9g, 0.187 moles, was added over a period of two hours during which time the temperature rose to 40°C. Aliquots were removed and the amount of unreacted isocyanate determined by infrared absorbance. When reaction of the isocyanate was complete the mixture was cooled and precipitated into diethyl ether. The product was isolated, dissolved in toluene, and reprecipitated three times. The final product was isolated and dried.

Acrylic copolymers were prepared by typical solution polymerization techniques. A 400g solution of monomers was prepared containing butyl methacrylate, ethyl acrylate, methyl methacrylate and in some cases 2-hydroxyethyl acrylate. Dicumyl peroxide, 5.6g, was added to the monomer solution and the resulting solution added dropwise to 100g of methylamylketone at reflux, 145°, over a 3 hour period. The resulting polymers had a molecular weight of approximately 3000 as determined by size exclusion chromatography. Acrylic copolymers were blended with the blocked isocyanate to give a stoichiometric ratio of isocyanate to hydroxyl. Catalysts, when used, were present at a level of 0.2-0.5% by weight.

FT-IR Analysis

Spectra were obtained using a Digilab FTS-15E Fourier Transform Spectrophotometer. A reference spectrum of the crystal was collected prior to application of the sample. A thin film was cast from solution onto a NaCl crystal and the solvent allowed to evaporate. The crystal was placed into a heated cell (Model No.018-5322 Foxboro/Analabs) which was interfaced with a DuPont 900 Differential Thermal Analyzer to provide a constant heating rate, typically 5°C/minute. Thirty scans at a resolution of 4 cm^{-1} were co-added to produce each interferogram while the temperature was increased from ambient to approximately 220°C at a rate of 5°C/minute. Following cure completion the spectra were computed and plotted as absorbance spectra.

The reaction was followed by monitoring the absorbances at 2256 cm^{-1}, 1738 cm^{-1}, and 1503 cm^{-1} as a function of time. Film thickness changes were compensated for by normalizing the isocyanate absorbance to a band which remains constant during the reaction. For these systems the band at 1446 cm^{-1} was used. The normalized absorbance, $\bar{A}(t,T)$, was then defined as

$$\bar{A}(t,T) = A(t,T)[A_{1446}(t,T)/A(t0)^{o}_{1446}]$$

where $A(t,T)$ is the absorbance of the band of interest, $A(t,T)_{1446}$ is the absorbance at 1446 cm^{-1} at time t and temperature T and $A(t=0)^{o}_{1446}$ is the absorbance of the 1446 cm^{-1} band at time zero. The normalized absorbance as a function of time and temperature was then used to obtain the fractional conversion curve.

Initially the conversion, $F(t,T)$, is assumed to be zero and the point at which the absorbance stabilizes after reaction corresponds to complete reaction with $F(t,T)$ equal to one. Fractional conversion as a function of temperature and time is calculated by:

$$F(t,T) = [A_f - \bar{A}(t,T)]/[A_f - A_o]$$

where A_o is the normalized initial absorbance, $A(t,T)$ is the normalized absorbance at temperature T and time t, and A_f is the normalized final absorbance. Kinetics parameters were determined using a Nelder-Mead Simplex minimization procedure described elsewhere [9].

DMA

A DuPont 990/981 Dynamic Mechanical Analyzer system was used in assessing changes in the mechanical behavior of the coatings during cure. Woven fiberglass braides (Potter Industries, Hasbrouck, N.J.), 1"x0.5", were used as inert substrates for the blocked isocyanate containing coatings. Samples were applied uniformly to fiber-glass braid pre-mounted horizontally in the instrument clamps. Relative modulus and energy dissipation of all samples were recorded as a function of time at a heating rate of 5^oC/minute under a dry nitrogen purge of 5 l/minute. Samples were scanned from room temperature to 230^oC, cooled, and rerun from approximately -80^oC through 230^oC. Experimental details unique to kinetics studies by DMA have been described elsewhere [4,8,9].

High Performance Size Exclusion Chromatography

The instrument used in this study was the Waters Associates Model 150C ALC/GPC. The instrument was operated at 40^oC, with Burdick and Jackson distilled in glass tetrahydrofuran (THF) as the eluting solvent. The sample column bank consisted of two 50 cm Varian Instruments MicroPak® TSK gel columns (TSK 2000H and TSK 3000H).(MicroPak® is a registered trademark of Varian Associates, Inc., Palo Alto, CA). The flow rate was set at 1.0 ml/min.

RESULTS AND DISCUSSION

Effect of Catalysts on Deblocking of Methyl Ethyl Ketoxime (MEKO) Blocked T-1890

The initial and final spectra obtained during the deblocking of T-1890 blocked with MEKO are shown in Figure 1. Several major spectral changes occur during the deblocking reaction. Absorbances at 1738 cm^{-1} due to urethane carbonyl functionality and the other at 1503 cm^{-1} due to urethane N-H functionality associated with the blocked isocyanate reactant decrease as the reaction progresses. A large absorbance at 2256 cm^{-1} due to isocyanate functionality formed during the deblocking increases during the reaction.

The changes in the normalized absorbance of these bands during deblocking are shown in Figure 2. Deblocking begins near

110°C, reaches a maximum rate near 170°C, and approaches completion around 200°C. Comparison with the normalized isocyanate absorbance of the original T-1890 indicates nearly quantitative deblocking of the MEKO blocked adduct.

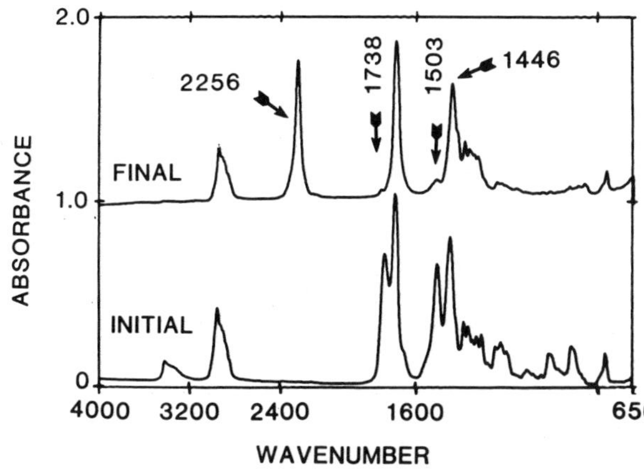

Figure 1 Initial and Final Spectra During Deblocking of T-1890/ MEKO.

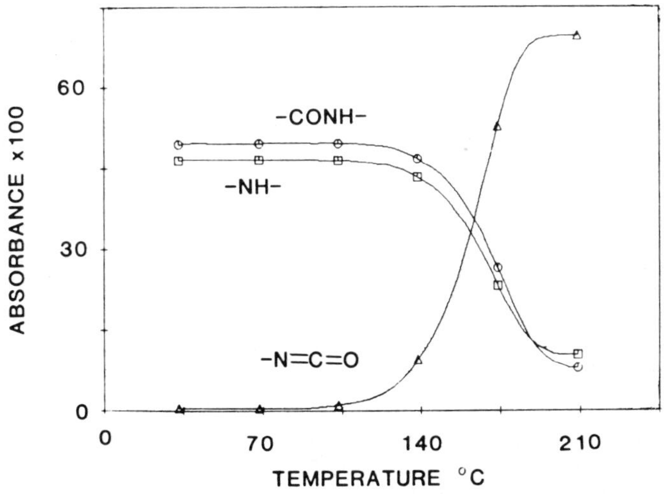

Figure 2 Absorbance of Functional Groups During Deblocking.

Reaction kinetics parameters determined from the Nelder-Mead Simplex minimization procedure are given in Table I and are in good agreement with previously reported results for this system [10]. Figure 3 shows that the conversion curve generated from the calculated kinetics parameters agrees closely with the experimental conversion curve.

TABLE I

Effect of Catalyst on Deblocking of T-1890/MEKO

Catalyst	n	E	lnA
-	1.0	23.2	20.6
TBAC	1.0	21.5	18.7
DBTDL (0.2%)	1.1	19.6	16.8
DBTDL (0.5%)	1.8	29.2	30.5
DABCO	0.7	20.6	18.5

The addition of catalysts to increase the cure response in blocked isocyanate coatings is well known. While it is often assumed that they function by catalyzing deblocking, little work has been done to support this assumption. Several reported catalysts were investigated to determine their effect on deblocking kinetics. The conversion curves for isocyanate formation in the presence of tetrabutylammonium chloride, TBAC, 2,2,2-bicyclooctane, DABCO, and dibutyltin dilaurate, DBTDL are shown in Figure 4. TBAC has no effect on the rate of deblocking. DABCO acts as a mild catalyst for deblocking, lowering the temperature at which deblocking begins and increasing the conversion throughout the reaction. While a preliminary study [3] indicated the presence of side reactions catalyzed by DABCO leading to lower isocyanate absorbance at the completion of reaction, these results indicate a nearly quantitative deblocking.

DBTDL does change the course of the deblocking reaction. The initial portion of the conversion curve for DBTDL catalyzed deblocking reaction, is identical to the uncatalyzed case. At

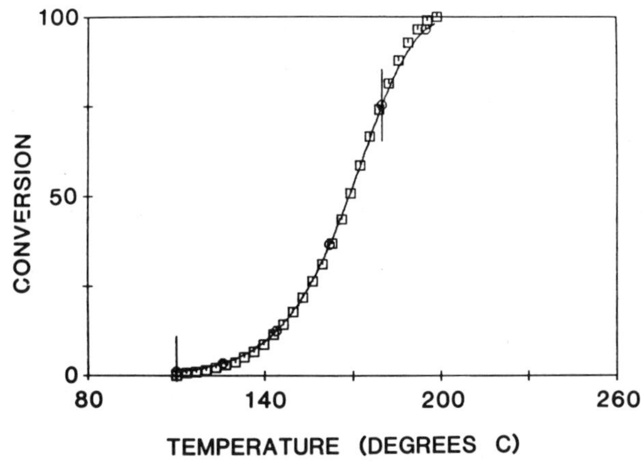

Figure 3 Experimental Data and Calculated Conversion Cure for T-1890/MEKO
□ Experimental Data
⊖ Calculated Curve.

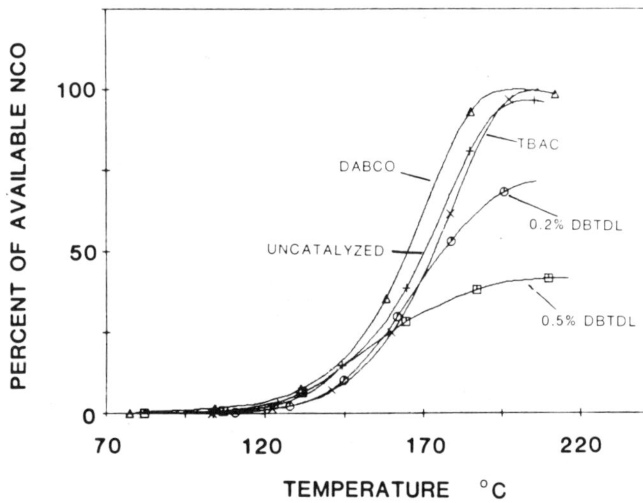

Figure 4 Effect of Catalysts on Deblocking.

higher temperatures the concentration of free isocyanate becomes less than in the absence of catalyst. Increasing DBTDL concentration decreases the total NCO released during deblocking. This decreased isocyanate concentration can be explained by the DBTDL catalyzed reaction of free isocyanate groups with the remaining urethane linkages to form allophanate structures.

Scheme I

Extended trimerization of free isocyanate groups would result in decreased -NCO absorbance and increased absorbance of the normalizing band and may account for some of the decrease in isocyanate absorbance. Volatilization of the T-1890 caused by catalyzed decomposition of the central isocyanurate ring would not explain the results due to the compensating loss of absorbance of the normalizing band.

While difference spectra obtained for the initial portion of the DBTDL catalyzed deblocking reaction are virtually identical to the uncatalyzed case, spectra obtained in the latter stages show marked differences as shown in Figure 5. Absorbances at 1637 cm^{-1} and 1659 cm^{-1} are observed in the presence of catalyst which do not appear in the uncatalyzed sample. One of these bands is believed to be due to amide structures resulting from the DBTDL catalyzed Beckman rearrangement of the free oxime, as shown in Scheme II, to give N-substituted amides which on further reaction with free isocyanate would give products stable over the temperature studied.

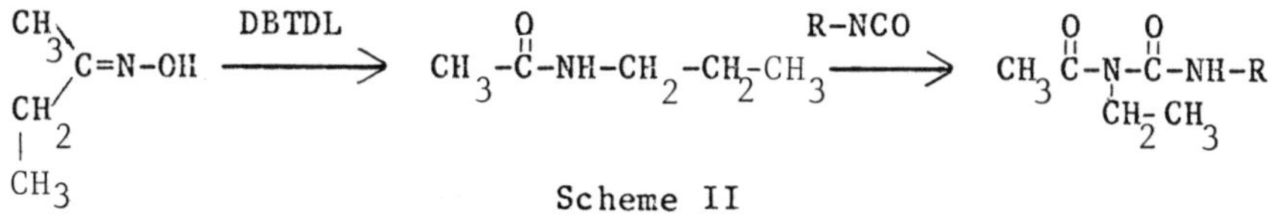

Scheme II

The other band observed in the presence of DBTDL catalyst may be due to allophanate structure described previously.

Absorbance curves for the N-H functionality and C=O functionality throughout the catalyzed deblocking are shown in Figure 6. The deblocking continues as evidenced by the continued decrease in N-H and C=O even while the isocyanate does not continue to increase. At higher temperatures, as side reactions become more important, no increase in absorbance is observed at 2256 cm^{-1} even though the deblocking continues as evidenced by the decrease in absorbance at 1503 cm^{-1} and 1738 cm^{-1}.

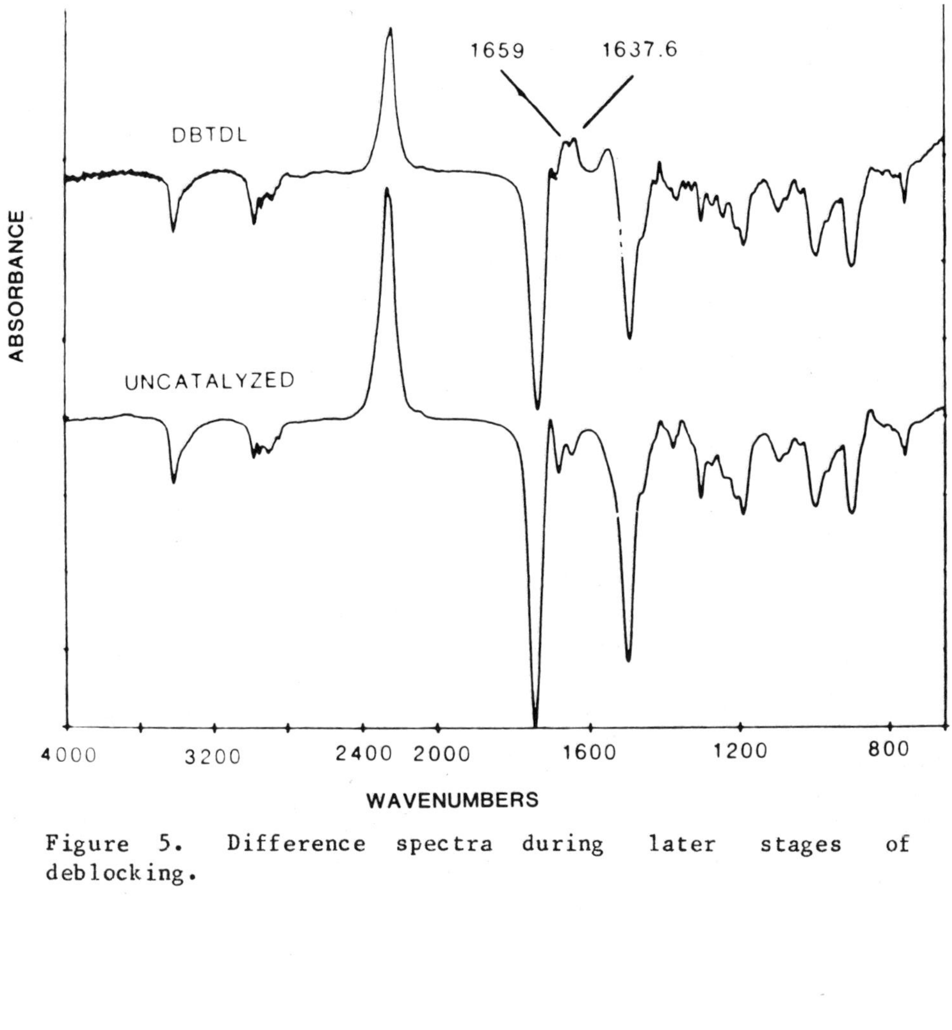

Figure 5. Difference spectra during later stages of deblocking.

Formation of allophanate structure is supported by size exclusion chromatography characterization of the deblocked materials. Curve A of Figure 7 is the chromatogram of T-1890 following one temperature scan. This compares with chromatogram B, the end product following the deblocking of T-1890/MEKO with no catalyst. The product of deblocking is nearly identical to T-1890 after a temperature scan. The addition of DBTDL drastically changes the deblocked product composition as shown in chromatogram C. More high molecular weight product is observed supporting a side reaction between free isocyanate and the still unblocked material leading to allophanate groups. In addition close examination of the chromatogram reveals that the major peak corresponds to a slightly higher molecular weight than in the uncatalyzed sample. This is consistent with the proposed deblocking, Beckman rearrangement of the MEKO, and reaction with isocyanate to form stable products.

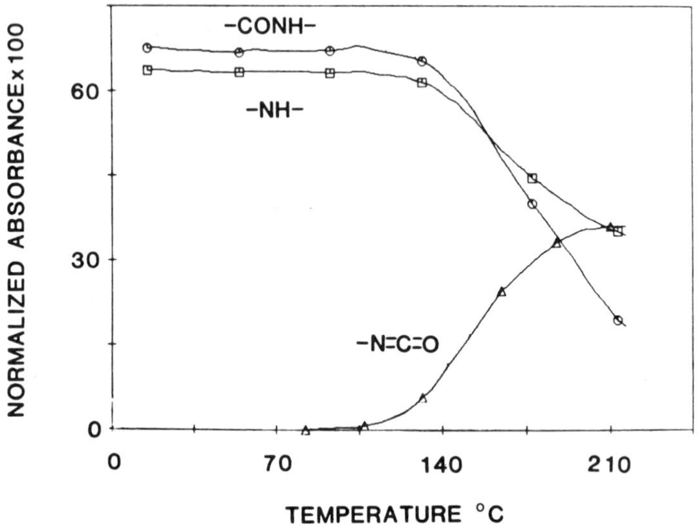

Figure 6. Absorbance of functional groups during DBTDL catalyzed deblocking.

Figure 8 shows the isocyanate absorbance as a function of temperature during the heating of unblocked T-1890. Uncatalyzed material shows no loss of absorbance while in the presence of catalyst a significant loss of isocyanate absorbance begins near 130°C. The slow linear increase in absorbance throughout the uncatalyzed curve is a result of the faster rate of broadening of the normalizing peak at 1446 cm^{-1} relative to the isocyanate absorbance resulting in an apparent increase in isocyanate absorbance.

Cure of Blocked Isocyanate Containing Acrylic Coatings

The concentration of free isocyanate functionality during cure of an actual polymer system containing blocked isocyanate and hydroxyl is much more complicated. Free isocyanate is formed during beblocking and consumed during the cure reaction with hydroxyl functionality. The concentration of isocyanate during cure rises during the early stages of reaction as deblocking dominates, plateaus as the deblocking is completed and the reaction with hydroxyl functionality dominates. The curve cannot be analyzed quantitatively using our kinetics approach but does

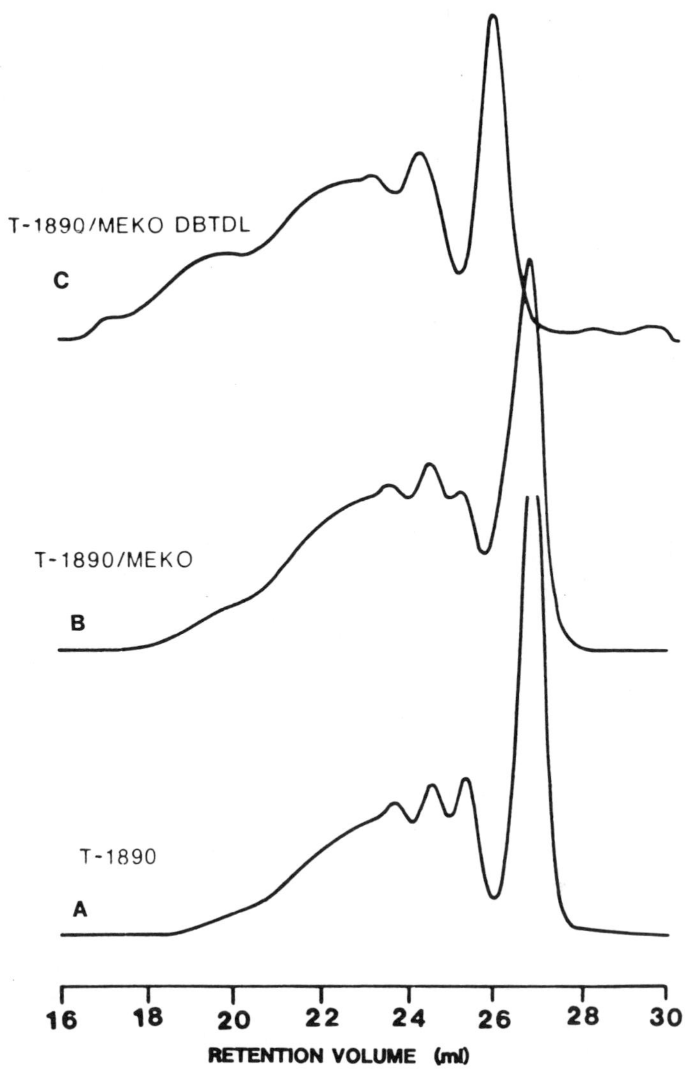

Figure 7 HPGPC Analysis of Heated Samples.

give qualitative information concerning the cure chemistry.

The concentration of free isocyanate functionality during the cure of an actual polymer system is more complex as shown in Figure 9 for a stoichiometric ratio of total -NCO to hydroxyl functionality. The onset of deblocking occurs at the same temperature as the neat blocked isocyanate, near 90°C. In the initial stages of the reaction, deblocking proceeds faster than cure and the absorbance of free isocyanate rises. Near 170°C the rate of deblocking and cure equalize and the isocyanate concentration reaches a maximum. As the cure proceeds the isocyanate concentration decreases due to the completion of the deblocking coupled with an increased rate of reaction of isocyanate with hydroxyl.

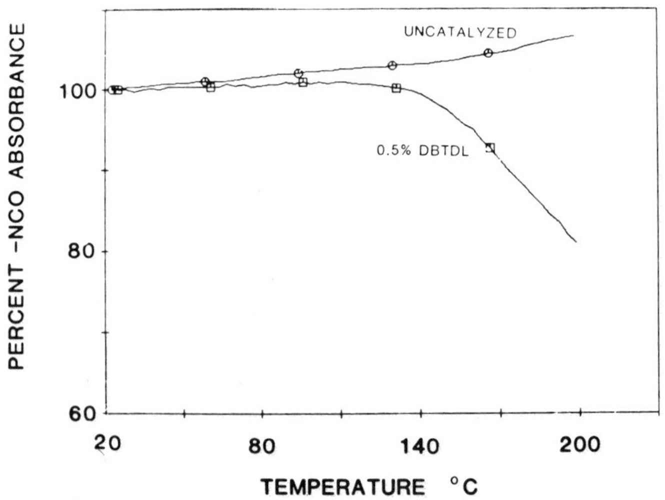

Figure 8. Isocyanate absorbance during heating of T-1890.

The relative modulus measured by DMA, shown in Figure 9, as a function of temperature represents the physical development of cure. Although the deblocking reaction begins near 90°C a modulus increase is not observed until about 170°C. A plot of -NCO absorbance for T-1890 deblocking in the presence of a nonfunctional acrylic is included in Figure 9 to represent the free isocyanate levels expected in the absence of hydroxyl functionality. The difference between this latter curve and the curve obtained in the presence of hydroxyl functionality gives a measure of isocyanate functionality which has deblocked and reacted with the acrylic polymer to form crosslinks. As crosslinks forms the gel point is reached and the modulus begins

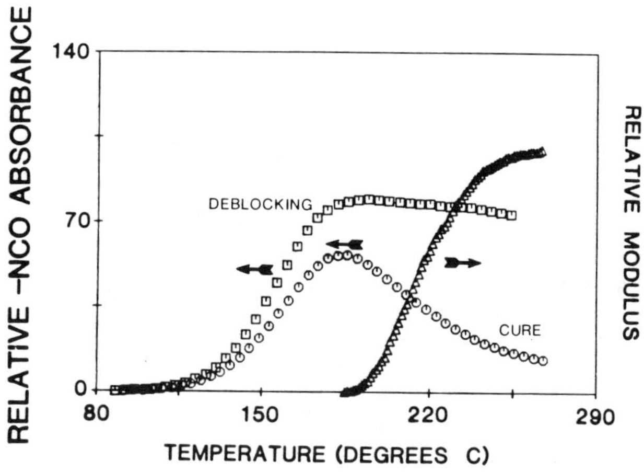

Figure 9 Cure of Uncatalyzed Hydroxy Acrylic.

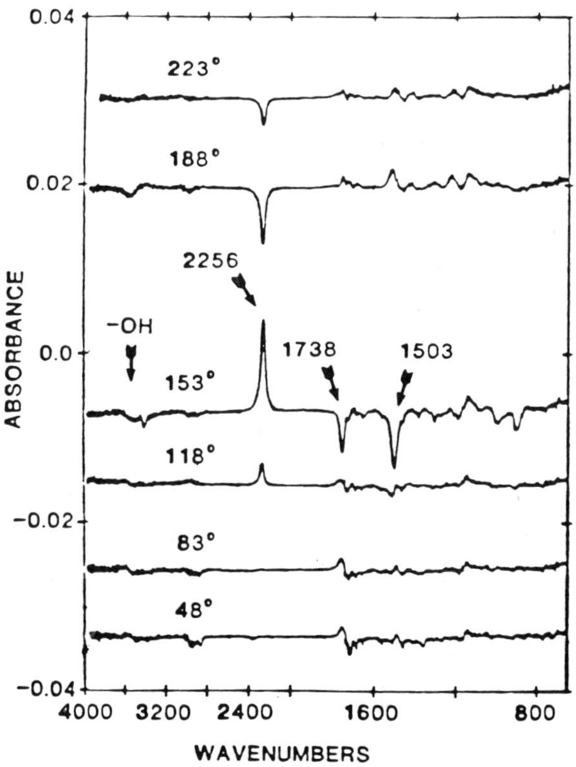

Figure 10 Differential Changes During Uncatalyzed Cure.

to increase. Based on the difference in absorbance between the deblocking curve and the cure curve, the gel point is reached when approximately 40% of the total available isocyanate functionality has formed crosslinks. The delayed cure, in this system, is not a result of a slow deblocking reaction, but rather a slow isocyanate-alcohol cure reaction.

Difference spectra during various stages of the cure reaction yield more information concerning the cure process. As shown in Figure 10, the initial difference spectra at 48°C and 83°C show only small changes, probably related to minor peak shifts and loss of residual solvents. At 118°C, slightly after the onset of deblocking, the difference spectrum shows a significant increase in the isocyanate functionality but no changes in the hydroxyl functionality. At 153°C, deblocking reaches a maximum and the isocyanate peak becomes quite large. Decreases in absorbance are found at 1503 cm^{-1} and 1738 cm^{-1} indicative of deblocking. At 188°C, where deblocking has ceased, a net loss of isocyanate functionality, which in combination with the decrease in hydroxyl absorbance near 3500 cm^{-1}, gives clear evidence for the cure reaction. The peaks at 1503 cm^{-1} and 1738 cm^{-1} have disappeared, demonstrating that the deblocking reaction is complete. As a temperature of 223°C the only appreciable change is a slight decrease in isocyanate functionality during the final phases of cure.

Comparison of the chemical and physical cure curves based on FT-IR and DMA, respectively, are shown in Figure 11 for the DBTDL catalyzed system. The onset of cure occurs at roughly 31% conversion of the isocyanate groups. Throughout the entire cure reaction, the isocyanate absorbance is very small in contrast to the uncatalyzed system where large isocyanate absorbances are observed. Difference spectra at various stages of the cure reaction are shown in Figure 12. At 124°C, a slight isocyanate absorbance is observed with corresponding decreases in absorbances at 1503 cm^{-1} and 1738 cm^{-1} due to decreasing concentrations of the blocked isocyanate. In addition, a loss of hydroxyl functionality is observed at 124°C demonstrating the occurrence of appreciable crosslinking at lower temperatures in the presence of DBTDL. The lack of isocyanate absorbance during the cure reaction is thus due to rapid consumption of free isocyanate in the cure reaction. In the presence of DBTDL catalyst the rate limiting step in the cure reaction is the deblocking step.

CONCLUSIONS

The use of dynamic temperature scanning FT-IR in combination with DMA has been shown to provide new insights into the cure mechanism of blocked isocyanate containing coatings by allowing

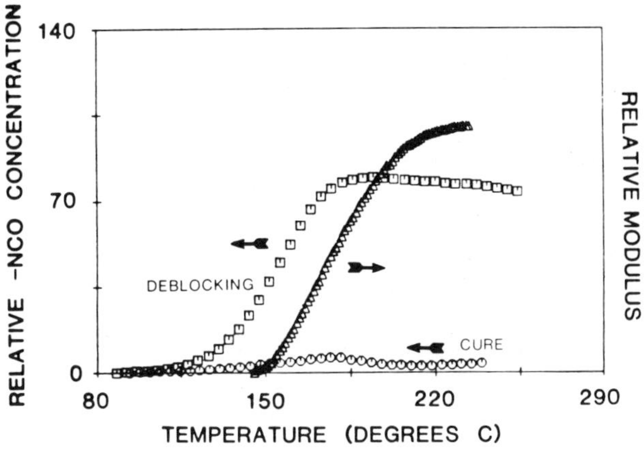

Figure 11. DBTDL Catalyzed Cure.

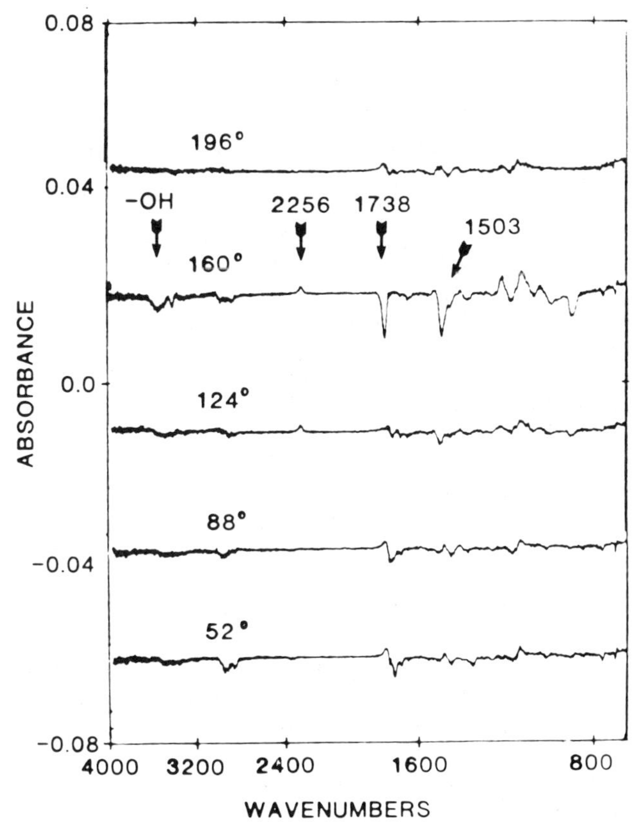

Figure 12. Differential Changes During DBTDL Catalyzed Cure.

the individual deblocking and crosslinking reactions to be observed during the cure process. In uncatalyzed systems it was found that the deblocking and subsequent crosslinking reaction occur at approximately equal rates.

Using the aforementioned techniques it was found that DBTDL does not catalyze deblocking of MEKO blocked T-1890 but does significantly catalyze the cure reaction between free isocyanate and hydroxyl functionalities resulting in a faster cure reaction. DBTDL was also found to catalyze side reactions in neat T-1890 leading to consumption of a portion of the total isocyanate functionality. During the cure reaction, in view of the extremely fast cure reaction, these side reactions are believed to be negligible.

REFERENCES

1. Z.Wicks, Prog. Org. Coatings, 3, 73 (1975).

2. Ibid., 9, 3 (1981).

3. G.M.Carlson, C.M.Neag, C.Kuo and T.Provder, in "Advances in Urethane Science and Technology, Vol.9", K.C.Frisch and D.Klempner, Eds., Technomic Publishing, Lancaster, Pa., (1984) p.47.

4. L.T.Phai, F.Viollaz, Y.Chamberlin, T.M.Lau and J.P.Pascault, Makromol. Chem., 185, 281 (1984).

5. V.Mirgel, Proc. XV Congress AFTPV, Cannes, (1982) p.173.

6. T.Provder, C.M.Neag, G.M.Carlson, C.Kuo and R.M.Holsworth, in "Analytical Calorimetry, Vol.5", P.S.Gill and J.Johnson, Eds., Plenum, New York (1984) p.377.

7. T.Provder, R.M.Holsworth, T.H.Grentzer and S.A.Kline, in "Advances in Chemistry Series, No.203", C.D.Craver, Ed., (1983) p.233.

8. T.Provder, R.M.Holsworth and T.H.Grentzer, ibid., (1983) p.77.

9. M.E.Koehler, A.F.Kah, C.M.Neag, T.F.Niemann, F.B.Malihi and T.Provder, in "Analytical Carorimetry, Vol.5", P.S.Gill and J.Johnson, Eds., Plenum, New York (1984) p.361.

10. P.I.Kordomenos, A.H.Dervan and J.Kresta, J. Coatings Tech., 54, NO687,43 (1982).

HYDROGEN BONDING IN NYLON 66 AND MODEL COMPOUNDS

Dana Garcia* and Howard W. Starkweather, Jr.

Central Research and Development Department
Experimental Station
E.I.DuPont de Nemours & Company
Wilmington, Delaware 19898

*Present address:
The B.F.Goodrich Company
Research & Development Center
9921 Brecksviller Rd., Brecksviller, OH 44141

INTRODUCTION

Ever since its discovery [1] almost a century ago, the hydrogen bond, its character, and effect on molecular structure and properties have been the topic of numerous studies [2,3]. The hydrogen bond is unquestionably an essential idea in explaining the structure of natural proteins and synthetic polyamides as well as small molecule association phenomena, intramolecular ring closure, and catalysis.

Although less extensively studied, the presence of a hydrogen bonded network in polyamides, such as nylon 66, has been known for some time [4-9]. One should nevertheless mention that although most of the existing literature refers to the NH---CO interaction as hydrogen bonding, early work by Cannon [10,11] points out, based on the trans configuration of the amide group, the polarizability of the nitrogen atom and the magnitude of the observed infrared frequency changes on dissociation, that the most favored interaction would be of dipole-dipole nature. Presently this issue remains unresolved. What is clear from the work of Trifan [4] and others [5-9] is the complete association of all amide groups in linear aliphatic polyamides in the crystal and in the amorphous phase at low temperatures. As a result, a number of physical [13] and mechanical properties [12] of these polymers have been attributed to this fully bonded motif and its high degree of retention at elevated temperatures.

The enthalpies (Δ H) of hydrogen bond breaking for a number of polyamides have been reported in the literature [4-6,10] and are in the range from 7 kcal/mol to about 12 kcal/mol. While entropy (Δ S) values have not been directly reported, extrapolation of existing data [6] yields values on the order of 26 e.u.

This paper presents the results of our study of hydrogen bond breaking in Nylon 66. The aim is to investigate the hydrogen bond breaking process, its associated thermodynamic parameters, and the extent to which it has taken place at temperatures close to the melting point.

EXPERIMENTAL

Nylon 66 samples were prepared by casting from a 2% solution of hexafluoroisopropanol directly on a 13 mm diameter NaCl disk. Prior to use the cast film samples were throughly dried in a vacuum oven at 120°C for at least 48 h.

N,N′-diacetylhexamethylenediamine (m.p.=125°C) was obtained from Dr. S.Mazur and N-butylacetamide (m.p.=-11°C) was purchased from Aldrich Chemical Co.

The samples were placed in a variable temperature Barnes infrared cell, taking care to minimize any exposure to moisture. A temperature controller was used with the thermocouple in intimate contact with the NaCl disks containing the sample. The error in the temperature reading was estimated at no more than 1°C. Prior to data collection the cell was equilibrated for at least 5 min at the desired temperature. Nylon 66 and the two monomeric amides showed good thermal stability in the temperature range of interest. On cooling back to room temperature, the initial spectrum was always regenerated.

The infrared spectra were run in a Nicolet 7199 FT-IR equipped with a liquid nitrogen cooled MCT detector. Typically 200-400 scans were averaged at 1.0 cm^{-1} resolution and ratioed against a previously run background of the empty cell obtained at the same temperature. Spectra were recorded at a number of temperatures between 27°C and 270°C.

The Nylon 66 samples had a peak melting temperature of 256°C with the melting process ending at 267°C and a 40% crystallinity [14], as determined by DSC.

RESULTS

The infrared spectra of solution cast Nylon 66 films at 27°C, 200°C and 260°C are shown in Figure 1. It is readily observable that fundamental spectral changes have occured on heating.

At 27°C, the infrared spectrum (Figures 1a and 1b) does not reveal the presence of any absorption above 3400 cm^{-1} indicative of non-bonded amide groups, in agreement with previous workers [4-12,15]. The bonded NH stretching mode is located at 3300 cm^{-1}.

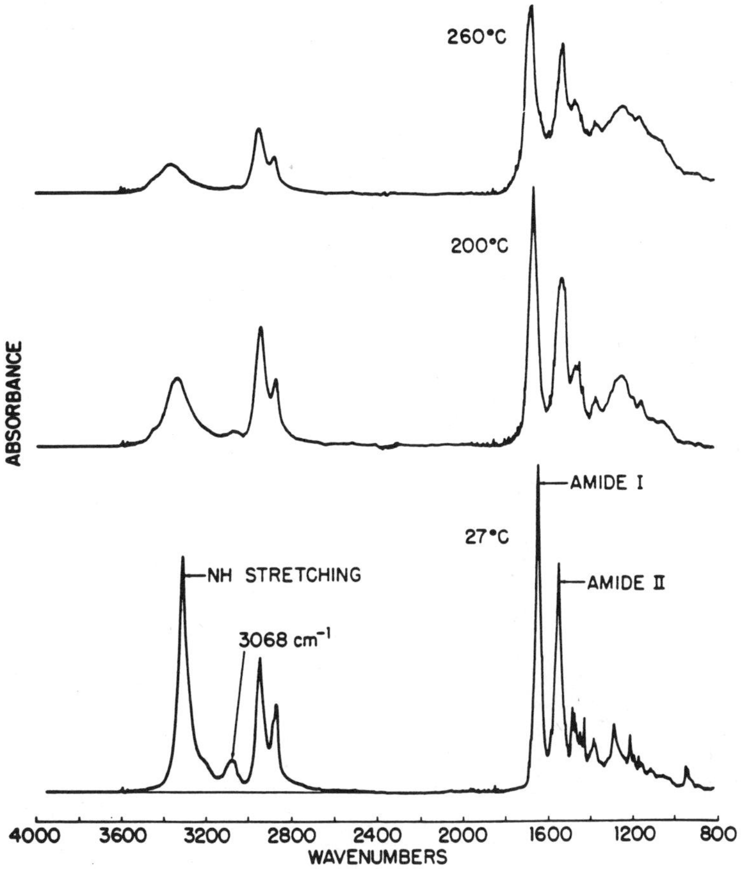

Figure 1a. Spectrum of Nylon 66 at 27°C, 200°C, and 260°C. (All three spectra are on the same absorbance scale.)

With increasing temperature, this band becomes progressively broader, shifts to larger wavenumbers, and develops an increasingly pronounced shoulder above 3400 cm^{-1}. In a similar fashion, the amide I (1634 cm^{-1}) and amide II (1541 cm^{-1}) bands broaden and shift to larger and smaller wavenumbers, respectively, though less dramatically than the NH stretching mode. The exact assignment and origin of the band at 3068 cm^{-1} is presently unresolved. The most likely assignments are the first overtone of the amide II band [10] or a combination amide I + amide II bands [7]. The former possibility is probably the more reasonable. It is based on the peak spectral position in a number of amides and polyamides [10], its temperature dependence, and the results of isotope substitution experiments [16].

The changes thus far described are consistent with the process of hydrogen bond breaking. Equivalently, in terms of dipole interactions, these same frequency and width changes were explained by Cannon [10,11] via increases in the amide-amide contact distance.

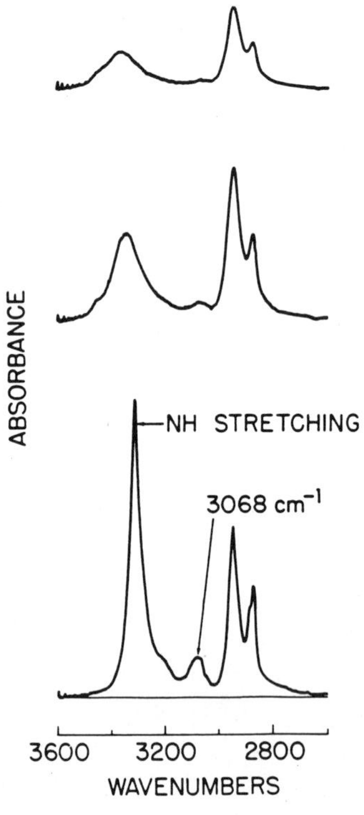

Figure 1b. Enlarged presentation of the spectra in the region of the NH stretching band.

The spectral region between 1470 cm^{-1} and 1000 cm^{-1} also exhibits broadening as the melting point is approached, eventually collapsing to a broad envelope as a result of the loss of coupling in the crystalline segments of the polymer [10,11].

Experimentally, a number of spectroscopic and nonspectroscopic methods [2,3,17] have been used to study the nature and dynamics of hydrogen bonded systems. Probably one of the most widely used methods has been infrared spectroscopy [17]. A molecule exhibiting hydrogen bonding alters its infrared spectrum in a highly dramatic fashion. Clear manifestations of hydrogen bonding are the large frequency shifts and the band broadening of the donor's hydrogen stretching vibration (ν NH 3300 cm^{-1}). Theoretical and experimental studies [2,3,17] have attributed these observed spectroscopic changes to a number of anharmonic coupling effects which could give rise to Fermi resonance, the possibility of a second potential well conductive to tunneling, and the structural disorder of nonequivalent hydrogen bonds; The band shape of the NH stretching vibration (ν NH) broadens even further and develops substantial asymmetry with increasing temperature, making quantitative analysis rather difficult (Figure 1). According to Romanowski [18], changing the temperatures by 100°C results in a tail on the order of 600 cm^{-1} on the low frequency side of the NH band.

Thus, the major problem one encounters in quantitative analysis are the uncertainty in band shape and the effect of overlapping bands. To overcome these problems, we have employed a curve analysis program in conjunction with a judicious choice of baseline (Figure 1). The curve analysis program (a detailed description can be found in reference 19), is a band fitting routine to Gaussian and Lorentzian shapes. From the total bonded and nonbonded NH stretching band, we have deconvolved the contributions of the 3068 cm^{-1} band and the CH stretching band below 3000 cm^{-1}. The low frequency shoulder on the 3300 cm^{-1} band was included in the calculation of the NH stretching area. We have not separated the nonbonded NH band from its bonded contribution; such a deconvolution is not needed for our data analysis.

The equilibrium between bonded and nonbonded amide can be represented according to Eq. (1), with an associated equilibrium constant K and thermodynamic parameters given by Eq. (2).

$$(\text{NH---OC})_{\text{bonded}} \quad (\text{NH})_{\text{free}} + (\text{O = C})_{\text{free}} \tag{1}$$

$$K = \frac{[\text{NH}]_{\text{free}}\,[\text{O = C}]_{\text{free}}}{[\text{NH---O = C}]_{\text{bonded}}} = \frac{C_o[1 - x_b]^2}{x_b} = \exp\left(-\frac{\Delta H}{RT} + \frac{\Delta S}{R}\right) \tag{2}$$

where C_o is the total concentration of amide groups and X_b the bonded amide fraction. In terms of experimentally measured quantities and remembering that essentially all amide groups are hydrgen bonded [4-12] at room temperature, the bonded amide fraction (X_b) becomes Eq. (3):

$$X_b = \frac{[A(T)/A_o - E_t/E_b]}{[1 - E_f/E_b]} \quad (3)$$

where A_o and A(T) are the integrated absorption band areas for the NH stretching mode (bonded and nonbonded) at room temperature and some higher temperature T, and E_f and E_b the integrated extinction coefficients for the free and bonded amide NH stretching mode. A similar data analysis can be found in the work of Schroeder and Cooper [6].

The E_f/E_b ratio for the NH stretching bands was calculated using the model compound data of MacKnight and Yang [5] for the value of E_f (4.47×10^3 l/mol.cm^2) and our experimentally determined value for E_b (1.44×10^4 l/mol.cm^2) according to their procedure [5]. The extrapolation from the model compound to the polymer will undoubtedly introduce some error in the value of the extinction coefficient ratio. Unfortunately, no direct-measurement could be made on the polymer because of the lack of an appropriate nonhydrogen bonding solvent. While we do not have an estimate of the error involved, the small changes in the integrated extinction coefficient observed by MacKnight and Yang [5] within a polyamide series indicates that the magnitude of the error introduced probably is not substantial. Furthermore, we checked the effect of varying the E_f/E_b ratio on the thermodynamic parameters of the hydrogen bond breaking process (Eqs. (2) and (3)) and observed approximately a 5-7% change in H and S for a 32% change in the E_f/E_b ratio.

The preceeding analysis also implicitly assumes that in the absence of chemical changes, in our case hydrogen bond breaking, the integrated area of the absorption band is independent of temperature. As long as the correct band shape has been integrated, this assumption is justified [20].

Figure 2 shows the bonded amide fraction (X_b) as a function of temperature. At temperatures below the onset of the melting transition (below 230°C by DSC), the bonded amide fraction is always higher than the experimentally determined crystalline content of the sample (40%), consistent with the belief that hydrogen bond breaking does not occur in the crystal. Collaborating evidence to this effect was obtained from temperature dependent X-ray diffraction studies by Starkweather and Jones [14]. They observed that the crystal interchain separation within the hydrogen bonded sheets is almost independent of temperatures up to 250°C. In addition, our own studies on N,N′-diacetylhexamethylenediamine, a compeltely crystalline material at room temperature, have shown no spectral changes up to the melting point (125°C), where we first start observing evidence for hydrogen bond breaking.

A van′t Hoff plot was constructed using the data in Figure 2 in conjunction with Eq. (2) (see Figure 3). The enthalpy (ΔH) and entropy (ΔS) of the hydrogen bond breaking process can thus be obtained from the slope ($-\Delta H/R$) and the intercept ($\Delta S/R - \ln C_o$) of the linear least squares line fitted through the experimental points (see Table 1).

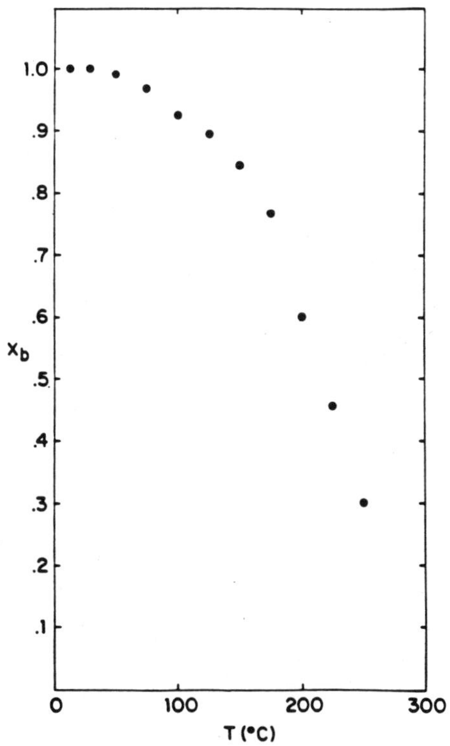

Figure 2. Plot of the total hydrogen bonded fraction (X_b) versus temperature (T) for Nylon 66.

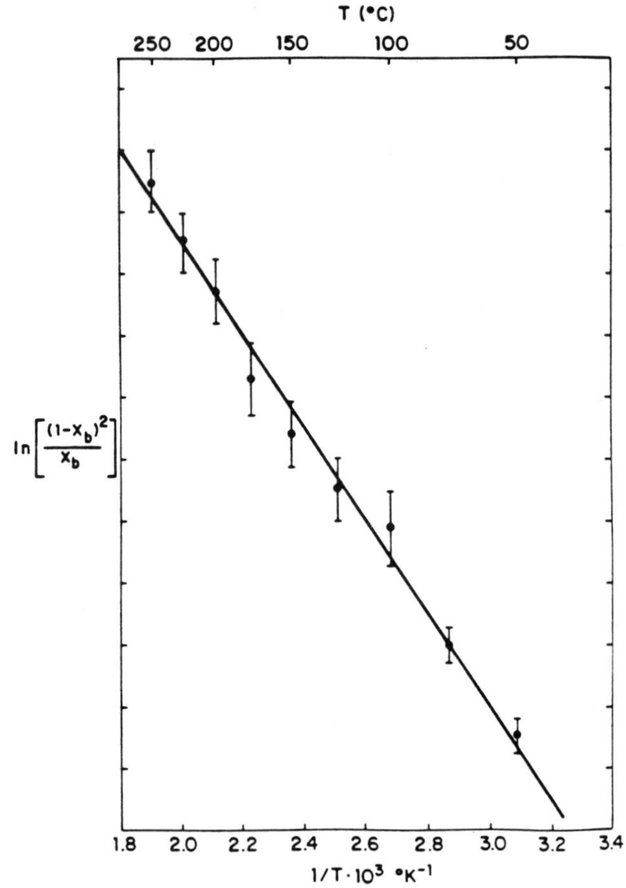

Figure 3. Van't Hoff plot for Nylon 66.

Our value of 14.83±0.70 kcal/mol of NH, for the enthalpy of hydrogen bond breaking in Nylon 66 is higher than the average value for polyurethanes and polyamides reported by Trifan and Terenzi [4] (8.36 kcal/mol) as well as MacKnight and Yang [5] (8.5-10.5 kcal/mol) for a number of polyurethanes or calculated by Cannon [10,11] from dipole-dipole interactions. Better agreement is found with the work of Schroeder and Cooper [6] which quotes values of 11-12 kcal/mol for semicrystalline polyamides such as Nylon 11 and Nylon 12.

Remembering from the preceeding paragraphs that we have established the absence of hydrogen bond breaking in the crystal below the melting temperature, we believe that a correct data analysis should include an approproate correction to account for the 40% crystalline content of our sample.

The bonded amide fraction (X_b^o) in the amorphous fraction of the polymer, assuming that E_b is the same for both the amorphous and the crystalline fractions, then becomes

$$X_b^o = \frac{A(T)/A_o - E_f/E_b}{1 - E_f/E_b} - 0.40 \qquad (4)$$

and the corrected equilibrium constant K_{corr} will be given by Eq. (5).

$$K_{corr} = \frac{C_o(0.6 - X_b^o)^2}{X_b^o} \qquad (5)$$

The van't Hoff plot using the corrected expression for the equilibrium constant (Eq. (5)) is shown in Figure 4 with its accompanying thermodynamic parameters in Table 1 (line 2).

The thermodynamic parameters (H and S) for the hydrogen bond breaking process in Nylon 66 (Table 1) are substantially higher than was expected for hydrogen bonds in model amides [2,21]. To determine if our values were singly characteristic of the polymer, we have studied N,N'-diacetylhexamethylenediamine and N-butylacetamide as diamide and monoamide model compounds.

In each case, a fully hydrogen bonded spectrum was obtained at low temperatures and the hydrogen bond breaking process was monitored above the compound's melting point. The data analysis was based on the NH stretching band using the procedure previously outlined (Eqs. (1) - (5)).

TABLE I

Thermodynamic Parameters for Hydrogen Bond Breaking in Nylon 66

Band Analysis	Crystallinity Correction	ΔH (Kcal/mol of NH)	$\Delta S/R - \ln C_o$ [a]	ΔS (e.u.)
NH stretch	no	14.83±0.70	14.45±0.80	33.29
NH stretch	yes	13.35±0.70	13.05±0.90	30.50

a. C_o = total concentration of NH groups[6] = $2\ \rho/M$; where ρ = density in g/cm^3 (ref. 29) and M = molecular weight of the repeat unit in g/mol.

TABLE II

Thermodynamic Parameter for Hydrogen Bond Breaking in Model Compounds

Compound	ΔH (kcal/mol of NH)	$\Delta S/R - \ln C_o$	ΔS (e.u.)	MP (°C)
N-burylacetamide[a]	7.15±.30	7.83±.40	19.81	-11
N,N'-diacetylhexamethyl-enediamine[b]	8.55±1.40	9.61±1.35	23.67	125

MP = melting point
a. $C_o = \rho/M$
b. $C_o = 2\ \rho/M$

Where C_o = total concentration of NH groups; M = the molecular weight and ρ = the density

The van't Hoff plots are shown in Figure 5 and the associated thermodynamic parameters in Table 2. We have also plotted the hydrogen-bonded amide fraction as a function of temperature for the two model compounds and for the amorphous component of nylon 66 (Figure 6). The large error bars associated with N,N'-di-acetylhexamethylenediamine are due to sample scattering effects and changes in the optical pathlength.

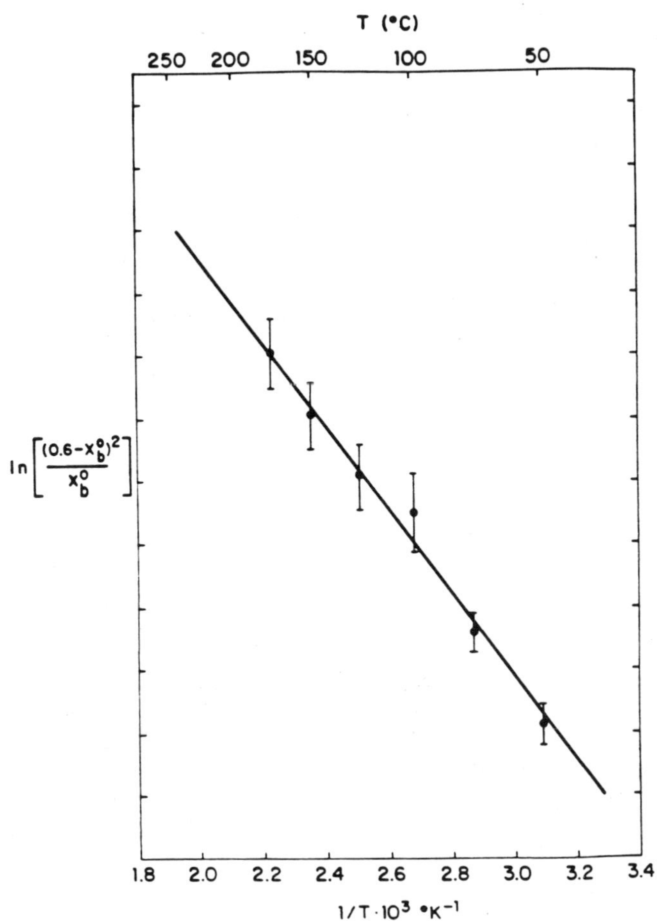

Figure 4. Van't Hoff plot using the crystallinity correction for Nylon 66.

DISCUSSION

When discussing hydrogen bonding, one has to consider both the structures of the bonded adducts and the nature of the surrounding medium. In the case of amide dimers in solution, these effects have been extensively investigated in the literature [2,21]. The effect of the surrounding medium (solvent) is usually

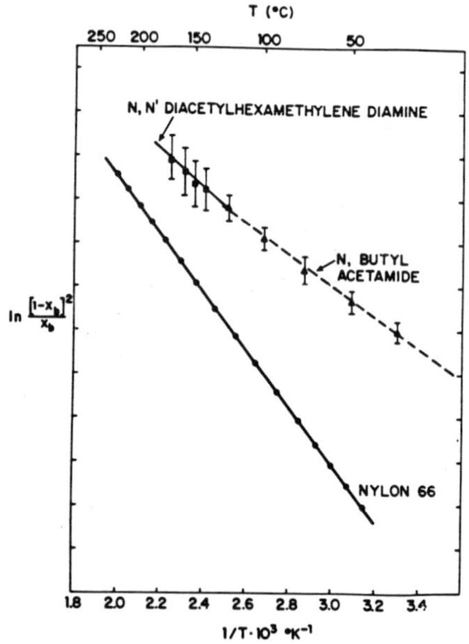

Figure 5. Van't Hoff plots for n-butylacetamide and N,N'-diacetylhexamethylenediamine.

explained in terms of the solvent's dielectric constant [21]. From a simple electrostatic model, the thermodynamic constants (ΔH, ΔS) are linearly proportional to the reciprocal of the dielectric constant of the solvent. Futhermore, the enthalpy and entropy are also highly dependent on the structure of the dimeric adduct [2,21]. If a cyclic structure is obtained, as has been proposed for ε-caprolactam, the magnitude of the thermodynamic parameters per hydrogen bond is higher than for a linear dimer such as N-methylacetamide. A strong dependence is also found with the magnitude of the hydrogen bond length [2,21].

In our study, the analysis is substantially complicated by the fact that we are not dealing with isolated molecules, but with a hydrogen bonded polymeric motif. This may explain the lack of information on amide model compounds in the neat state. We are thus forced to present a more qualitative picture of the hydrogen bond breaking process than we would have liked.

Comparing the data for Nylon 66 (Table 1, line 2) with that for amide model compounds (Table 2), it is easily seen that the polymer exhibits substantially higher enthalpy and entropy values. Furthermore, no difference within experimental error is observed between the monoamide and the diamide model componds (Table 2). This leads us to believe that no interaction is occuring between the adjacent amide groups in the polymer chain. Such a conclusion, of course, does not imply the absence of a cooperative effect in the transverse hydrogen bond chain direction. Thus, the obtained thermodynamic parameters for both polymer and model compounds are not characteristic of the breaking of a single isolated hydrogen bond, but reflect a multistep process which we have approximated via one equilibrium constant (Eqs. (1) and (2)). To completely describe our system, the equilibrium constant for each hydrogen bond breaking step would be needed. Unfortunately, the complexity of the spectra precludes any attempt to deconvolve each contributing species from bydrogen bonded dimer to hydrogen bonded polymer.

Faced with a similar problem in their study of N-methylacetamide, Loenstein and coworkers [26] resorted to an approximate description in which the bonded and nonbonded amide fractions were related to two equilibrium constants for dimerization and formation of higher order hydrogen bonded aggregates. Following on their procedure, we have attempted to model our system in a similar fashion. The results of our calculations for both Nylon 66 and model compounds revealed that a two equilibrium constant model is still inadequate to explain the complexity of the hydrogen bond breaking process. Our major concern when we undertook this calculation was the unusually high values of the entropies and enthalpies obtained using a single equilibrium constant model.

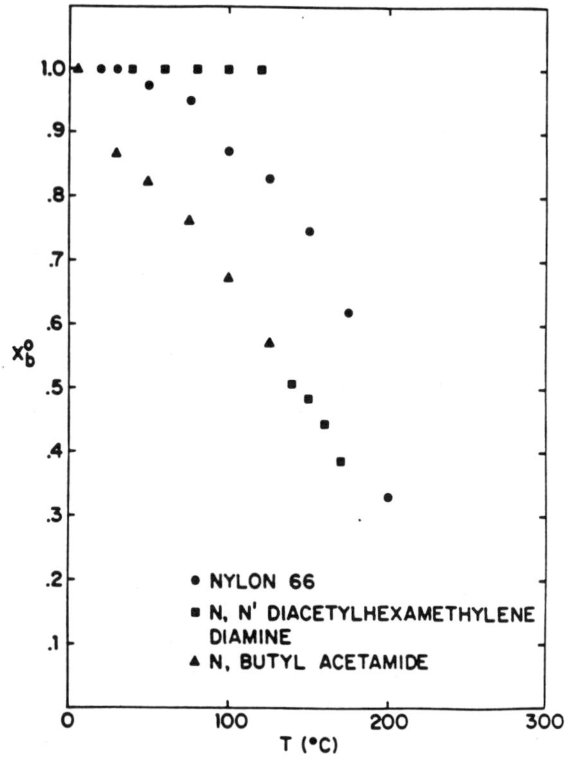

Figure 6. Plot of hydrogen bonded fraction (X_b^o) for Nylon 66 (amorphous component only), n-butylacetamide and N,N'-diacetylhexamethylenediamine as a function of temperature (T).

It soon became apparent that even when two equilibrium constants were employed, the thermodynamic parameters of both steps could not be lowered simultaneously by a substantial amount. As mentioned before, this may be a manifestation of the complexity of a hydrogen-bond breaking process. Furthermore, comparing our results with cited literature [2,21] values for the dimerization process of amides in solution may not be appropriate in view of the different adduct structures and dielectric constants of the media. Theoretical [27] and nonspectroscopic [28] studies of hydrogen bond breaking in small amides, yielded values of 8.6 kcal/mol-9.54 kcal/mol for the strength of the hydrogen bond, in good agreement with our own findings for model compounds. In terms of model compound-polymer analysis, we have to remember that the model compounds were studied in the melt where molecular mobility is high compared to the amorphous portion of the polymer below its melting point temperature. While in the melt, conformational rearrangement is not expected to require as much energy. In Nylon 66, to achieve a nonbonded conformation required overcoming a number of rotational energy barriers (Rotation around the C-C bond in ethane has a potential energy barrier of 2.80 kcal/mol [22]) about the chain bonds which will contribute to the enthalpy values. This line of reasoning is also consistent with the observed increase in entropy (Table 1, line 2 and Table 2).

Besides conformational contributions, we also expect differences in dielectric constant between the model compounds and Nylon 66. Although specific dielectric data were not available on the model compounds we studied, comparison with amides for which dielectric data exists [21] indicates that they exhibit higher dielectric constants than Nylon 66 which would be reflected in lower enthalpy and entropy values for the latter. When attributing effects to the dielectric constants of the medium, caution should be exercised, and we should remember that the dielectric constant is sensitive to the temperature and the degree of hydrogen bonding [21].

Changes in the hydrogen bond length are also reflected in the values of the thermodynamic parameters. The shorter the length, the stronger the bond will be. The hydrogen bond lengths in the crystal for N,N′-diacetylhexamethylenediamide [23] and Nylon 66 are 0.288 nm and 0.297 nm, respectively. While these values cannot be directly transfered to the melt and amorphous phase of Nylon 66 and remembering that a distribution of hydrogen bond lengths exists, there is no reason to believe that on the average the magnitude of the difference will be substantially greater than in the crystal. It is then unlikely that the enthalpy difference between N.N′-diacetylhexamethylenediamine and Nylon 66 is a result of bond length effects.

Regarding the degree of hydrogen bonding as a function of temperature, we have to conclude from our data (Figure 2) that below the onset of melting, only hydrogen bonds in the amorphous polymer fraction break. With the onset of melting and the collapse of the crystal structure, hydrogen bonds which had been present in the crystal also break. Extrapolation of our data above the melting point yields a hydrogen bonded fraction of 0.266 at 270°C and 0.166 at 300°C in agreement with the data of Cooper [6] on Nylon 11 and Nylon 12. These results are also consistent with the previously proposed idea [25] of a substantial degree of hydrogen bonding in the melt.

CONCLUSIONS

Hydrogen bond breaking in Nylon 66 and model compounds is a complex multistep process. Our results have shown that while no interaction is occuring between adjacent amide groups in the polymer chain, there is, nevertheless, a cooperative effect in the transverse hydrogen-bond direction. The thermodynamic parameters obtained via the one equlibrium contant approximation are not characteristic of the breaking of a single hydrogen bond, but reflect the complexity of our system. Furthermore, for Nylon 66 there are additional contribution to the enthalpy due to rotational barriers around the polymer chain which have to be overcome. We have also established that hydrogen bonds in Nylon 66 crystals do not break below the onset of melting and that any analysis of hydrogen bond breaking in semicrystalline polymers should include such a crystallinity correction.

ACKNOWLEDGEMENT

The authors thank Drs. A.D.English, S.Mazur, B.Matheson, and Prof. S.Krimm for many useful discussions, comments, and suggestions. The technical assistance of Mr. G.Wilson is also acknowledged.

1. W.Nerust, Z. Physik. Chem., 8, 110 (1981).

2. G.C.Pimentel and A.L.McClellan,"The Hydrogen Bond", Reinhold, New York (1960).

3. P.Shuster, G.Zundel and C.Samdorphy,"The Hydrogen Bond", North Holland, New York (1976).

4. D.S.Trifan and J.F.Terenzi, J. Polym. Sci., 28, 443 (1958).

5. W.J.MacKnight and M.Yang, J. Polym. Sci. Symposium, 42, 817 (1973).

6. L.R.Schroeder and S.L.Cooper, J. Appl. Phys., 47, 4310 (1976).

7. E.Bessler and G.Bier, Makcomol. Chem., 122, 30 (1969).

8. J.Dechant and R.Danz, Plaste Kautschuk, 19, 250 (1972).

9. A.Anton, J. Appl. Polym. Sci., 12, 2117 (1968).

10. C.G.Cannon, Spectrochim. Acta, 16, 302 (1969).

11. ibid., Mikrochim. Acta, 2-3, 555 (1955).

12. S.Furuhara, Y.Suzuki and S.Omiote, Kautschuk Gummi Kunststoffe, 30, 434 (1977).

13. C.Carfagna, M.Vacatello and P.Corradini, J. Polym. Sci. Polym. Chem. Edi., 15, 1 (1977).

14. H.W.Starkweather and G.A.Jones, J. Polym. Sci. Polym. Phys. Ed., 19, 467 (1981).

15. U.G.Gafunov and I.I.Novak, Zh.. Prikl. Spektr., 15, 690 (1971).

16. M.Beer, H.B.Kessler and G.B.Sutherland, J. Chem. Phys., 29, 1097 (1958).

17. T.M.Theophanides,"Infrared and Raman Spectroscopy of Biological Molecules", D.Reidel, Holland (1978).

18. H.Romanowski and L.Sobozyk, Chem. Phys. Lett., 58, 73 (1978).

19. J.S.Nelder and J.Mead, Computer J., 7, 308 (1965).

20. E.B.Wilson, J.C.Decius and P.C.Cross,"Molecular Vibrations", McGraw Hill, New York (1955).

21. S.N.Vinogradov and R.H.Lindell,"Hydrogen Bonding", Van Nostrand Reinhold, Holland (1971).

22. P.J.Flory,"Principles of Polymer Chemistry", Cornell University Press, Ithaca (1978).

23. M.Bailey, Acta Crystallogr., 8, 575 (1955).

24. Bernard-Houplain, Marie-Claude and C.Sandorfy, Can. J. Chem., 51, 3640 (1973).

25. H.W.Starkweather, J.F.Whitney, and D.R.Johnson, J. Polym. Sci. A, 1, 715 (1963).

26. H.Lowenstein, H.Lassen and A.Hvidt, Acta Chem. Scand., 24, 1687 (1970).

27. P.Hobza, F.Mulder and C.Sandorfy, J. Am. Chem. Soc., 104, 925 (1982).

28. M.Davies and H.Jones, Trans. Faraday Soc., 55, 1329 (1959).

29. M.J.Kohan,"Nylon Plastics", John Wiley and Sons, New York (1973).

COMBINATION OF DIFFUSE REFLECTANCE FT-IR SPECTROSCOPY, FOURIER SELF-DECONVOLUTION AND CURVE-FITTING FOR THE INVESTIGATION OF REACTING COALS

Peter R. Griffiths and Shih-Hsien Wang

Department of Chemistry
University of California
Riverside, CA 92521

INTRODUCTION

Coal is a remarkably complex mixture of polymeric materials with mineral inclusions. The organic macerals are believed to consist of a skeletal structure which is composed of clusters of condensed aromatic and hydroaromatic nuclei, joined by methylene, polymethylene, or ether linkages. This lattice is of high molecular weight and is believed to contain smaller, more hydrogen rich, molecules in the pores. The infrared spectrum of most coals is relatively featureless, consisting of several broad bands assignable to common functional groups such as CH_2, CH_3, C=O, etc., but giving little detailed information on the precise molecular environment of each group. Several workers [1-4] have shown that overlapping bands in coal spectra may be resolved by Fourier self-deconvolution [5]. We have assigned many of these features to functional groups in specific environments.

An example of the power of Fourier self-deconvolution (FSD) to resolve absorption features caused by the aliphatic C-H streching modes of a medium volatile bituminous coal is shown in Figure 1. Also shown in this figure is the fourth derivative of the spectrum computed by fitting a small region of the spectrum to a seventh order polynomial and calculating the derivative of this function [7]. Each feature in the fourth derivative spectrum was also observed in the deconvolved spectrum. Since the derivative spectrum was not calculated using a Fourier transform, the possibility that the features observed in the deconvolved spectrum are due to truncation artifacts (side-lobes) is small. The fact that all features are reproducible indicates that they are not due to noise in the original spectrum. It is noteworthy that second derivative spectra of coals which had been reported previously [8,9] did not show as many spectral features as the spectra shown in Figures 1b and 1c.

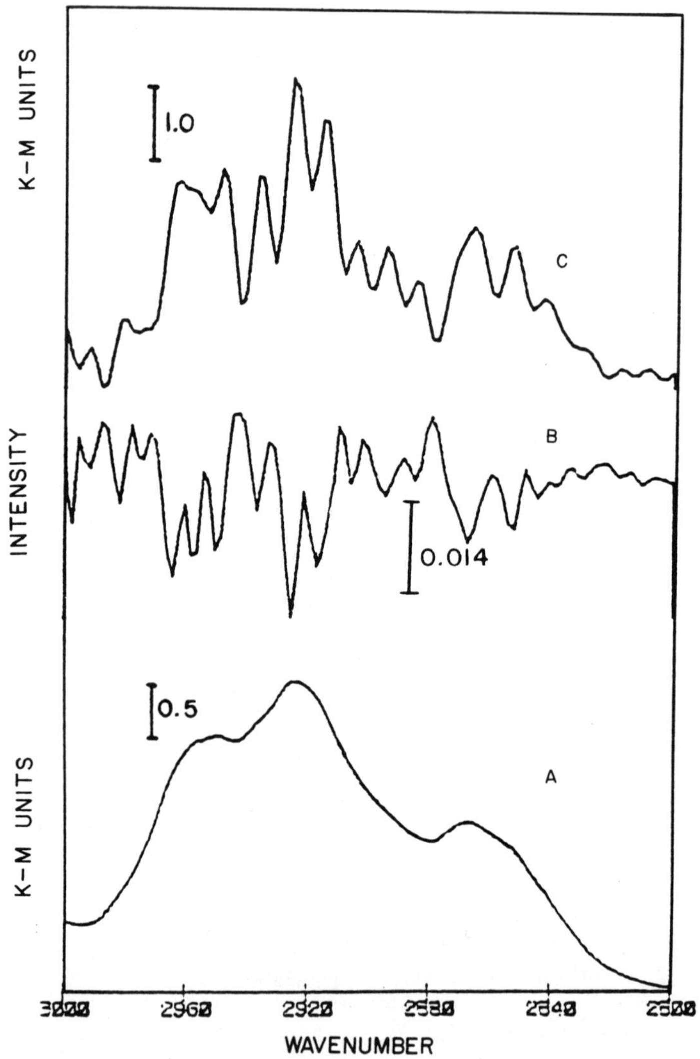

Figure 1 (a) Diffuse reflectance spectrum of medium volatile bituminous coal PSOC 990 in the region between 3000 and 2800 cm^{-1}. (b) Fourth derivative of this spectrum. (c) Deconvolved spectrum computed with $\gamma' = 10\ cm^{-1}$

Quantitative analysis of samples from an even-order derivative spectrum is, of course, possible and has been achieved in infrared spectrometry [10]. Nevertheless for this approach to be useful, the bands in the derivative spectrum must be well resolved. In the case of the infrared spectra of coals, this criterion is never met.

Spectra which have been subjected to FSD should be more useful for quantitative analysis than derivative spectra since the areas of each band are identical before and after deconvolution. This can be proven mathematically, since the first point in the Fourier domain array (the intensity of which is directly proportional to the integrated area of all peaks in the spectrum) is multiplied by unity in the deconvolution operation. Unless all spectral features are completely resolved, however, the effect of overlap by neighboring bands still precludes the accurate determination of each absorbing species by a simple measurement of band intensities. We have therefore attempted to fit spectra which had been subjected to FSD using a least-squares curve-fitting routine, and to determine the areas of each peak in the deconvolved spectrum from the parameters giving the least-squares best fit.

Previously we had attempted to curve-fit the spectra of coals in several spectral regions using techniques which have been described by Painter et al. [8,9], but we found that a given spectrum could be fit to approximately the same degree of accuracy by many different combinations of bands. Through the application of FSD, not only could the number of bands included in the curve-fitting program be limited to those observable in the deconvolved spectrum, but the initial estimate of their absorption frequencies could also be obtained from this spectrum.

Three parameters are usually varied for each spectral feature, i, in order to obtain the best fit. These are the peak absorbance, A_i^o, full width at half-height, $2\gamma_i$, and the wavenumber corresponding to the maximum absorbance, $\bar{\nu}_i^o$. If poor initial estimates of these parameters are made, it is easy to obtain a "best-fit" which is totally incorrect. To avoid this possibility, the values of $\bar{\nu}_i^o$ which are entered into the program were obtained directly from the deconvolved spectrum, and none of the $\bar{\nu}_i^o$ values were allowed to change during the first set of iterations. After the best fit had been obtained in this manner, all three parameters (A_i^o, γ_i, and $\bar{\nu}_i^o$) were permitted to vary. In this way we believed that the result would be a meaningful best-fit to the deconvolved spectrum. The procedure of only permitting the number of bands in the fit to equal the number observed in the deconvolved and fourth derivative spectra and not allowing the $\bar{\nu}_i^o$ values to vary initially from the values obtained directly from the deconvolved spectrum gives us much more confidence in the validity of the fit.

Besides the three parameters discussed above, one other parameter can also be changed in many curve-fitting programs, including the one used in this work. This parameter controls the band shape [11,12]. In the work described in this paper, the band shape is assumed to be a linear combination of a Lorentzian and a Gaussian profile, i.e. the absorbance $A(\bar{\nu})_i$ of any component band at wavenumber, $\bar{\nu}$, is given by:

$$A(\bar{\nu})_i = \alpha A_i^o \frac{\gamma_i^2}{\gamma_i^2 + (\bar{\nu} - \bar{\nu}_i^o)^2} + (1 - \alpha)\, A_i^o \exp - \left[\frac{(\ln 2)\gamma_i}{(\bar{\nu} - \bar{\nu}_i^o)} \right]^2 \quad (1)$$

where α=1 for a pure Lorentzian profile and α =0 for a pure Gaussian profile.

It has been reported by Solomon et al. [13,14] that the entire mid-infrared spectrum of coal can be fit by the sum of 26 Gaussian bands, and that Gaussian bands give a significantly better fit than Lorentzians. However, many of these 26 bands can be further resolved by FSD. For example, whereas Solomon et al. required only four bands in the aliphatic C-H stretching region (3000-2800 cm^{-1}), thirteen bands are resolved by FSD as shown in Figure 1. We have therefore investigated how the apparent shapes of bands which are resolved by FSD change as the extent of deconvolution increases, and the results will be reported in this paper.

EXPERIMENTAL

All manipulations were performed on the diffuse reflectance (DR) spectra of neat powdered coals which had been ratioed against the spectrum of powdered KCl and converted to the Kubelka-Munk (K-M) function, $f(R_\infty)$. The intensity of bands in the K-M spectrum should be proportional to sample concentration and the band shapes should be equivalent to transmission spectra plotted linearly in absorbance.

Spectra were measured using a Digilab Model 296 interferometer equipped with optics for DR spectrometry designed and constructed in our laboratory [15,16]. One thousand successive scans, measured with a data acquisition rate of 20 kHz, were signal-averaged to obtain each spectrum (acquition time about 8 minutes per spectrum). All spectra were measured at 4 cm^{-1} resolution for archival purposes, but it was rarely found necessary to use more than half of the Fourier domain array during Fourier self-deconvolution.

All coals were obtained from the Pennsylvania State University/ Department of Energy coal bank. Samples were ground for 2 minutes in a Wig-L-Bug grinder (Crescent Manufacturing Co., Chicago, IL). During the investigation of air oxidation of coals at elevated temperature, about 0.5 g of the powdered coal was placed on a glass Petri dish which was in turn placed in an oven. Aliquots weighing approximately 60 mg were removed at the desired intervals and transferred directly to a sample cup for measurement of the DR spectrum.

The programs for curve-fitting and FSD were written by D.J.Moffatt at the National Research Council of Canada(NRCC) and supplied to us by Dr. D.G.Cameron. The FSD program was modified slightly to permit deconvolution to be specified directly in terms of γ', the amount by which the half-width of each band is decreased. We prefer this parameter to the more commonly used "resolution enhancement parameter", k, given by:

$$k = \frac{\gamma_i}{\gamma_i - \gamma'}$$

since γ_i is generally different for all bands in the spectrum.

RESULTS AND DISCUSSION

Deconvolved spectra can be fit remarkably well. The result of fitting the deconvolved spectrum shown in Figure 1 (computed with $\gamma'=10\ cm^{-1}$) with thirteen components is illustrated in Figure 2. Each of the thirteen components is shown together with the difference spectrum between the sum of these bands and the deconvolved spectrum. In Figures 3 and 4, the corresponding data are shown for the same original spectrum subjected to smaller amounts of deconvolution (γ'=8.5 and 7.0 cm^{-1}, respectively). The values of $\bar{\nu}_i^o$ for each component band were the same as those found when $\gamma'=10\ cm^{-1}$.

Although curve-fitting techniques have been applied previously to the infrared spectra of coals in this region, the original spectrum was always fit without reducing the widths of the bands. The number of bands in the most reliable investigations was estimated from the second derivative of the spectrum [8,9]. Fewer component bands were observed in the second derivative spectrum of a given coal than we observed in either the fourth derivative spectrum or the deconvolved spectrum computed with $\gamma'=10\ cm^{-1}$. In addition, we found that unless some information is included in addition to the number of features observed in these spectra, it is possible to fit the spectrum equally well by many different combinations of parameters. For example, the result of using a completely different set of parameters to the ones used for Figure 4 in order to fit the spectrum computed with $\gamma'=7\ cm^{-1}$ is shown in Figure 5. The standard deviations of the difference spectra shown in Figures 4 and 5 are almost identical.

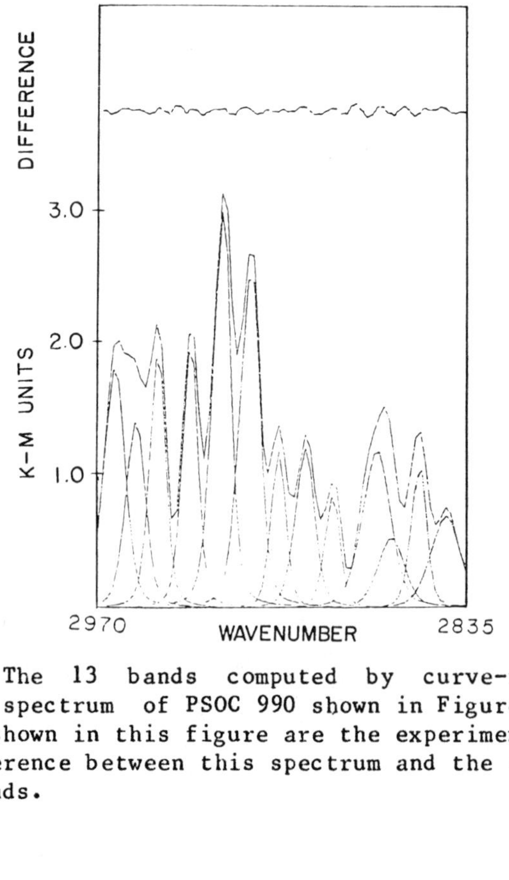

Figure 2. The 13 bands computed by curve-fitting the deconvolved spectrum of PSOC 990 shown in Figure 1, with α - 0.30. Also shown in this figure are the experimental spectrum and the difference between this spectrum and the sum of the 13 component bands.

The values of A_i^o and γ_i used for Figure 5 were then changed to $A_i^o(\gamma_i/\gamma_i-3)$ and (γ_i-3), respectively, and the values of $\bar{\nu}_i^o$ were left unchanged. The result of subtracting the synthetic spectrum computed using these parameters from the experimental spectrum computed with $\gamma'=10\ cm^{-1}$ is shown in Figure 6. Comparison of the difference spectra in Figures 2 and 6 illustrates the superiority of the parameters used for Figure 2, even though the difference spectra in Figures 4 and 5 were very similar.

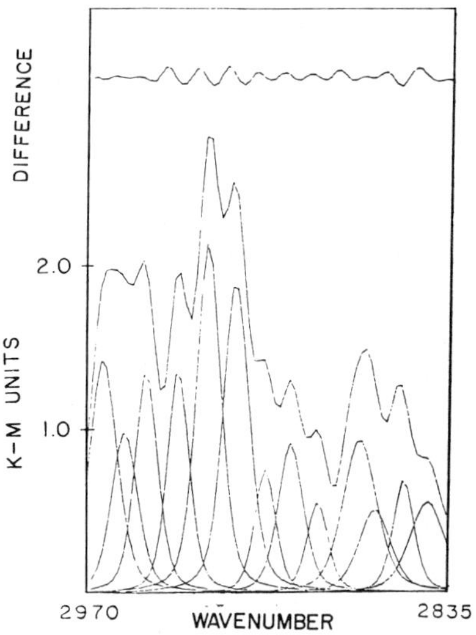

Figure 3 Result of curve-fitting the spectrum of PSOC 990 computed with $\gamma' = 8.5\ cm^{-1}$. In the region between 2970 and 2835 cm^{-1}. Values of ν_i^o for each band were taken from the data shown in Fig. 2, and A_i and γ_i were allowed to vary; $\alpha = 0.50$ for each band.

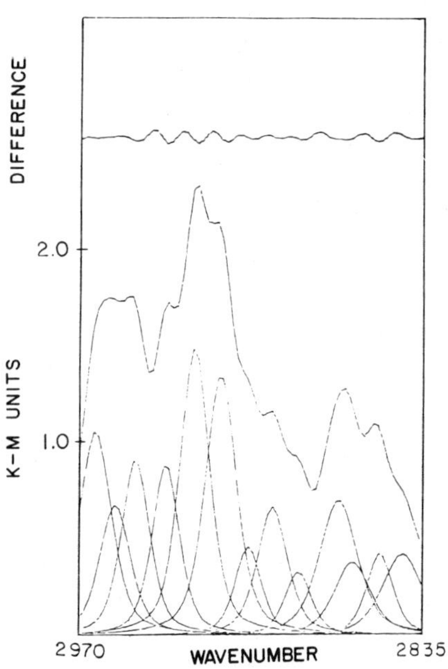

Figure 4 Corresponding data to Figure 3, but for deconvolution with $\gamma' = 7.0\ cm^{-1}$, $\alpha = 0.60$ for each band.

During this work it was found that the bandshape required to obtain the best fit changed as the value of γ' was altered. The greater the extent of deconvolution, the greater the proportion of Gaussian component $(1-\alpha)$ in Eq. (1) that was required to obtain the best fit. This finding should not be unexpected for samples as complex as coals. Each of the component bands shown in Figure 2 has been assigned to aliphatic C-H groups in different molecular environments [3], yet none is due to a single molecule. There is a range over which each molecule containing one of these functional groups will absorb, and it is not unlikely that there will be a normal probability distribution of peak wavenumbers among this ensemble of molecules. Thus even though the profile of each band in the spectrum of an individual molecule may be Lorentzian, the Gaussian distribution of peak wavenumbers will impart a considerable fraction of a Gaussian profile for the ensemble of different molecules found in coal.

The shape of the component bands in coal spectra can probably be better approximated as a Voigt profile, i.e. the convolution of a Lorentzian and Gaussian shape, rather than the linear combination shown in Eq. (1). However Voigt profiles are very hard to simulate [17-19], and the linear combination should be an adequate approximation. The result of deconvolution of a Voigt profile in the Fourier domain using an exponential weighting function, $\exp(2\pi\gamma' x)$, will be to progressively decrease the Lorentzian component of the profile as γ' is increased (after the reverse transform has been computed) leaving in the limit a purely Gaussian shape.

In practice the values of $(1-\alpha)$ required to give the best fit for spectra deconvolved using γ'=7.0, 8.5, and 10 cm^{-1} were 0.40, 0.50, and 0.70, respectively. Even for the spectrum of a pure compound, it is probable that the fraction of the Gaussian profile required to give a good fit will increase as γ' is increased since it has been found previously [11,12] that the spectra of pure compounds are better fit by including a small amount of a Gaussian contribution.

The question of whether peak areas are maintained during deconvolution was also addressed. The areas of each component band shown in Figures 2,3, and 4 are listed in Table 1. It can be seen that the average standard deviation for the thirteen peaks is about 3.5%, and that there is no obvious trend in the way that the peak areas vary during deconvolution. We are therefore confident that the approach described in this paper gives reliable quantitative information.

TABLE I

Band Parameters Computed by Curve-fitting in the Region Between 2970 and 2835 cm^{-1} for the Spectrum of PSOC 990.

Original Position ($\bar{\nu}_i^o$)	Width (γ_i)	A_i^o (with γ' = 10.0 cm^{-1})	Area (with γ' = 10.0 cm^{-1})	A_i^o (with γ' = 8.5 cm^{-1})	Area (with γ' = 8.5 cm^{-1})	A_i^o (with γ' = 7.0 cm^{-1})	Area (with γ' = 7.0 cm^{-1})
2963.95	14.48	1.88	20.54	1.42	21.48	1.13	22.32
2956.34	14.14	1.46	14.76	0.98	14.07	0.73	13.67
2948.45	13.41	2.00	16.63	1.34	16.69	0.98	16.53
2936.78	13.23	2.08	16.41	1.35	16.27	0.95	15.67
2925.61	13.83	3.09	28.85	2.12	28.65	1.59	28.83
2915.46	14.19	2.64	26.98	1.88	27.15	1.44	27.39
2904.79	13.13	1.22	9.31	0.75	8.79	0.49	8.05
2895.15	14.14	1.25	12.59	0.91	13.00	0.71	13.45
2885.02	13.25	0.88	6.94	0.55	6.60	0.35	5.84
2868.87	16.09	1.23	18.28	0.93	17.90	0.75	18.05
2863.26	15.72	0.60	8.45	0.50	9.22	0.41	9.41
2852.21	13.25	0.93	7.28	0.68	8.20	0.46	7.51
2843.07	16.56	0.72	11.48	0.55	11.24	0.45	11.46

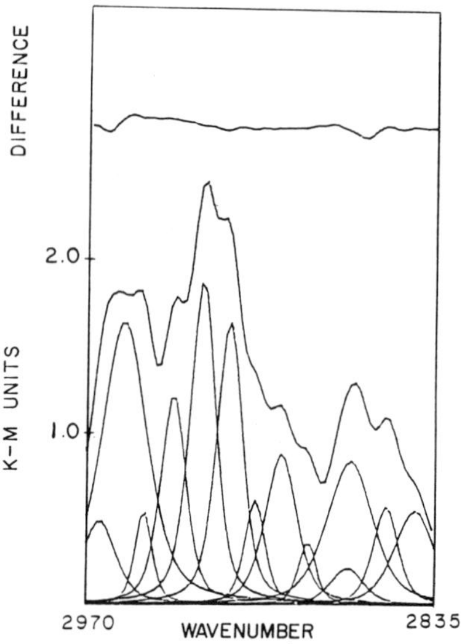

Figure 5 The result of fitting the spectrum shown in Figure 4 (γ' = 7.0 cm^{-1}), but allowing $\bar{\nu}_i^o$ to vary as well as A_i^o and γ_i; α is still equal to 0.60.

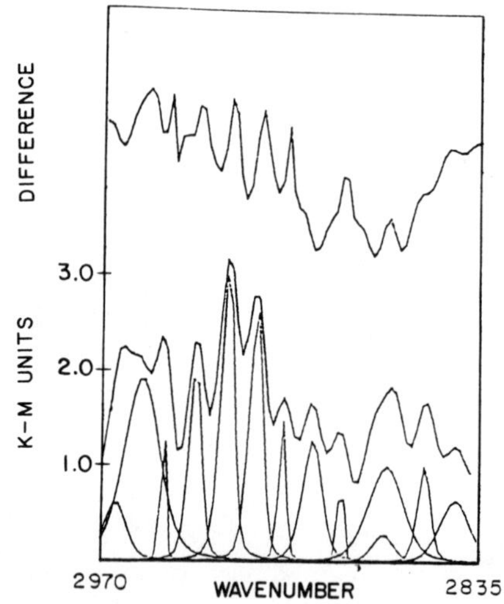

Figure 6 The result of attempting to fit a deconvolved spectrum computed with γ' = 10 cm^{-1} using the same $\bar{\nu}_i^o$ values used for Figure 5, and changing the values of A_i^o and γ_i to $(A_i^o\gamma_i/\gamma_i-3)$ and (γ_i-3), respectively. The value of α was changed to 0.30 to be consistent with the data shown in Figure 2.

One other useful finding was discovered when these numerical techniques were applied to the study of band parameters of a series of spectra measured during the 300°C air oxidation of a medium volatile bituminous coal. The deconvolved spectra in the aliphatic C-H stretching region are shown in Figure 7. Each of the deconvolved spectra was fit independently using the procedure described in this paper. The $\bar{\nu}_i^o$ values for most of the bands were remarkably constant. One band, however, showed a monotonic variation of $\bar{\nu}_i^o$ during oxidation, see Figure 8.

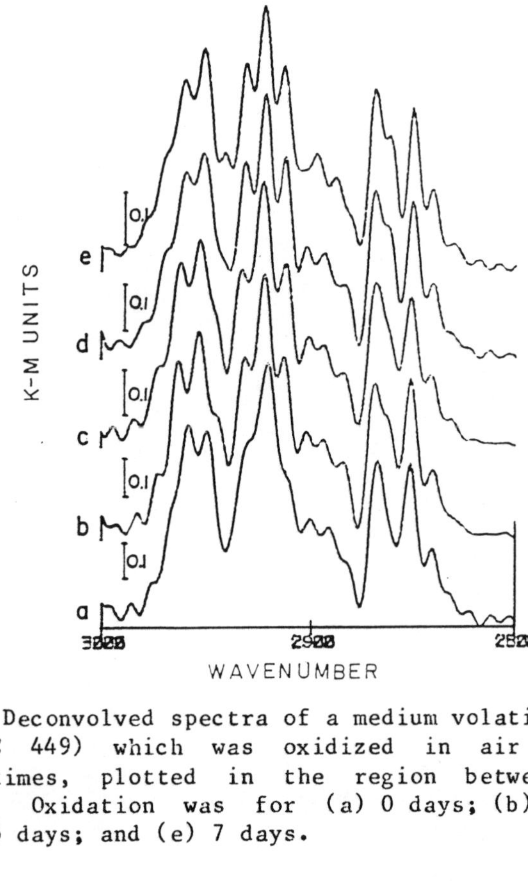

Figure 7. Deconvolved spectra of a medium volatile bituminous coal (PSOC 449) which was oxidized in air at 150°C for different times, plotted in the region between 3000 and 2800 cm^{-1}. Oxidation was for (a) 0 days; (b) 1 day; (c) 3 days; (d) 5 days; and (e) 7 days.

We believe that the reason for this behavior is that this band is still composed of two unresolved components. The higher frequency component is thought to be the symmetrical C-H stretching mode of methyl group attached to alkyl chains (CH_2-R) and the

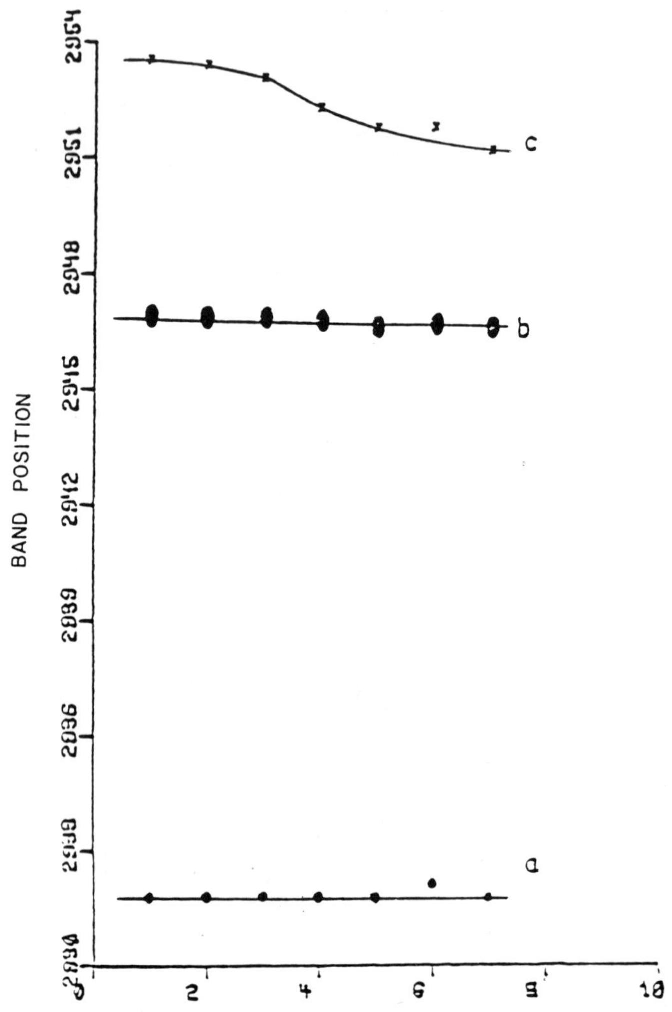

Figure 8 Variation with time of oxidation of the best-fit values of $\bar{\nu}_i^o$ for three C-H stretching bands which were resolved from a single band by FSD.

lower frequency component is thought to be the corresponding vibration of methoxy groups [3]. As the oxidation proceeds, the concentration of CH_2-R groups decreases and the concentration of CH_2-O- groups increases. If these bands are separated by less than the instrumental resolution, they cannot be resolved by FSD. From the data shown in Figure 8, it would appear that the CH_2-R band is located near 2954 cm^{-1} and the CH_2-O- band is located near 2951 cm^{-1}. Thus even though these two bands cannot be resolved, FSD still permits important qualitative information on each to be obtained through the proper application of curve-fitting routines.

CONCLUSION

It is apparent that Fourier self-deconvolution gives important quantitative information on powdered polymeric materials. We believe that through the combined use of diffuse reflectance infrared spectrometry, Fourier self-deconvolution, and curve-fitting, detailed mechanistic and kinetic data on coal and other complex organic mixtures will be obtained.

ACKNOWLEDGEMENT

This paper was prepared with the support of the U.S.Department of Energy, Grant No. DE-FG22-82PC50797. However, any opinions, findings, conclusions, or recommendations expressed herein are those of the authors and do not necessarily reflect the views of DOE.

REFERENCES

1. P.R.Griffiths, S-H. Wang, P.W.Yang, D.E.Henry and I.M.Hamadeh, ACS Div. Fuel Chem. Preprints, 28, 27 (1983).

2. E.L.Fuller, Jr., N.R.Smyrl, R.W.Smithwick and C.S.Daw, ACS Div. Fuel Chem. Preprints, 28, 44 (1983).

3. P.R.Griffiths and S-H. Wang, Fuel, 64, 229 (1985).

4. P.B.Tooke and A.Grint, Fuel, 62, 1003 (1983).

5. J.K.Kauppinen, D.J.Moffatt, H.H.Mantsch and D.G.Cameron, Appl. Spectrosc., 35, 271 (1981).

6. S.-H. Wang, Ph.D. Dissertation, Ohio University, Athens, Ohio (1984).

7. P.C.Gillette, D.Kormos, M.K.Antoon and J.L.Koenig,"Selected Computer Programs with Application to Infrared Spectroscopy", Digilab Users Group, Cambridge, MA (1979).

8. P.C.Painter, R.W.Snyder, M.Starsinic, M.M.Coleman, D.W.Kuehn and A.Davis, in "Coal and Coal Products: Analytical Characterization Techniques", E.L.Fuller, Jr., Ed., Am. Chem. Soc. Symp. Ser. 205 (1982) p.47.

9. P.C.Painter, R.W.Snyder, M.Starsinic, M.M.Coleman, D.W.Kuehn and A.Davis, Appl.Spectrosc., 35, 475 (1981).

10. J.N-P.Sun, P.R.Griffiths and C.A.Sperati, Spectrochim. Acta, 39A, 587 (1983).

11. R.N.Jones, Appl. Opt., 8, 597 (1967).

12. R.N.Jones, Pure Appl. Chem., 18, 303 (1969).

13. a) P.R.Solomon, ACS Div. Fuel Chem. Preprints, 24, 184 (1979). b) Ibid., in "Coal and Coal Products: Analytical Characterization Techniques", E.L.Fuller, Jr., Ed., Am. Chem. Soc. Symp. Ser. 205 (1982) p.77.

14. M.P.Fuller and P.R.Griffiths, Anal. Chem. 50, 1906 (1978).

15. M.P.Fuller and P.R.Griffiths, Appl. Spectrosc., 34, 533 (1980).

16. S.R.Dryson, J. Quant. Spectroc. Radiat. Transfer, 16, 611 (1976).

17. R.J.Noll and A.Pires, Appl. Spectrosc., 34, 351 (1980).

18. A.Klim, J. Quant. Spectrosc. Radiat. Transfer, 26, 537 (1980).

USE OF CURVE ANALYSIS TO ANALYZE OVERLAPPING BANDS IN THE INFRARED SPECTRA OF POLYMERS

B. Jasse

Laboratoire de Physicochimie Structurale et
Macromoléculaire, Ecole Supérieure de Physique
et de Chimie Industrielles de Paris
10, rue Vauquelin 75231 Paris Cedex 05, France

ABSTRACT

Derivative spectroscopy and Fourier self-deconvolution methods used to enhance the apparent resolution of spectra were applied to overlapped regions of the infrared spectra of three polymers: (i) the methylene rocking band of polyethylene in the 750 to 700 cm^{-1} range; (ii) polystyrene in the 600 to 500 cm^{-1} range which is sensitive to the conformational structure; and (iii) an epoxy resin in the range 3200 to 2700 cm^{-1} where an absorption band relative to the epoxy group at 2984 cm^{-1} allows to study the curing reaction. Fourier self-deconvolution was found to give the best results to estimate peak positions and half-widths. The curve analysis of the different spectra was done using deconvolution results as starting values.

INTRODUCTION

Infrared spectra of polymers very often consist of peaks which are extensively overlapped making quantitative analysis difficult. Furthermore, changes in band shape or frequency of the overlapping peaks can occur when the polymer is submitted to specific treatments such as stretching, heating or blending. The techniques commonly used as factor analysis and difference spectroscopy are not reliable in such conditions and curve fitting of individual bands to a composite spectrum appears as a very attractive method to separate the different absorption bands. However, this method requires assumptions regarding the number of component present, their wavenumbers and widths, data which are frequently not available.

The first method proposed to increase the apparent resolution of a spectrum was the use of second and fourth derivatives in frequency space, whereas recent approaches of this problem are focused on calculation in Fourier space. A schematic diagram of the methods presently usable in a routine manner is shown hereafter.

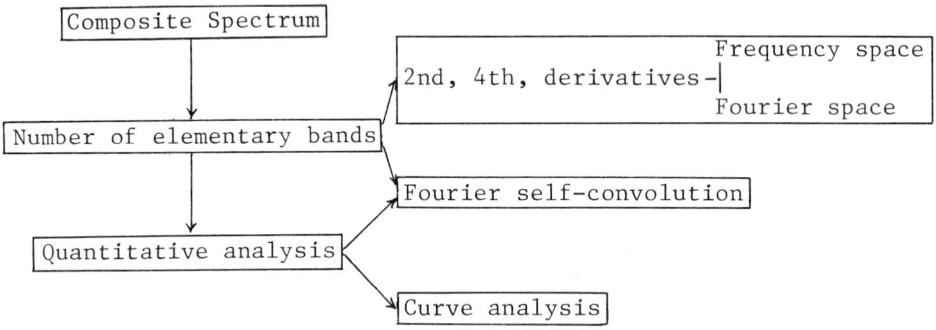

Diagram for the analysis of overlapping absorption bands.

In the present paper we describe the application of the different methods allowing to increase the apparent resolution of infrared spectra to the analysis of overlapped regions in the infrared spectra of three polymers: (i) polyethylene methylene rocking bands in the 750-700 cm^{-1} range; (ii) polystyrene in the 600-500 cm^{-1} region which is sensitive to the conformational structure; (iii) an epoxy resin in the range 3200-2700 cm^{-1} where an absorption band relative to the epoxy group at 2984 cm^{-1} allow study of the curing reaction.

B. THEORETICAL BACKGROUND

A. Derivative Spectroscopy

Derivation techniques have been proposed for many years to analyze spectroscopic data but only with the recent development of digitized spectra has the method been developed. A detailed study of this technique was recently published by Maddams et al. [1-3]. A very simple approach to the calculation of derivative of digitized spectra has been proposed by Butler and Hopkins [4]. If the ordinate values are measured at points n and n + α, then

$$y^{1}\left(n + \frac{\alpha}{2}\right) = y(n + \alpha) - y(n)$$

The complete derivative of the spectrum is obtained by repeating the process sequentially in steps of one data point. Recently,

Koenig and co-workers [5] used a seventh-order method of undetermined coefficient function. Higher order derivatives are readily calculated by repeating the process.

The limit of resolution of two Lorentzian peaks depends on intensity ratio and separation of the two peaks. Different examples of separation of two overlapping Lorentzian peaks can be found in references 1,8 and 9.

In practice, the limiting factor in derivation in frequency space is the noise level, which increases rapidly with each order of derivation. A substantial amount of smoothing is often necessary [5].

An alternative route to derivative spectroscopy is the use of Fourier space calculations [6] which permit the user to selectively eliminate the noise component and retain only the useful information.

The method involves two steps: application of the derivative weighting function and application of a truncating or smoothing weighting function to the Fourier transform of the spectral region under study.

A spectrum $E(\nu)$ and an interferogram $I(x)$ are a pair of Fourier transform; that is

$$E(\nu) = \int_{-\infty}^{\infty} I(x)\exp(i2\pi\nu x)\, dx = \mathcal{F}\{I(x)\}$$

$$I(x) = \int_{-\infty}^{\infty} E(\nu)\exp(-i2\pi\nu x)\, d\nu = \mathcal{F}^{-1}\{(E(\nu))\}$$

where $\mathcal{F}$ and $\mathcal{F}$ are the Fourier and inverse Fourier transforms, respectively.

The n-th derivative of $E(\nu)$ is then given by:

$$d^nE(\nu)/d\nu^n = \int_{-\infty}^{\infty} (i2\pi x)^n\, I(x)\, \exp(i2\pi\nu x)\, dx$$

which is the complex Fourier transform of the interferogram apodized by a complex function $A_n(x)=(i2\pi x)^n$.

For even order derivatives the apodization function is real and even and is given by: $A_n(x)=(-1)^{n/2}(2\pi x)^n$.

In practice, due to noise, it is necessary to smooth the spectrum before the derivatives are computed. Boxcar apodization B(x) is the best smoothing function and the derivatives are then calculated using the apodization function $A(x)=B(x)A_n(x)$.

B. Fourier Self-Deconvolution

Another method for increasing the apparent spectral resolution of overlapping bands is deconvolution in Fourier space [7]. In this technique, the inverse Fourier transform of the system to be deconvolved $I(x) = \mathcal{F}^{-1}[E(\nu)]$ is multiplied by a function $D(x)/\mathcal{F}^{-1}[E_o(\nu)]$ where D(x) is an apodization function and $E_o(\nu)$ the intrinsic line shape function of the spectrum $E(\nu)$. The self-deconvoluted spectrum $E'(\nu)$ is then the Fourier transform of the new interferogram $I'(x)$. The division by $\mathcal{F}^{-1}[E_o(\nu)]$ results in the self-deconvolution and D(x) determines the line shape function $\mathcal{F}[D(x)]$. In practice the intrinsic line shape is approximated by a Lorentzian line

$$E_o(\nu) = \frac{\sigma/\pi}{\sigma^2 + \nu^2} \qquad \mathcal{F}^{-1}\{E_o(\nu)\} - \exp(-2\pi\sigma|x|)$$

where 2σ is the width at half-height of $E_o(\nu)$. The efficiency of the operation is given by the parameter $K=2\sigma/\Delta\nu_{1/2}$ where $\Delta\nu_{1/2}$ is the half-width of $\mathcal{F}[D(x)]$.

An important point is that the integrated intensity is not modified by the self-deconvolution process [6]. This means that quantitative information can be obtained from the spectra with enhanced resolution.

C. Curve Analysis

The number of components, their wavenumbers and widths being estimated, it is usually a simple task to obtain a good fit between a synthesized spectrum consisting of a sum of individual bands and an observed composite spectrum. The main problem is the choice of mathematical function most suitable for characterizing the observed band shapes. In a review of curve fitting procedures and their limitations, Maddams [8] has pointed out that the evidence in favor of the rather general applicability of the Lorentzian shape is very strong. The various computing routines for curve fitting are all based on least squares refinement procedures.

EXPERIMENTAL

Infrared spectra were obtained on a Nicolet 7199 FT-IR spectrometer the software routines of which carry out all the previous mathematical treatments: DR_1 and DR_2 software routines for obtaining 1st and 2nd derivatives in frequency space, respectively. Calculation is based on Butler and Hopkins method [4]. DERIVE-FTN routine generates derivatives in Fourier space and IRD-CON. FTN routine performs self-deconvolution, a Bessel function being used as apodization function.

The curve analysis program CAP was used to analyze the spectra with Lorentzian band shape. This program allowed handling of up to 26 peaks in the manual fitting mode. Initial position of the different peaks were taken from self-deconvolution. After a manual adjustment, the final fitting step was performed using the computer.

RESULTS AND DISCUSSION

A. Methylene Rocking Region of Polyethylene

This PE mode which splits in two absorption bands at 731 and 720 cm^{-1} due to intermolecular interaction in crystalline state was chosen to check the reliability of self-deconvolution in quantitative analysis. The spectrum, its second and fourth derivatives and the result of self-deconvolution are shown in Figure 1. One can visually detect an assymmetry on the high wavenumber side of the 720 cm^{-1} band. This broadening can be interpreted in terms of the superposition of contribution of all trans sequence of various sequence length [10]. In the second derivative the presence of an additional band on the high frequency side of the 720 cm^{-1} can be inferred from the asymmetry of the positive side lobes. No information can be deduced from the fourth derivative due to increasing noise, and only the two main bands are observed. In the self-deconvolved spectrum, a shoulder around 725 cm^{-1} is clearly detectable. Results of curve analysis of the original and self-deconvolved spectrum are given in Table 1. As expected three bands are necessary to get a good fit. The error is very small for wave numbers and area percentage of the three bands between the two methods of analysis. The spectral resolution in the self-deconvolved spectrum has been improved by a factor of about 2. However, in self-deconvolution process the increase in apparent resolution is balanced by a higher RMS

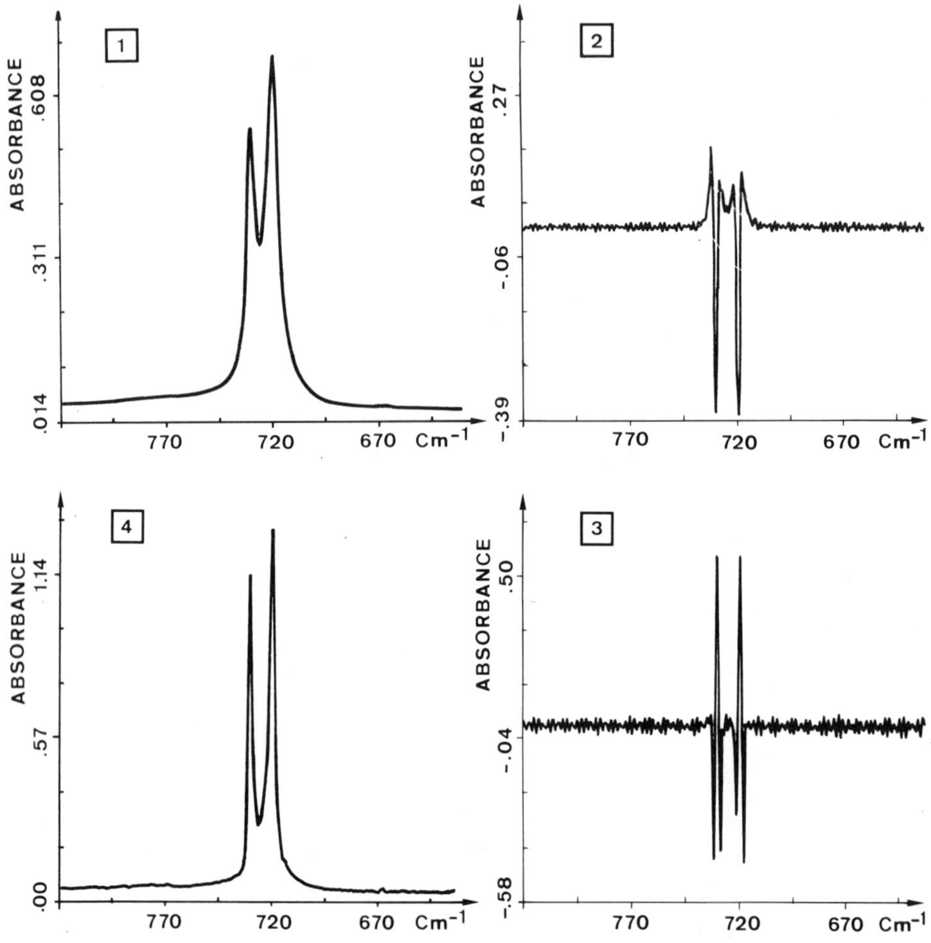

Figure 1 Polyethylene (1) spectrum in the 645 to 820 cm^{-1} range; (2) second derivative (DR2); (3) fourth derivative (DR2 x DR2); (4) self-deconvolution (VFO = 3 cm^{-1}; K = 2.5)

TABLE I

Curve analysis of the 670 to 770 cm^{-1} range of the PE infrared spectrum and the self-deconvoluted spectrum (resolution: 2 cm^{-1}; number of scans: 100)

	Band wave number (cm^{-1})	Band half-width (cm^{-1})	Peak height (arbitrary units)	Area of band (arbitrary units)	% area
Spectrum	719.81	12.60	302	806	60.2
	725.68	14.24	31	93	6.9
	730.05	9.64	214	439	32.8
Self-deconvolution	719.83	6.10	298	642	59.2
	724.50	7.9	27	74	6.8
	730.05	3.94	266	367	33.9

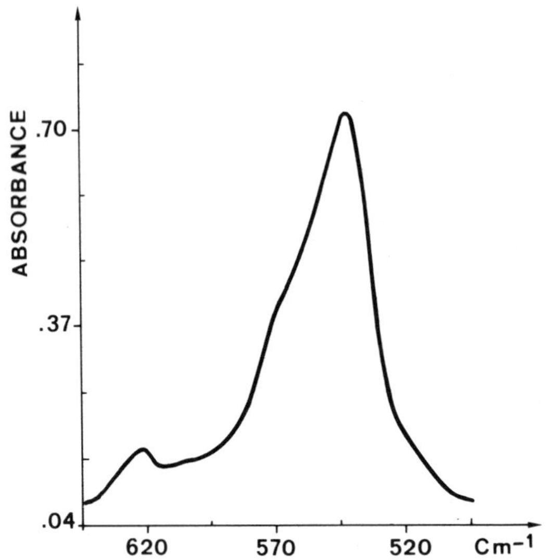

Figure 2 Atactic polystyrene spectrum in the 495 to 645^{-1} range.

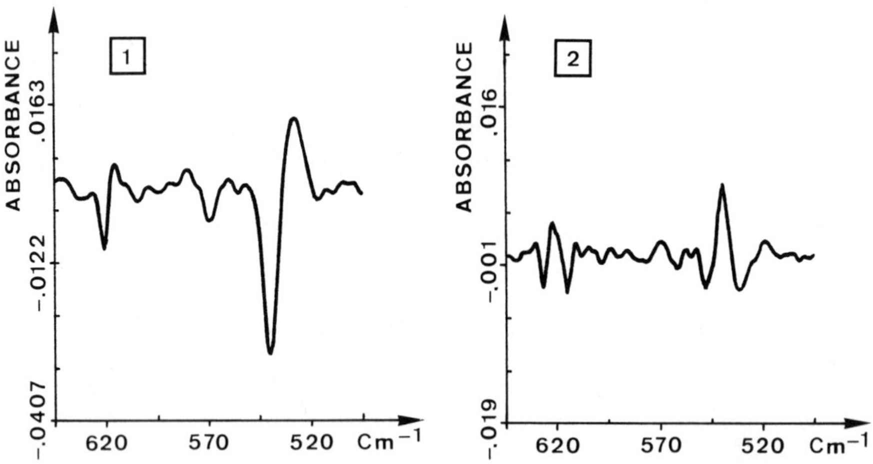

Figure 3 Atactic polystyrene. (1) second derivative (DR2); (2) fourth derivative (DR2 x DR2)

difference between real and synthesized spectra due to increased noise and curve analysis is more readily achieved on the original spectrum. Experimental error in the area is around ±1%.

B. Polystyrene

It is interesting to know what kind of information on conformational structures can be deduced from the vibrational spectrum of atactic polystyrene (PS). Previous studies on PS model molecules (11,12) clearly pointed out that the 16b out-of-plane mode of the ring, which is observed in the 600-500 cm^{-1} range, is particularly sensitive to the conformation of the aliphatic chain. It was possible to conclude that the wavenumber of the vibration is stable at 540 cm^{-1} when a sequence of four or more <u>trans</u> conformation carbon-carbon bonds occurs. When <u>gauche</u> conformations are present, the vibration of the benzene rings in the neighborhood of the gauche links is shifted towards higher frequencies and observed around 550 cm^{-1}.

As shown in Figure 2, four bands are visually detectable in atactic PS spectrum, as two peaks at 622 and 540 cm^{-1} and two shoulders around 565 and 515 cm^{-1}. The 622 cm^{-1} band is assigned to the 6b mode of the ring, which is insensitive to the conformational structure.

Second and fourth derivatives in frequency space gives no more information about the number of bands due to increased noise (see Figure 3). On the other hand, derivatives in Fourier space and self-deconvolution (Figure 4, a,b and c) indicate that eight absorption bands are present. Curve analysis (Table 2 and Figure 4, d) shows that the 600-500 cm^{-1} region of atactic PS spectrum is constituted of three major absorption bands at 568, 555 and 540 cm^{-1}. The 540 cm^{-1} band corresponds to <u>trans</u> segments while the 555 cm^{-1} band has to be assigned to <u>gauche</u> (TG and GG) conformational structures.

As far as the 568 cm^{-1} band is concerned, the assignment is more questionable. A strong sharp band is observed at 566 cm^{-1} in crystalline isotactic PS; however, as shown in the self-deconvolved spectra of amorphous and crystalline isotatic PS (Figure 5), the 566 cm^{-1} absorption band is not present in the amorphous polymer and is obviously characteristic of crystalline arrangements. Such structures are more unlikely expected to occur in the atactic polymer and the best assignment for the 568 cm^{-1} band is a combination between the 9b in-plane mode of the ring, which is observed around 235 cm^{-1} and a skeletal deformation mode which appears around 329 cm^{-1}. The vibrational analysis of PS model compounds supports such an assignment [11,13].

TABLE II

Curve analysis of the 500 to 650 cm^{-1} range of the PS infrared spectrum. (resolution: 2 cm^{-1}; number of scans = 1.000)

Band wave number (cm^{-1})	Band half-width (cm^{-1})	Peak height (arbitrary units)	Area of band (arbitrary units)
515.31	23.70	10	46
540.79	20.70	262	1049
554.07	25.94	142	722
568.59	23.54	88	404
590.98	27.54	22	118
605.19	20.50	8	32
622.17	16.48	35	113
632.21	26.90	14	73

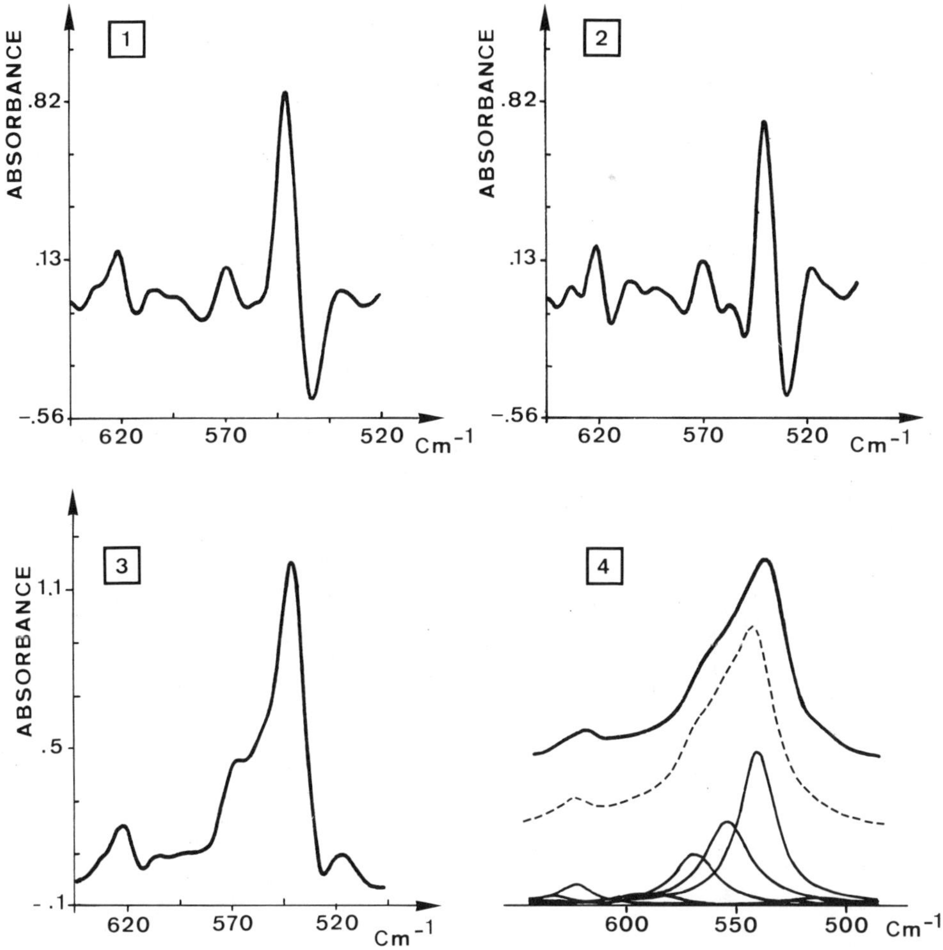

Figure 4 Atactic polystyrene. (1) second derivative (DERIVE. FTN; VFO = 2, VF1 = 0.6); (2) fourth derivative (DERIVE.FTN; VFO = 3, VF1 = 0.6); (3) self-deconvolution (VFO = 15 cm^{-1}; K = 1.6); (4) Curve analysis (CAP).

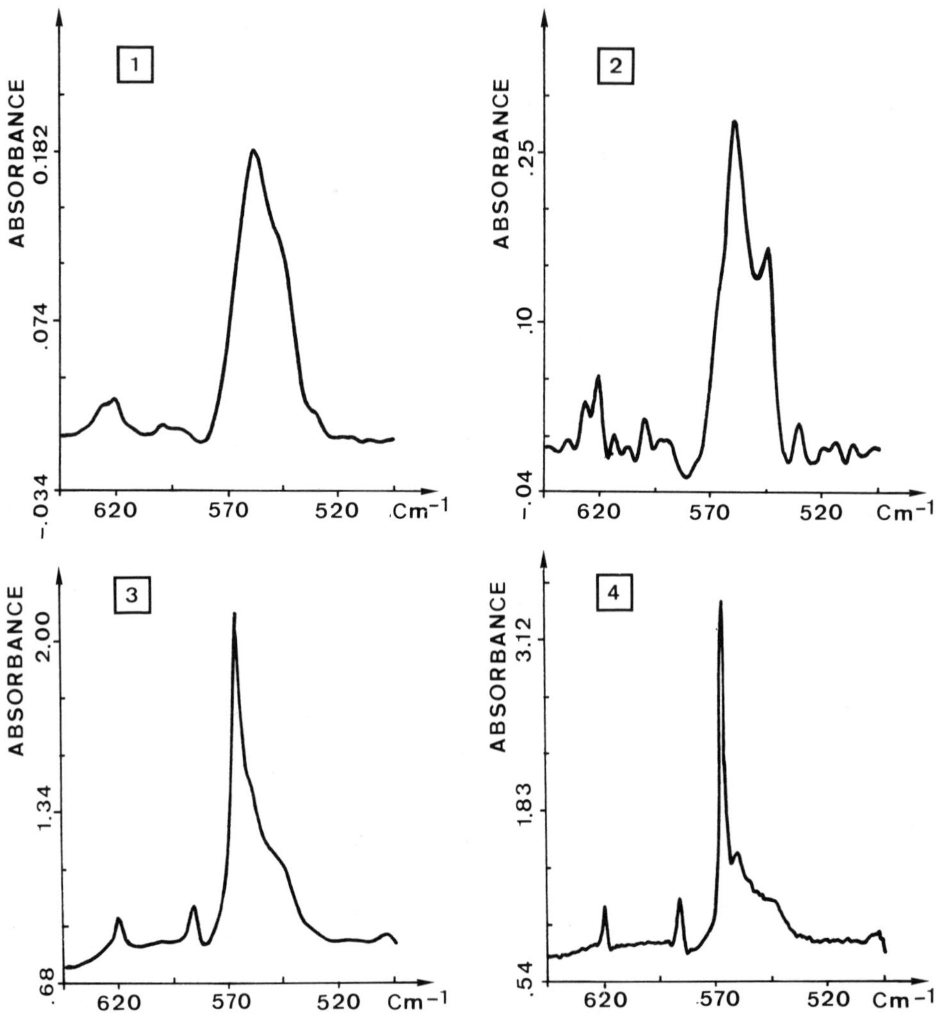

Figure 5 Isostactic polystyrene. (1) spectrum of amorphous i.PS; (2) self-deconvolution of amorphous i.PS; (VFO = 3 cm^{-1}; K = 1.8; (3) spectrum of crystalline i.PS (VFO - 3 cm^{-1}; K = 2).

In conclusion, the new methods of analysis mentioned above produce a new insight into the 630 to 500 cm^{-1} range of the PS spectrum. Instead of the four absorption bands visually detectable, eight bands are present, two of them being related to the out-of-plane mode of the benzene ring (ν_{16b}) which is sensitive to the conformational structure of the aliphatic chain. Furthermore, it is impossible to distinguish the TG isotactic and GG syndiotactic conformations.

C. Epoxy Resin

The resin studied (Lopox 152) is based on tetraglycidyl-4,4′-diaminodiphenylmethane and 4,4′-diaminodiphenyl sulfone. The 3200-2700 cm^{-1} range was analyzed to follow the change in intensity during the curing reaction of the 2984 cm^{-1} band which is related to the epoxy group. Fourth derivative and self-deconvolution of the spectrum of a prepolymer are shown in Figure 6. Eleven absorption bands are clearly detectable. Examination of the spectrum obtained at the end of the curing process, when all the epoxy groups have disappeared, shows that a weak additional band exists at 3008 cm^{-1}, corresponding to an aromatic ring mode.

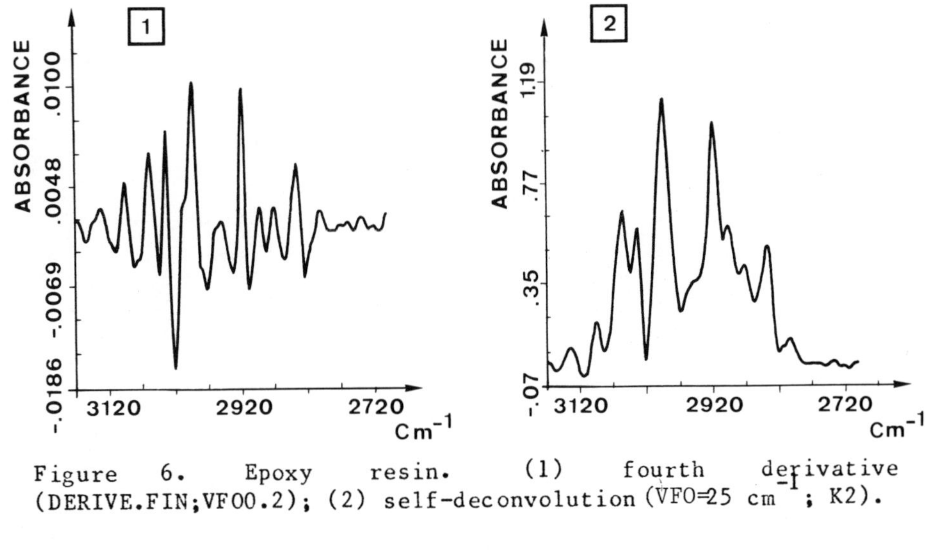

Figure 6. Epoxy resin. (1) fourth derivative (DERIVE.FIN;VF00.2); (2) self-deconvolution (VFO=25 cm^{-1}; K2).

In the uncured polymer only a very small shoulder appears on the high frequency side of the self-deconvoluted 2994 cm^{-1} peak. However a noticeable shoulder exists at 3008 cm^{-1} in the second and fourth derivatives in Fourier space, and although the intensity of this band represents only 6% of the 2994 cm^{-1} peak, it remains detectable. Curve analysis (Figure 7) was done using twelve absorption bands. All these bands were assigned with the help of model molecules. This example points out the interest in considering spectra obtained at different stages of reaction for the analysis of reacting systems. Weak overlapping bands can be enhanced or revealed by the disappearance of reacting groups and new bands can also appear.

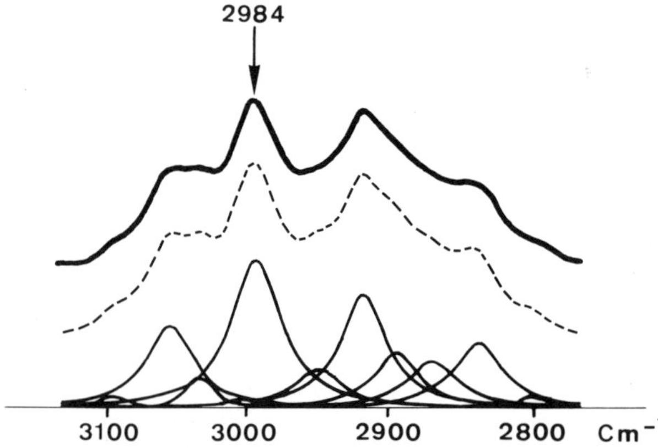

Figure 7. Epoxy resin. Curve analysis (CAP).

CONCLUSION

The examples presented in this paper show that, among the different techniques now available with commercial FT-IR software routines, Fourier space derivatives and self-deconvolution represent a useful tool for improved analysis of overlapping bands. All these programs, which are fast and very easy to run, can be used in a routine manner and provide starting estimates of the parameters to be used in the curve analysis process. A general criterion which should always be applied is that the peaks obtained should be readily explicable in terms of probable chemical groups and for this purpose, the use of model compounds is of a very useful help.

REFERENCES

1. W.F.Maddams and W.L.Mead, Spectrochim. Acta, 38A, 437 (1982).

2. S.Hawkes, W.F.Maddams, W.L.Mead and M.J.Southon, Spectrochim. Acta, 38A, 445 (1982).

3. W.F.Maddams and M.J.Southon, Spectrochim. Acta, 38A, 459 (1982).

4. W.L.Butler and D.W.Hopkins, Photochem. Photobiol., 12, 451 (1970).

5. P.C.Gillette, J.B.Lando and J.L.Koenig, Appl. Spetrosc., 36, 401 (1982).

6. J.K.Kauppinen, D.J.Moffatt, H.H.Mantsch and D.G.Cameron, Anal. Chem., 53, 1454 (1981).

7. Ibid., Appl. Spectrosc., 35, 271 (1981).

8. W.F.Maddams, Appl. Spectrosc., 34, 245 (1980).

9. W.F.Maddams and P.B.Tooke, J. Macromol. Sci. Chem. Ed., A17, 951 (1982).

10. P.C.Painter, M.M.Coleman and J.L.Koenig, "The Theory of Vibrational Spectroscopy and its Application to Polymeric Materials", Wiley and Sons, New York (1982) p.355.

11. B.Jasse, A.Lety and L.Monnerie, J. Molecul. Struct., 18, 413 (1973).

12. B.Jasse and L.Monnerie, J. Molecul. Struct., 39, 165 (1977).

13. Ibid., J. Phys. D: Appl. Phys., 8, 863 (1975).

APPLICATION OF CURVE FIT AND DECONVOLUTION TO POLYMER ANALYSIS

Patricia B. Roush, Robert W. Hannah and John P. Coates

Perkin-Elmer Corporation
901 Ethan Allen Highway
Ridgefield, CT 06877

Alan Bunn

ICI P & P, Ltd. Co.
Wilton
England

H.A. Willis

Computer Enhanced Spectroscopy
Wiley and Sons
England

INTRODUCTION

A great wealth of information about a material is found in its infrared spectrum. For many years infrared was used not just for identifying componds but also for obtaining semi-quantitative information about structural units. In more recent years other instrumental techniques, such as NMR, have been developed and have superseded infrared spectroscopy in the quantitative determination of structural information. Now, however, because of the availability of high signal-to-noise infrared spectrometers, and the computers associated with the spectrometers, structural information, which could not be accessed previously, can be extracted from the infrared spectrum. Techniques such as deconvolution and curve fit analysis provided by the computer, allow more structural information to be made visible in the infrared spectrum than was once thought possible. In the deconvolution process broad complex bands can be mathematically narrowed to yield an improvement in the observation of the overlapped component bands. The results obtained are often similar to those obtained from samples analyzed at low temperatures.

This paper will deal with two examples of deconvolution applied to polymer analysis with supporting data from a curve fit analysis. The first example deals with ethylene propylene copolymers and shows the enhancement of the infrared spectrum to give information about the length of the methylene sequences in the polymer. In the second example the technique is used to provide information about the crystallinity of polypropylene samples.

EXPERIMENTAL

The ethylene propylene copolymers used in this study were obtained from ICI, England. Carbon-13 NMR (C-13) spectra were obtained by Dr. Alan Bunn of ICI Research. Results of the C-13 analyses are given in Table 1 for four of the samples. These represent the number of respective sequences of CH_2 groups normalized to 100 carbon atoms.

Table 1

Sample	$(CH_2)_3$	$(CH_2)_5$	$(CH_2)_{n>7}$
1	100	0	0
2	91	9	0
3	82	11	6
4	78	8	14

Infrared spectra were obtained for thin self supporting films using the Perkin-Elmer Model 1800 Fourier Transform Spectrophotometer. Relevant operating conditions were 4 double beam scans at 2 cm^{-1} resolution, weak Beer-Norton apodization, and double sided interferograms. These conditions yield a peak to peak signal-to-noise in excess of 1500 which satisfies the requirements for the amount of resolution enhancement to be done. Spectra were stored on disk at 1 cm^{-1} data interval for the full range 4000 to 450 cm^{-1}.

In order to improve the observability of the overlapping bands in the CH_2 rocking region near 730 cm^{-1}, mathematical resolution enhancement was performed. The process is identical to the Fourier self-deconvolution [1-3] except that the mathematical operations are performed in the spectral domain rather than in the Fourier domain.

The Fourier self-deconvolution technique involves the enhancement of the high resolution components of a spectrum in the Fourier, or time domain by multiplication with an exponential function. The degree of the enhancement is related to the power of the exponent. This process is limited, however, by a dramatic increase in noise and spectral artifacts caused by the truncation of the transformed data. These are in part, compensated by the application of an apodization function. The net result is that the time domain data is multiplied by a combined exponent/apodization function. To observe the effects in the spectral domain, it is of course, necessary to perform a second Fourier transformation. Typically the complete procedure is repeated a number of times until the desired result is achieved. A multiplication in the time domain is equivalent to a convolution in spectral space. Therefore, an alternative approach to the Fourier self deconvolution procedure, which was adopted in this paper, is to perform a Fourier transform on the combined mathematical enhancement function and to convolve the result of this transformation with the spectrum in spectral space. The advantages are obvious in terms of convenience and time saving. Furthermore, this approach has been incorporated into an interactive graphics based routine which enables the user to appreciate the effects of the enhancement within a few seconds.

An independent iterative curve fit analysis was also performed on the overlapping CH_2 rocking band. The program used provided the means to vary the contribution of Gaussian and Lorentzian functions of individual bands plus the ability to specify the line width at half height. The true heights of the individual component bands were not assumed and were evaluated in the iterative process by comparison of the intensities in the synthesized result to the intensities in the original spectrum. Each iteration involved an examination of the residue between the calculated and real spectrum and a readjustment of the parameters. Although the curve fit analysis can be performed independently, information obtained from second derivative or deconvolved spectra can be used as initial inputs into the curve fit procedure. In addition, the several mathematical approaches were used as checks against one another.

In the second part of our study, two samples of isotactic polypropylene were used. Sample A was prepared by heating the polypropylene to 170°C and then quenching it in liquid nitrogen. Sample B was allowed to slowly cool from the melt temperature. The infrared spectra were obtained on thin self supporting films using the Perkin-Elmer Model 1800 FT-IR under similar operating conditions as for the ethylene propylene copolymer samples.

In order to ascertain if more information about the degree of crystallinity of the sample could be obtained, several of the bands between 1350 and 750 cm^{-1} were deconvolved using the spectral deconvolution technique described above.

RESULTS AND DISCUSSION

The random ethylene propylene copolymer produces a complex band envelope in the 730 cm^{-1} region, such as that shown in Figure 1, and this is assigned to the CH_2 (methylene) rocking vibration. It is well known that the highly coupled CH_2 rocking frequency is shifted appreciably depending on the CH_2 sequence length. Based on model compounds, the assignments given in Table 2 have been determined.

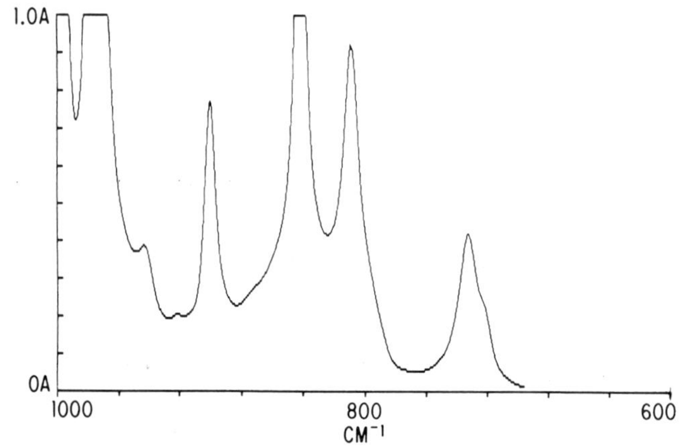

Figure 1. Statistical ethylene/propylene copolymer.

Table 2

Methylene Sequence Length	Approximate Band Position cm^{-1}, in random EP Copolymers
1	815
2	750
3	735
4	730
5	720

In theory, it should be possible to determine the presence of the various sequence lengths and to quantitatively determine the number of the respective sequences per 100 carbon atoms as is done by C-13 NMR. In practice the band overlap is such that the number of C3 sequences may be distinguished from the total of the C5 and longer sequences, but it was not possible to distinguish among the C5 and longer sequences. It would appear that the number of C3 sequences and the total number of C5 and longer sequences may be determined. However, additional problems exist which make the quantitative measurement uncertain. As the number of long sequences increases, the likelihood of crystalline areas occuring may also increase. It has been recognized by Willis and others [4] that the half widths of absorption bands of crystalline polymers are narrower than for amorphous polymers. The half band widths are indeed changing in the set of polymers being described here and whether this is due to crystalline effects remains to be determined. It is quite clear that the various components contributing to the band envelope near 730 cm^{-1} have different band widths. Furthermore, these band widths are changing in a manner which is not completely predictable. The net result of this is to increase the uncertainty in the quantitative measurement of the concentrations of the methylene sequence lengths. Quite obviously, more work needs to be done on additional samples.

Spectra of several polymers are shown in figures 2 through 8. In each case, the spectrum of the original data and the deconvolved spectrum are given. A summary of the measured band positions and half band widths in the deconvolved spectra is given in Table 3 for several polymer samples.

Table 3

Sequences	Band Positions		Half Band Widths 732 cm^{-1}	720 cm^{-1}
1. 100 C3	732.2	719.8	5.1	6.2
2. 91C3, 9C5	732.4	720.4	5.8	5.4
3. 78C3, 8C5, 14C7	732.1	720.2	6.2	4.9
4. 82C3, 11C5, 6C7	732.4	721.2	5.8	6.8
5. Some C3, Most C5 and longer	729.9	719.9	4.0	4.0
6. Some C3, Most C5 and longer	729.7	720.5	4.0	4.8
7. C3 and longer	730.1	719.9	5.1	5.0

Note: C3, C5, etc = $(CH_2)_3$, $(CH_2)_5$, etc.

The NMR results for the first four samples are given in Table 1. The infrared spectrum of sample 1 is shown in Figure 2. This sample is said to have all $(CH_2)_3$ sequences. However, there clearly is a very weak band at 719.8 cm^{-1}, and this must be assigned to a very low concentration of longer CH_2 sequences. In addition, a third weak band near 739 cm^{-1} was observed in the deconvolved spectrum and its presence was confirmed in the results of the curve fit analysis of the band as shown in Figure 9. This band was observed in several of the deconvolved spectra. Its assignment is uncertain, but based on infrared data for model componds, it is possibly related to the presence of terminal n-propyl groups.

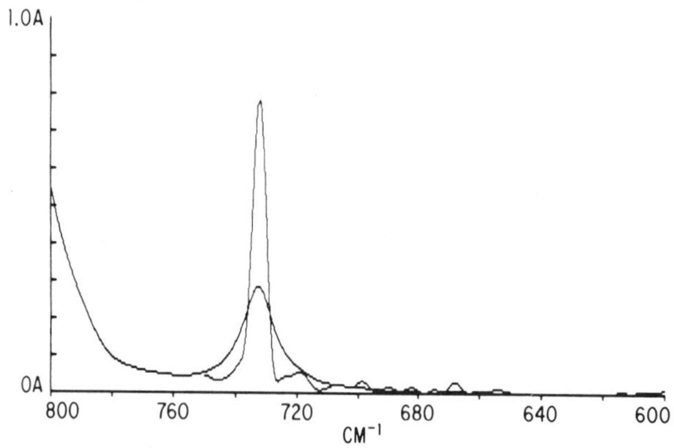

Figure 2. Ethylene propylene, random copolymers. Sample 91/27; all C3 sequences. Original and deconvolved.

The spectrum of sample 2 is shown in Figure 3. The deconvolved spectrum clearly shows the higher proportion of $(CH_2)_5$ sequences. Samples 3 and 4 shown in Figures 4 and 5, have still higher proportions of C5 and longer sequences and the ratio of band intensities for the 730 to 720 cm^{-1} bands qualitatively agrees with the NMR data. However, while the NMR data indicates substantial number of C7 and longer sequences, the 720 cm^{-1} band has not been resolved further. This might have been predicted since the position of the $(CH_2)_7$ rocking vibration is only 2 cm^{-1} lower than the band due to $(CH_2)_5$ sequences. In view of the half widths, which are probably between 8 and 12 cm^{-1}, the possibility of observing two bands by deconvolution is not possible. Curve fit analysis for sample 3, Figure 9, also showed only the two major bands at 732 and 720 cm^{-1}. The very weak band at 750 cm^{-1} may be assigned to $(CH_2)_2$ sequences. This band may also be observed in several of the deconvolved spectra.

The last three samples have different spectra and exhibit different behavior during deconvolution from the first four samples. Spectra are shown in Figures 6 through 8. The half widths of the deconvolved bands are consistently less, and the intensity of the lower frequency band at 720 cm^{-1} is greater than the 732 cm^{-1} band. This is consistent with a higher proportion of longer sequences but in the absence of NMR data to calibrate the behavior of these samples, it is not possible to completely interpret the data. Nevertheless, the observed narrowing of the 732 and 720 cm^{-1} bands, and the greater intensity of the 720 cm^{-1} band

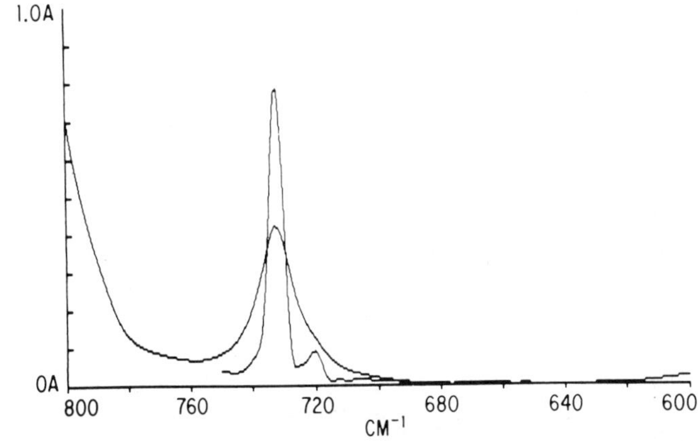

Figure 3 Ethylene Propylene, Random Copolymers Sample EFL200; C 3 and 9 C 5 sequences Original and deconvolved.

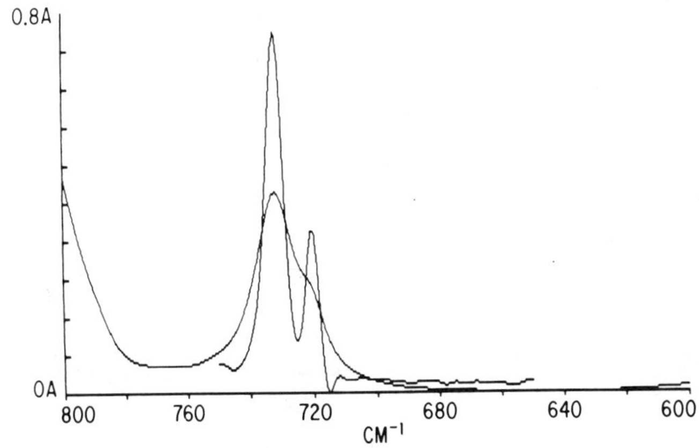

Figure 4 Ethylene Propylene, Random Copolymers Sample XS 1121; 78 C 3, 8 C 5, and 14 C 7 and longer.

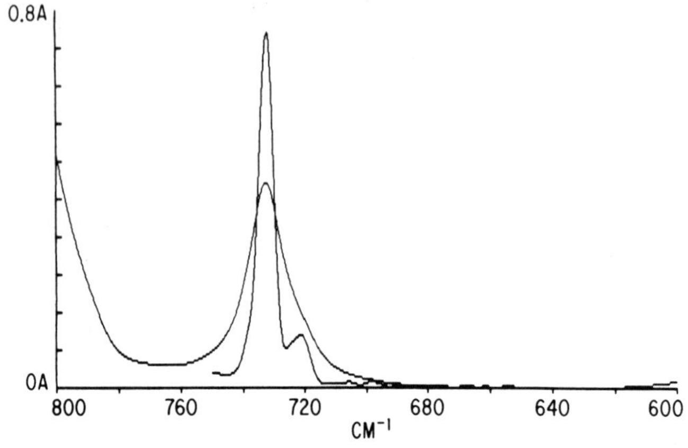

Figure 5 Ethylene Propylene, Random Copolymers Sample PP304/27; 82 C3, 11 C5, and 8 C7 and longer.

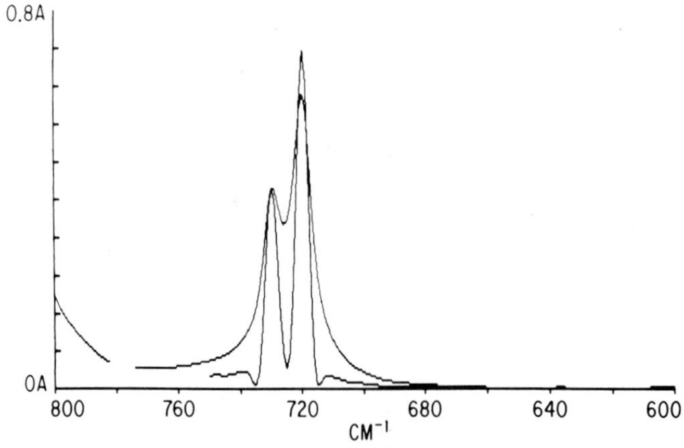

Figure 6 Ethylene propylene, Random Copolymers Sample 80/167, C3 and longer.

assigned to the longer sequences suggests more longer sequences with a resultant higher ordering in a larger proportion of the unit cells. Whether this implies a more crystalline polymer or not requires more investigation.

From the results seen in the ethylene-propylene copolymer samples, it appeared that spectral deconvolution might give some useful information about the dgree of crystallinity in a sample.

For many years it was thought that the infrared spectrum could be used to directly measure the amount of crystallinity in a polymer. What actually was being observed was not necessarily crystalline material, but the presence of a particular rotational conformer found in the polymer molecule that can exist in a crystalline form. Often a measurement of the trans to gauche ratio was made and was equated to the crystalline to amorphous ratio in the sample. However, although trans conformer did exist in the crystalline form, it was also present in the amorphous form as well. Clearly then in a polymer sample containing any degree of crystallinity, there would be overlapping bands due to the amorphous and crystalline forms present in the sample. The more crystalline bands were observed to be narrower because of higher order in the crystalline regions. The amorphous bands were broader due to the number of different environments about the particular group. Since the crystalline bands are expected to be narrower and sharper than the corresponding overlapping amorphous bands, it should be possible to effectively separate the broad amorphous bands and the narrow crystalline bands using deconvolution.

Isotactic polypropylene is a pure head-to-tail polymer having either an all -d or an all -l configuration. In the crystalline form the long-chain molecules assume the form of three-fold helices. The crystallographic unit cell is monoclinic and contains four chains. The observed infrared spectrum for crystalline isotactic polypropylene is quite simple and can be interpreted in terms of a single helix containing three monomer units. This indicates that the intermolecular interactions in crystalline polypropylene are relatively weak, and also points out how different values for crystallinity of a sample can be obtained using different analytical techniques. For example, a sample could have had enough time to form helices on cooling from the melt, and thus give a high infrared crystallinity, yet not have had enough time to form a three-dimensionally ordered polymer network. Thus a lower crystallinity would be measured by X-ray techniques.

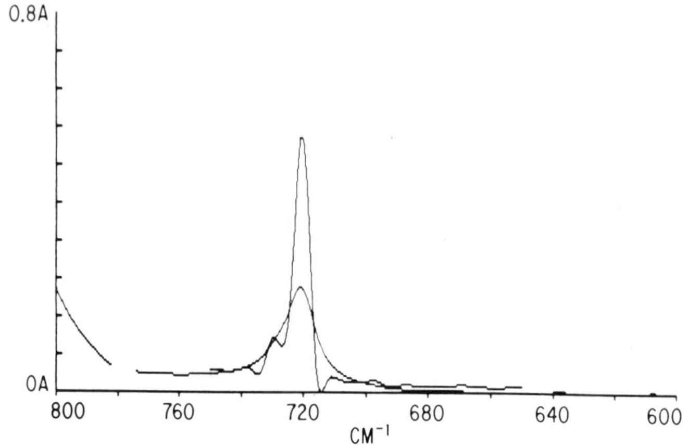

Figure 7 Ethylene Propylene, Random Copolymers
Sample 80/170, C3 and longer plus some C2.

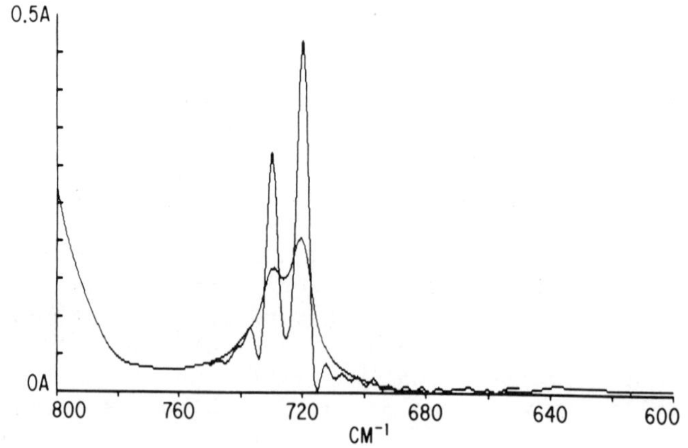

Figure 8 Ethylene propylene, Random Copolymers
Sample 80/142, C3 and longer.

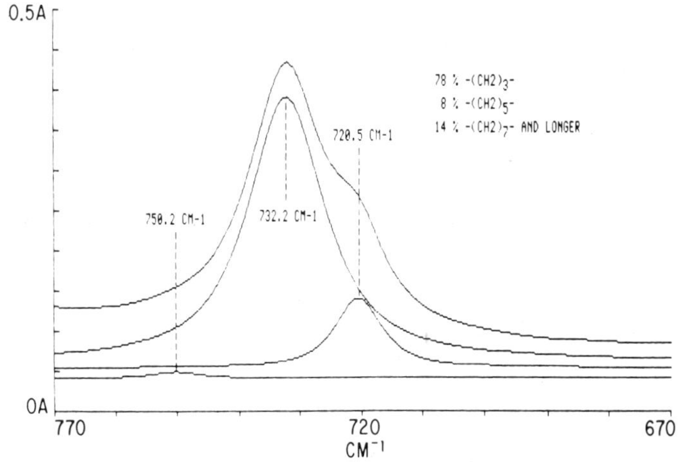

Figure 9. Ethylene/propylene copolymer.

The infrared spectra of crystalline and amorphous polypropylene show very little differences in the methyl and methylene stretching and bending modes. The major differences in the spectra are noticed in the longer wavelength region below about 1400 cm^{-1}, where strongly coupled highly conformation sensitive bands are observed. The locations of the bands found in crystalline polypropylene are given in Table 4 (5). In particular the bands at 1167, 997 and 841 cm^{-1} are characteristic of the threefold helical structure found in the crystalline polypropylene. These bands are either absent or very weak in amorphous polypropylene.

Table 4

Frequency, cm^{-1}	Intensity	Frequency, cm^{-1}	Intensity
1377	s	1103	w
1360	m	1045	w
1329	w	997	s
1303	w	972	s
1297	w	940	w
1255	w	899	w
1219	w	841	s
1168	s	809	m
1153	sh		

The band at 972 cm^{-1}, which is due to the methyl rocking vibration is often used as an internal standard to measure the degree of crystallinity in polypropylene samples, as it is reported to show very little change between the crystalline and amorphous forms [6]. The spectra of two samples of isotactic polypropylene are shown in Figure 10 and 11. The spectra were normalized to give the same intensity for the 973 cm^{-1} band, to remove differences in sample thickness. The intensities and half band widths for samples A and B are given in Table 5. By comparing the intensities of the 1168, 998 and 840 cm^{-1} bands, it is clear that sample B contains more crystalline character.

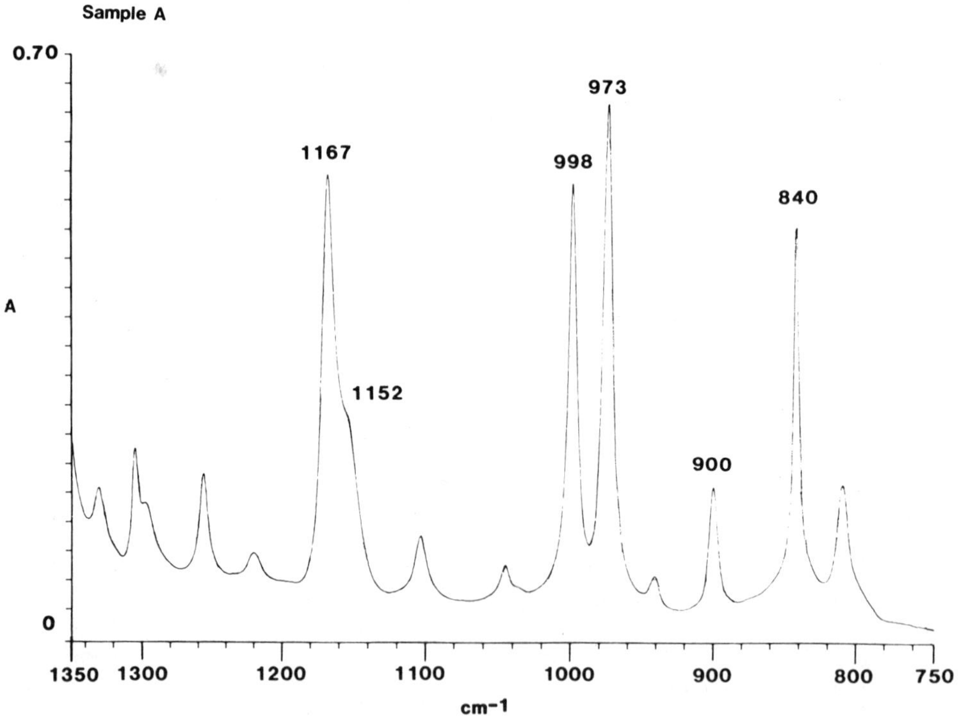

Figure 10. Isotactic polypropylene.

Table 5

Band Position, cm^{-1}	Sample A Intensity Absorbance	Sample A Half-Band Widths, cm^{-1}	Sample B Intensity Absorbance	Sample B Half-Band Widths, cm^{-1}
1168 *	0.543	*	0.595	*
1152	0.228	sh	0.208	sh
997	0.513	7.3	0.557	6.5
973	0.638	8.3	0.638	7.5
841	0.480	4.6	0.603	3.6

*; Shoulder at 1152 cm^{-1} prevents accurate measurement.

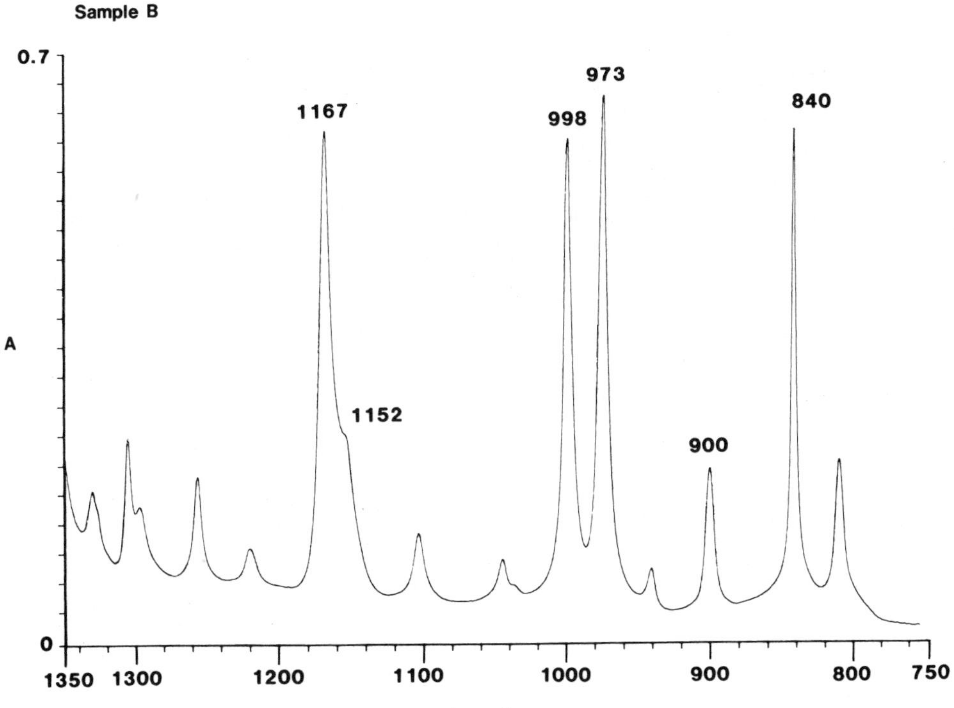

Figure 11. Sample B.

Using the band at 973 cm^{-1} as the internal standard a spectral subtraction was performed. The difference spectrum, shown in Figure 12 clearly shows that sample B has more crystalline character. In the difference spectrum the band at 973 cm^{-1} appears as a derivative-looking feature. This indicates that there is a frequency shift between the two bands and may also indicate the presence of more than one band under the 973 cm^{-1} band. The derivative-looking band at 973 cm^{-1} is surprising, since the 973 cm^{-1} band has been reported to not change in amorphous or crystalline samples [6].

Several regions in the spectra of samples A and B were deconvolved. Figure 13 shows the original data and the deconvolved spectra for the region around 1170-1150 cm^{-1}. Since the original bands were approximately the same width, they were enhanced using the same enhancement factors.

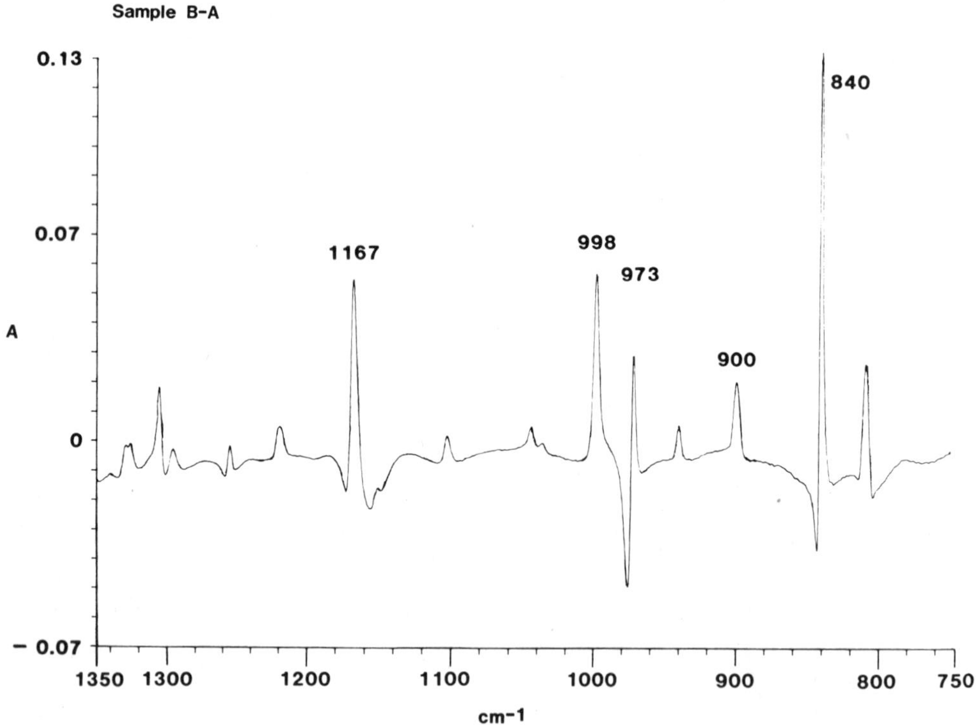

Figure 12. Sample B - A.

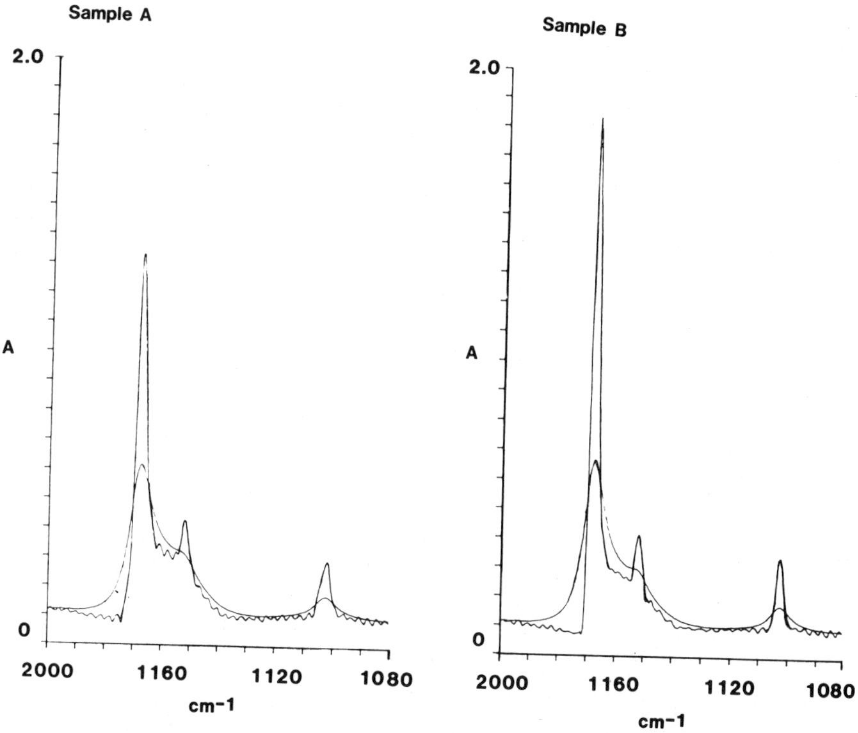

Figure 13. Sample A (left) and Sample B (right).

The band at 1168 cm^{-1} reduces to a 3-4 cm^{-1} half-band-width and the shoulder at 1152 cm^{-1} clearly appears. The 1168 cm^{-1} band in sample B is much more intense than in sample A, which is consistent with sample B containing more crystalline form. The small band centered at 1152 cm^{-1}, which is due to the methyl wagging, appears more intense in sample A than in sample B. However, from the deconvolved spectra there appears to be at least a third broader band in this region, which could be affecting the intensity of the 1152 cm^{-1} band.

The 1020-960 cm^{-1} region of sample A was deconvolved to a greater extent than sample B because the half-band-widths of these bands in sample A were wider in the original data. The resulting deconvolved spectra superimposed on the original data are shown in Figure 14. In this case the half-band-widths of the 998 cm^{-1} band are the same for both sample A and B and the intensity of these two bands is only slightly different. This is likely due to the fact that sample A was deconvolved more than sample B to produce this result. The band originally at 973 cm^{-1} is resolved in both samples to a sharp band at 972 cm^{-1} with a half-band-width of 2.5 - 3 cm^{-1} and a shoulder located at approximately 975 cm . The shoulder is more intense in sample A than sample B. This band at 973 cm^{-1} is reported to not vary in crystalline and amorphous samples and is used as an internal standard band in determining crystallinity.

The data here indicate that the band at 972 cm^{-1} is more intense in the more crystalline sample and the band at 975 cm^{-1} is more intense in the more amorphous sample. This result agrees with the difference spectrum in Figure 12 which indicates a frequency shift in the 973 cm^{-1} band of sample A and B. The band at 973 cm^{-1} is complex and could not be resolved any further from this data.

The band at 841 cm^{-1} was also deconvolved. The original spectra and superimposed deconvolutions are shown in Figure 15. In both cases the half-band-width was reduced to approximately 2 cm^{-1} using the identical deconvolution. The 841 cm^{-1} band, which is due to methine rocking, is more intense in sample B, which is again consistent with sample B having more crystalline character than sample A. A summary of the band positions and half-band-widths after deconvolution is given in Table 6.

From the deconvolution of various bands in the isotactic polypropylene samples it is apparent that several bands reported to be constant in amorphous and crystalline samples are changing more intense in amorphous samples. The 973 cm^{-1} band shows the most variation in that it is not one band but a sharp intense band at 972 cm^{-1} that increases in intensity with more crystalline samples and a smaller, broader band centered at 975 cm^{-1} that increases in intensity with more amorphous samples. The deconvolution techniques provided a way to better observe these bands. Whether the factor used in deconvolution can be used to quantitate the level of crystallinity or not requires further investigation.

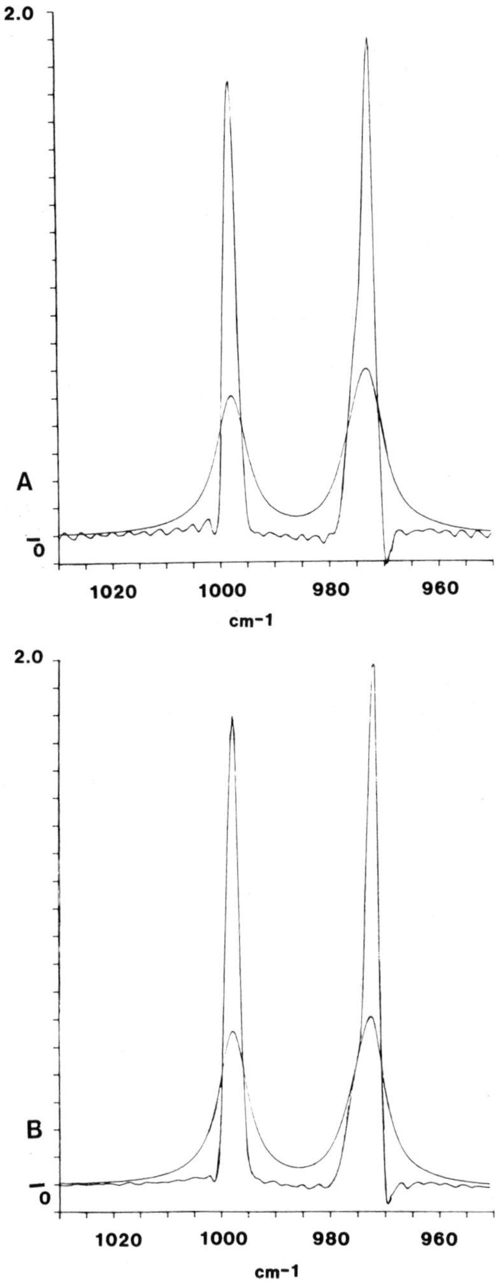

Figure 14 Sample A; top Sample B; bottom.

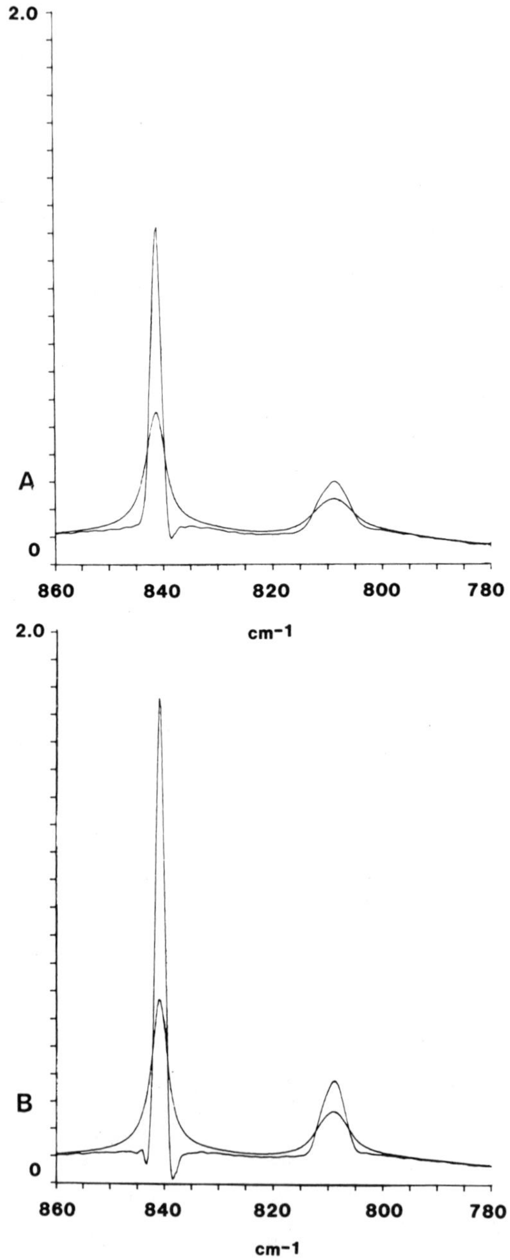

Figure 15 Sample A; top Sample B; bottom.

Table 6

Band Position cm^{-1}	Sample A Intensity Absorbance	Half-Band Width, cm^{-1}	Sample B Intensity Absorbance	Half-Band Width, cm^{-1}
1168	1.297	4.2	1.823	3.1
1152	0.343	-	0.326	-
998	1.691	3.8	1.740	3.8
972	1.855	3.0	1.979	2.5
841	1.184	2.3	1.750	2.0

CONCLUSION

The large amount of structural information inherent in the infrared spectra of polymers can be accessed by the use of spectral enhancement rountines such as spectral self-deconvolution. The additional application of a curve fit analysis can be used to support the validity of the enhanced data and, at times, can be used to determine the presence of broad bands that are not adequately indicated by the enhancement.

REFERENCES

1. W.E.Blas and G.W.Halsey, "Deconvolution of Absorption Spectra", Academic, New York (1981).

2. J.K.Kauppinen, D.J.Moffatt, H.H.Mantsch and d.G.Cameron, Appl. Spectrosc., 35, 271 (1981).

3. J.K.Kauppinen, D.J.Moffatt, D.G.Cameron and H.H.Mantsch, Appl. Aptics, 20, 20 (1981).

4. H.A.Willis and M.E.A.Cudby, "Structural Studies of Macromolecules by Spectroscopic Methods", K.J.Ivin, Ed., John Wiley and Sons, New York (1976).

5. I.J.Grant and I.M.Ward, Polymer, 6, 223 (1965).

6. D.O.Hummel, "Infrared Analysis of Polymers, Resins and Additives an Atlas", Vol.I, Wiley Interscience, New York (1971).

APPLYING VECTOR SOFTWARE CONCEPTS TO THE QUANTITATION OF POLYMER SYSTEMS

J.A. Miller and R.J. Obremski

Beckman Instruments, Inc.
Irvine, CA 92713

ABSTRACT

In the application of today's FT-IR systems to polymer science, the ease and the success of that application oftentimes is not the result of a benefit purely derived from Fourier transform infrared. Often the benefit to the analysis results from the ability to use some new accessory or the ability to apply an analytical concept to better advantage in conjunction with FT-IR. So too can there be new benefits and new improvements to methods through software and applied mathematics. Such is the basis for this description of vector concepts applied to the quantitation of polymer samples.

INTRODUCTION

For nearly four and a half decades, since the earliest infrared instruments, infrared analysis has contributed strongly to quantitation and control within the polymer science. At times, adequate answers have resulted merely from ratioing pairs of absorbance bands to measure relative amounts within copolymer systems or within polymer blends. In more traditional spectrophotometric quantitation, absorbance bands within an analyte spectrum are chosen for selectivity, sensitivity, and a linear response of intensity versus changing concentration. Measurements of these bands' net absorbances, relative to a calibration made across known concentration changes, have been the means, therefore, of producing a quantitative answer from an infrared spectrum. Employing the relationship of the Beer-Lambert law (A=abC) enabled the analyst to determine concentration using simple mathematics. When more than one analyte is to be determined a selective absorbance band for each analyte is chosen, and the measured, multiple net-absorbances are fitted to the simultaneous solution of multiple equations. With the addition of computational capabilities, and by applying matrix algebra to the solution of the multiple equations, computerized infrared speeds a complex analysis and permits quantitative schemes requiring only one careful calibration and the storage of essential factors in a calibration matrix. Mathematics employing matrix concepts have, therefore, become common to infrared quantitative software, and there has been a growing tendency to apply other useful forms of mathematics. A more rigorous treatment of all these fundamentals is not intended here, since there are numerous texts to which one may refer [1-4].

FT-IR QUANTITATIVE ANALYSIS

These same traditional quantitative schemes have been carried forward into FT-IR instrumentation. Also, because quantitation in the infrared region is routinely practiced, an FT-IR without quantitation software would be considered incomplete. The fact is that FT-IR systems will benefit all areas of science bringing quantitative abilities to work in unusual sampling situations--as within hyphenated technologies, or in time-dependent studies, and in all applications demanding high signal-to-noise and speed. Speed is not always a consideration in the development of a quantitative method, but sample throughput (analyses per minute) may be important to a given experiment.

Applying multicomponent quantitative analysis methods directly to problem mixtures can save considerable time and speed throughput, because time-consuming separation steps can be avoided or minimized. But too often, today's analyst may lack complete familiarity with the capabilities of multicomponent quantitative analysis using matrix concepts or those concepts extended through the use of vectors. Our work actually began with applying vector mathematics to dispersive infrared data. The targeted use was always intended for FT-IR data, for as you will see, the Fourier transform plays an important part in our derivation of a vector. Vector concepts have been a natural extension of the application of mathematics to infrared analysis, and one reason for the pursuit of vectors has been for analytical speed.

As stated earlier, a complete definition of the fundamentals of multidimensional vector analysis [5-7] is beyond the scope of this paper, but suitable references will be listed to each topic area mentioned. The fitting of spectral information to vector space has been under investigation within our areas of product research for a number of years, and other groups have published vector concepts obviously from parallel but totally independent studies [8,9]. The mathematical basis is similar in nature to factor anlysis applied to spectra, and a number of papers have appeared defining applications of factor analysis [10-13].

In the broadest sense, these approaches all have a common thread in that there is good reason to deal with spectral information as a mathematician would deal with large number sets. An excellent example of a mathematical assessment of benefit to the analyst would be in the use of principal factor analysis of a quantitative calibration matrix to determine how many variables (components) are actually in evidence.

A spectroscopist is always attempting to learn as much as possible from the available spectrum, and the spectrum is the only available source of information. In progressing from analog systems to digital systems, the benefit may have been perceived as one of new control or enhanced speed associated with a computer. the real and perhaps unperceived value is that now the analyst has access to digitized numbers defining all the spectral information in subtle detail. This detail resides as well in the interferogram or transforms of the spectrum.

VECTOR QUANTITATIVE ANALYSIS

Many facets of a vector quantitative analysis are similar to the traditional form of a quantitative method. Data is assessed in absorbance. While each spectrum within the FT-IR system resides as a %T versus wavenumber data file (Figure 1), the first step in the vector program is to convert the %T file to an absorbance file (Figure 2) using the familiar relationship of A=-log T.

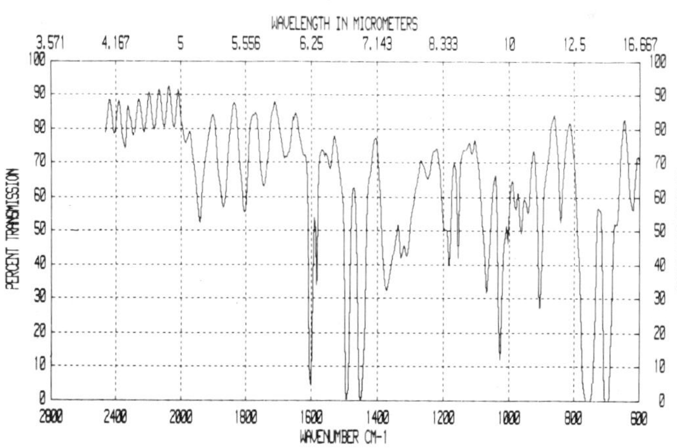

Figure 1. Polystyrene spectrum (%T).

This conversion places the data in a form that is proportional to concentration. The next step involves a Fourier transform of the absorbance spectral file to an absorbance transform (Figure 3).

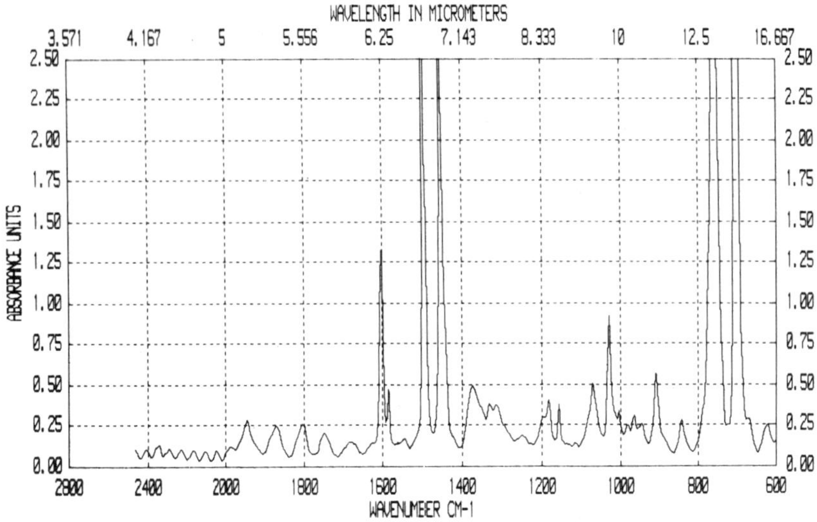

Figure 2 Polystyrene Spectrum (Abs.).

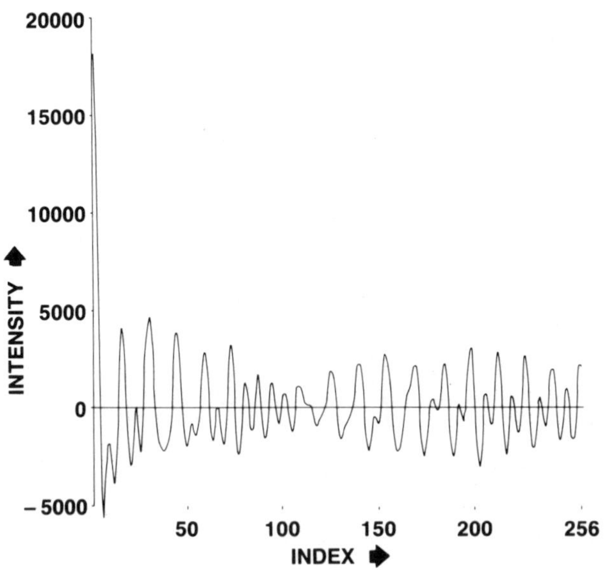

Figure 3 Fourier Transform of Absorbance Spectrum.

This preserves the property of being proportional to concentration, but places the data in a form such that each data point contains information regarding the entire spectrum and the overall function is an orthogonal series. The entire absorbance spectrum may be transformed, or only selected segments of the spectrum.

Table 1. Absorbance Transform Table

Index	Intensity	Index	Intensity
1	18179	6	-4246
2	13456	7	-2630
3	3462	-	-
4	-4068	-	-
5	-5822	256	2101

The data points within the absorbance transform are tabulated in memory as a series of discrete data points, Table 1. They may be plotted graphically, Figure 4. The fact that this data series represents an orthogonal function benefits us, in that we can use segments of this transform to build vector representations of varying lengths. These are referred to as "vector index ranges".

As in a traditional method development, absorbance bands will be carefully selected for each analyte within the mixture to minimize overlap and secure selectivity and purity for each analyte's measurement; hence the use of selected segments of the spectrum. Spectral segments are selected to bracket obvious changes occurring due to concentration changes in the calibration standards.

When these segments are expressed in vector form, net-absorbances or peak areas are not measured; rather, all the spectral information within a given segment of the spectrum is assessed. So, band shapes, areas, widths, intensities, even the most subtle changes in a segment's character are available for assessment in the vector representation.

Within the absorbance transform data, some terms do a better job than others in defining all the subtleties of the changing spectral information. Different vector index ranges may be chosen and quickly built into P-matrix [14,15] form as trial calibrations for the method (see Figure 4). The best selections of vector indices and of spectral segments are assessed by observing the effects seen upon the slopes of the calibration curves and by submitting known mixture for analysis. Obviously, the steeper the slopes for each analyte, the better the sensitivity and selectivity for changing concentrations. Good agreement in the analyses of known samples is the final test for any quantitative method.

Incorporating the vector information in P-matrix mathematics provides speed in the ultimate concentration determinations, but matrix concepts provide avenues for additional benefits. The associated set of mutually orthogonal eigenvector and scaler eigenvalues are used to determine the number of principal components varying within the matrix. This approach has been demonstrated [16] to be of value in polymer science and in polymer blending for following reactions to confirm whether unwanted (reactions) components have resulted.

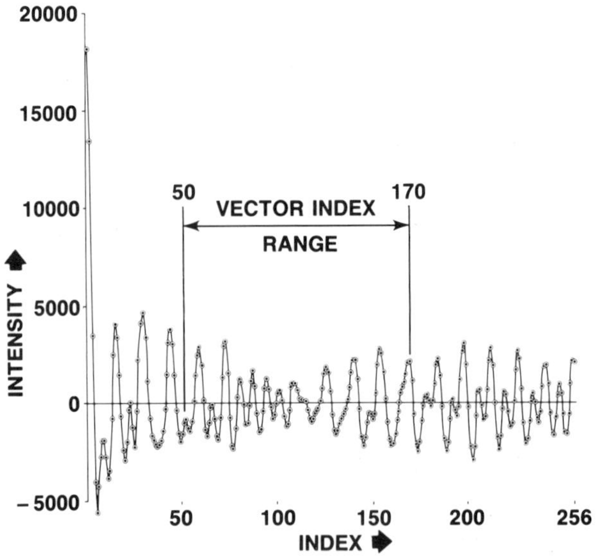

Figure 4. Absorbance transform.

In addition, when the analysis is overspecified by including more standards than the number of components in the mixture, a least-squares linear regression is used to calculate a P-matrix. Then, the P-matrix is used to recompute the concentrations of all the components within the standards and the calculated concentrations are compared to the actual concentrations.

A final benefit to the polymer analyst (within this all-inclusive approach of vector expressions and applied matrix mathematics) is realized when the matrix is constructed in a manner independent of sample pathlength [17]. This approach to matrix work enables the analyst to build the matrix with standards of known concentrations, but of undetermined pathlengths. After the matrix is constructed, another standard of known concentrations is submitted for assay, and this analysis is used to correct the matrix for indeterminate pathlength.

Therefore, polymer films (evaporated or merely quickly pressed to an acceptable range of thickness) may be easily accommodated in calibration and sample analysis.

EXPERIMENTAL SECTION

Apparatus

All spectra were prepared using a Beckman Model 2100 FT-IR with a DTGS detector. Spectra were averages of 64 scans at 4 cm^{-1} resolution employing only boxcar truncation. No other forms of data manipulation were applied to the data before using the vector quantitation software.

Sample Preparation

Pressed films of poly(vinyl acetate) (PVA) and poly(vinyl chloride) (PVC) copolymers were obtained for these trials. Transmission spectra were run on these films as received. Table 2 summarizes the various film thicknesses and PVA percentages for these samples.

Table 2. Sample Thicknesses and PVA%

Film Tag	Estimated Thickness (mils)	% Vinyl Acetate Range
A_{1+2}	2.5 - 3	1.5 - 2.0
B_{1+2}	2.25 - 3	4 - 6
C_{1+2}	~2	10 - 11
D_{1+2}	1.25 - 2	13.5 - 14.5
E_{1+2}	1.25	11 - 12
F_{1+2}	~2	~9
G_{1+2}	3 - 4	Ø

Duplicate films had been pressed from each copolymer, and one set was used for calibration; the other set used for testing the method as unknowns. A second set of spectra were prepared from small samplings cut from these films. These samplings were dissolved in THF and very thin films were cast upon KBr plates. After the THF had evaporated, these cast films were used to calibrate a second method to evaluate performance on weak spectra and on spectra exhibiting scattering and sloping baselines; the usual problems associated with this means of sample preparation.

RESULTS AND DISCUSSION

Figure 5 illustrates a typical spectrum for one of these films, as recieved. It is obvious that these film thicknesses will not permit the use of the carbonyl absorption (1735 cm^{-1}) to be used in assessing the level of PVA. This particular sample represents the mid-range of PVA percentages from this group, and the carbonyl absorption is off scale.

A region of second choice would be surrounding the C-O stretching at 1030 cm^{-1}. Figure 6 views the spectrum from 1100 cm^{-1} to 800 cm^{-1} for one of these copolymer samples plus that for the PVC polymer. Certainly this region and the absorption at 1030 cm^{-1} could be used even in traditional quantitative schemes. Dissolution of the samples and running them as solution in fixed path cells would be an acceptable approach. Other means could employ ratioing tehniques, or if the films exhibited interference fringes, these could be used to determine film thickness.

These films did not exhibit fringe patterns, and it was our intent to use the full force of this software to determine the PVA and PVC component percentages in as direct a fashion as possible. Figure 7 displays the pair seen in Figure 6, but with a third sample spectrum added, and Figure 8 shows this region with still another added.

The changing absorbance at 1030 cm^{-1} is obvious and there is an associated, diminishing change seen in the band at 965 cm^{-1}. And as expected, the baselines for these varying film thicknesses are also moving up and down -- especially in the higher wavenumbers near 1100 cm^{-1}. This would normally make it difficult to quantitate with good precision using net absorbance measurements or even band areas.

Using the indeterminate pathlength concept for adjusting the P-matrix, three standards are used in the calibration. Figure 9 shows the menu for establishing a calibration and Table 3 lists the selected standards and their percentages input for the calibration.

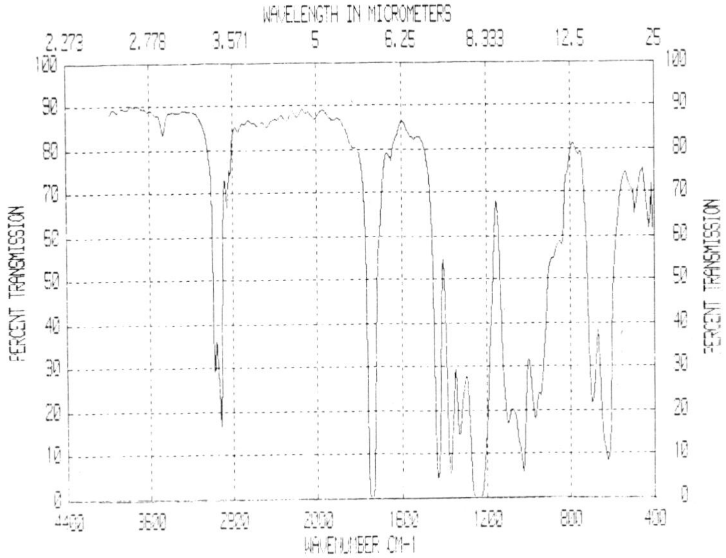

Figure 5 Typical Copolymer Spectrum.

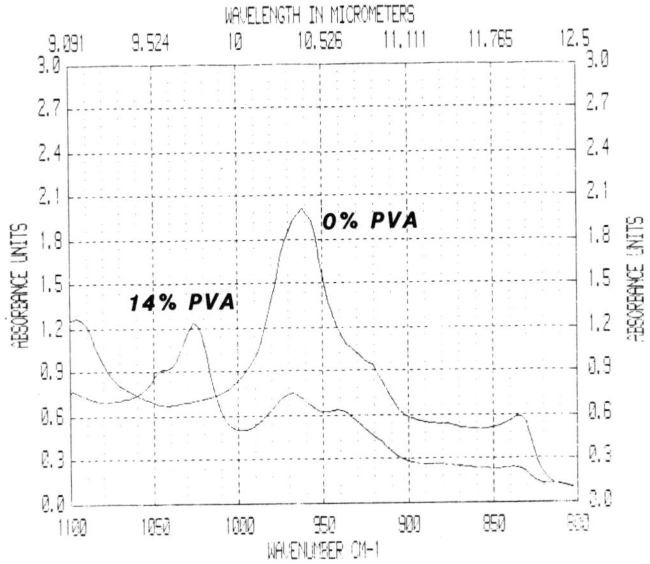

Figure 6 C-O Stretch Region (Abs.).

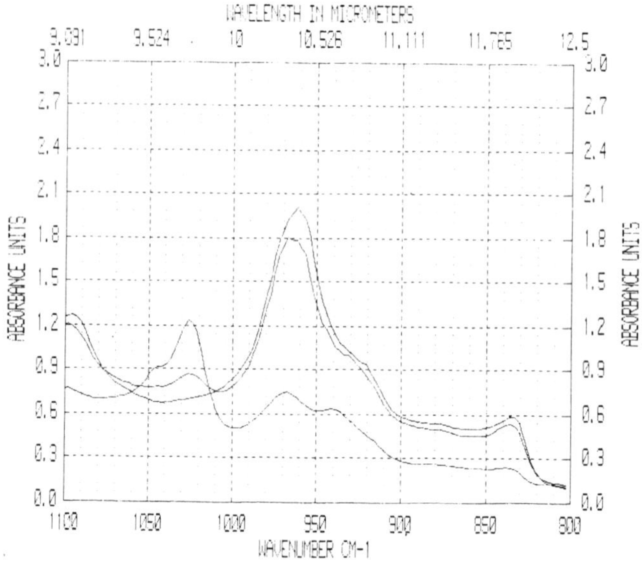

Figure 7 Additional Standard.

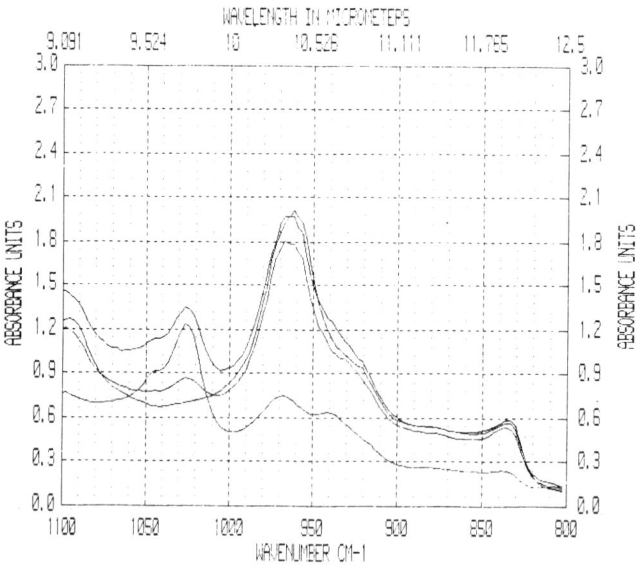

Figure 8 Four Film Samples.

In the work with the pressed films and with the thinner, cast films, all calibration trials were run using the analysis range of 1150 to 800 cm^{-1} noted in Figure 9. In this particular trial (Figure 9) the absorbance transform for that analysis range was assessed beginning with index 140 of the transform. That point and the next thirty indices were to be used to construct a vector for each standard. These vectors were then built into a P-matrix and adjusted for pathlength. Several trials were actually attempted to determine which starting index would produce the best calibration. Table 4 lists the variations noted in the slope of the calibration curves for these trials with varying starting indices.

Table 3.

Standard number	1	
Spectrum name	A2 FILM	[DISK1]
Run new scan	[NO]	
Component	concentration	units
1	1.7500	PER CENT
2	98.500	PER CENT
Standard number	2	
Spectrum name	D2 FILM	[DISK1]
Run new scan	[NO]	
Component	concentration	units
1	14.000	PER CENT
2	86.000	PER CENT
Standard number	3	
Spectrum name	F2 FILM	[DISK1]
Run new scan	[NO]	
Component	concentration	units
1	9.0000	PER CENT
2	91.000	PER CENT

Table 4. Calibration Curve Slope Variation for Film Samples

Trial	Starting Index	Slope of Component 1	Slope of Component 2
1	125	1.000	0.3744
2	140	1.000	0.9886
3	145	1.000	0.5228
4	135	1.000	0.5855

From this tabulation it is easily seen that the steepest slope is associated with the second calibration trial.

Table 5 presents a printout of the p-matrix comparison of recomputed values versus actual values for this calibration. Note too, the number of components was assessed using principal factor analysis, and only two variables were indeed found changing within the matrix of eigenvectors. To test accuracy for this calibration, the duplicate set of samples from the group were submitted for analysis against this calibration.

```
BUILD OR MODIFY AN ANALYSIS

Analysis name               FILM4D2  [DISK1]
Lock                        0000
Number of components         2
Number of standards          3
Indeterminate pathlength    [YES]
Normalize to weight/volume  [NO]
Starting index              140
Analysis range(s)           [1]
                            1150 to  800 cm-1

Additional information:
No additional information
```

Figure 9. Analysis conditions.

Table 5. Vector Quantitative Analysis
(Results of Analysis No.1 A)

Component	Concentration	Units
1	1.5817	%
2	98.418	%
Residual	1.1822	-

Component 1 refers to the PVA content, because that correlates to the order in which concentration values were entered in the set up menu, refer to Table 3. The values are normalized to 100%; this is a requirement of the indeterminant pathlength adjustment.

Also seen in this printout is a term referred to as a residual. This number indicates how well the vector of the unknown fit the spatial plane of the calibration vectors. In a two component mixture, such as in these copolymers, the vectors for each component would define a two dimensional plane in space. The unknown´s vector should fall upon that same plane; however, in any analysis there are factors present which may lift the sample vector off that plane. These factors are: noise, intermolecular actions (band shifts, etc.), contaminants, and background shifts. Another way of defining the residual is that it represents the portion of the unknown's vector which cannot be expressed as a linear combination of the standard vectors. The residual value reflects how well the sample (real world) compares to the calibration plane (ideal world), and is not a numerical value for error. It is obvious that a calibration set up should be composed of standard mixtures that best represent a production or real world analysis. The lower the residual number, the better the samples' vector fit the plane of the calibration. When these residual numbers exceed (typically) 5.0, they indicate a problem with the sample -- i.e. too much noise, a contaminant, etc.

Table 6 itemizes the results from the analysis of all the available duplicates in this group of samples, used to test this calibration.

Table 6. Summary of Results in Test of This Calibration

Sample	Known %PVA	Computed %PVA	Computed %PVC	Residual
A	1.5 - 2.0	1.577	98.423	1.182
B	4.0 - 6.0	5.113	94.887	2.517
C	10.0 -11.0	9.221	90.779	3.036
D	13.5 -14.5	13.361	86.639	1.672
E	11.0 -12.0	11.925	88.075	1.261
F	9	9.872	90.128	1.827
G	0	0.171	99.829	10.756

In all instances except for C and G above, the computed values fall in the expected ranges, and the accompanying residual numbers are low. For sample C, the residual number is slightly higher, indicating some factor within that assay had pulled that vector slightly away from the calibration plane, but the values are still close to the expected range. In sample G, however, that represents an entirely different sample type. Sample G was not a copolymer but was poly(vinyl chloride). The high residual number for G is another confirmation of a difference between the copolymer calibration and this sample. As in a real sample analysis, this would have flagged a suspect sample and prompted closer scrutiny of the sample's spectrum.

To estimate precision, a single sample from the group was run in five replicate determinations. Table 7 lists the results for repeats of sample B.

Table 7. Precision of Replicate Analysis

Analysis	%PVA	%PVC	Residual
1	5.078	94.922	1.427
2	5.023	94.977	1.698
3	5.124	94.876	1.587
4	5.148	94.852	1.645
5	5.048	94.952	1.473
Average %	5.084	94.916	1.566
Std. Dev.	0.0516	0.0516	0.114

In the second half of this work, thinner films from solution evaporation on KBr plates were evaluated. Figure 10 illustrates two of the spectra used in the calibration. Most of the films yielded spectra with flat baselines similar to the 14% PVA spectrum in Figure 10.

The worst case of a sloping spectrum was encountered in the 2% PVA spectrum in this figure. To show the ability of vector mathematics to accommodate baseline problems, we selected to use this 2% PVA spectrum in the actual calibration for these samples.

Also, no expansion techniques were applied to any of these weak spectra. The differences in absorbance values were, therefore, very small numerically, as can be seen in Figure 12. This represents the region used for building the vectors and the three spectra used for calibration. Calibration was again pursued using indeterminate pathlength correction.

Calibration trials were tried at several different starting indices, and the best clibration slope was found at a starting index of 50. Whereas the calibration for the thicker films was achieved with a starting index of 140, these thin film spectra exhibit slightly varying data changes very close to the baseline. The lower index data points (near index 50) within these absorbance transforms must better describe these subtle undulations.

Such thin films and weak spectra would not be expected to yield comparable results to the thicker films and more optimum absorbances. The results were surprising in how close this calibration came to achieving reasonable values. Time did not permit duplicates of these cast films, but those films which were not used in the calibration were used as unknowns to test this set up.

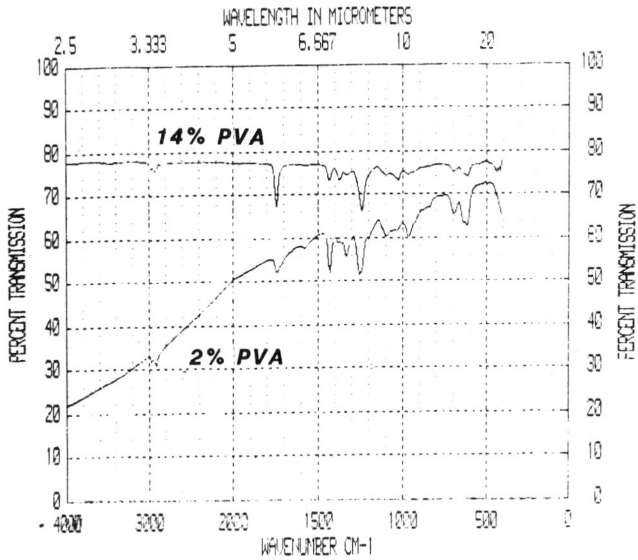

Figure 10 Evaporated Film Spectra.

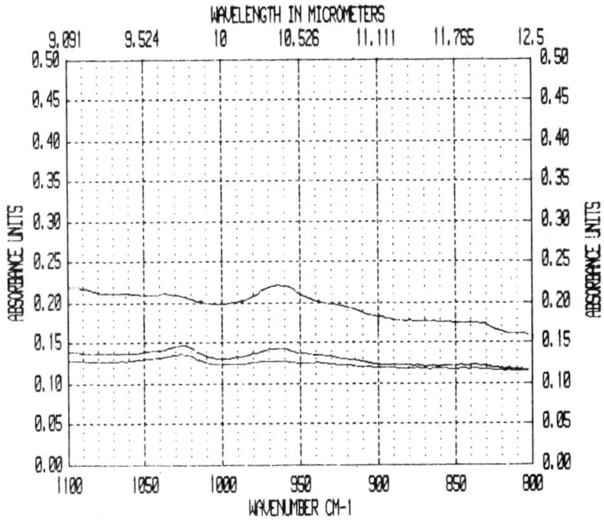

Figure 11 C-O Stretch Region (Abs.).

Table 8 details the results from samples B, C and E; samples A, D and F were used in this weak spectra calibration.

Table 8. Results in Evaluating Thin Film Calibration

Sample	%PVA	Found %PVA	%PVC	Residual
B	4 - 6	3.491	96.509	11.07
C	10 -11	11.628	88.372	13.59
E	11 -12	12.024	87.976	11.44

Now the residual values are much higher than in the more optimum sampling instance, but the values are within one percent of the expected range for PVA.

The residual values are most obviously indicating a larger effect of noise in these measurements. The signal levels within the spectra (Figure 12) are at absorbances between 0.10 to 0.25 A, with difference between sample readings in the third decimal. Noise has to be the most suspect cause for moving these sample vectors from the calibration plane.

CONCLUSIONS

Quantitation within the polymer sciences can be enhanced through the use of vector concepts in multicomponent infrared analysis. These benefits may be summarized as:

1. Vectors derived from absorbance transforms of spectra permit quantitative assessments even of subtle differences between spectra.

2. Selection of the optimum indices within the transform can overcome baseline and noise problems associated with the sample or sampling technique.

3. Vector representations may by expressed in multidimensional matrix form and benefit therefore from matrix concepts in:

   a. Computational speed across complex mixtures.
   b. Indeterminate pathlength samplings.
   c. Principal factor analysis.

ACKNOWLEDGEMENT

The authors wish to thank B.F.Goodrich Canada, Inc., Niagara chemical Plant, Niagara Falls, Ontario, for supplying the polymer samples used in this study.

REFERENCES

1. R.P.Bauman,"Absorption Spectroscopy", Wiley, New York (1962).

2. G.L.Clar, Ed.,"Encyclopedia of Spectroscopy", Reinhold, New York (1960).

3. M.M.Willard, L.I.Merritt,Jr. and J.A.Dean,"Instrumental Methods of Analysis", Fifth Ed., Van Nostrand, Princeton, N.J. (1974).

4. J.Haslam, J.A.Willis and D.C.M.Squirrell,"Identification and Analysis of Plastics", Second Ed., Iliffe Books, London (1972).

5. S.Lang,"Linear Algebra", Addison-Wesley, Reading, Ma. (1966).

6. G.Hadley,"Linear Algebra", Addison-Wesley, Reading, Ma. (1961).

7. G.W.Stewart,"Introduction to Matrix Corporations", Academic, New York (1973).

8. G.W.Small, G.T.Rasmussen and T.L.Isenhour, Appl. Spectrosc., 33, 444 (1979).

9. D.R.Morgan, Appl. Spectrosc., 31, 404 (1977).

10. W.F.Maddams, Appl. Spectrosc., 34, 245 (1980).

11. P.C.Gillete, J.B.Lando and J.L.Koenig, Anal. Chem., 55, 630 (1983).

12. P.C.Gillette and J.L.Koenig, Appl. Spectrosc., 36, 535 (1982).

13. Ibid., Appl. Spectrosc., 36, 661 (1982).

14. H.H.Harmon,"Modern Factor Analysis", N. of Chicago Press, Chicago (1967).

15. J.L.Koenig and M.J.M.Rodriguez, Appl.Spectrosc., 35, 543 (1981).

16. M.K.Antoon, L.D'Esposito and J.L.Koenig, Appl. Spectrosc., 33, 351 (1979).

17. B.R.Loy, R.W.Chrisman, T.A.Nyquist and C.L.Putzig, Appl. Spectrosc., 33, 638 (1979).

FT-IR STUDIES OF IONOMERS

Paul C. Painter, B.A. Brozoski* and M.M. Coleman

Polymer Science Program
Materials Science and Engineering
The Pennsylvania State University
University Park, PA 16802

* Experimental Station
E.I. DuPont de Nemours Company
Wilmington, DE 19898

INTRODUCTION

Perhaps the most widely accepted model of ionomer superstructure is the multiplet-cluster concept advanced by Eisenberg [1,2]. This framework has dominated the interpretation of many vibrational spectroscopic studies to the extent that specific infrared bands and Raman lines have been separately assigned to multiplets or clusters. We were no exception to this general rule and our initial FT-IR studies of the sodium and calcium salts of ethylene-methacrylic acid copolymers were interpreted in terms of Eisenburg's model [3,4]. Specific sharp bands in the spectra of quenched films or films held at elevated temperatures (in the range 70 to 130°C) were assigned to multiplets. Broad bands observed upon annealing these films for extended periods of time at room temperature were assigned to clusters. However, as a result of subsequent work on the salts of other group I and group II elements, we were forced to reevaluate these somewhat simplistic initial assignments [5]. In addition, even trace amounts of water can be shown to have a profound effect on the spectrum [6]. Because our publications in this field are, in effect, a series of successively modified interpretations, we will take this opportunity to present an overview of our work, with emphasis on the sensitivity of infrared spectroscopy to the local order found in these systems.

RESULTS AND DISCUSSION

Our initial work was based on studies of sodium and calcium ionomers [3,4]. In considering the evolution of our interpretation, it is useful to examine some of these initial results and then consider the modifications that were forced by the acquisition of additional data. Figure 1 compares the infrared spectrum of the sodium ionomer of an ethylene-methacrylic acid copolymer quenched from the melt to the spectra of the same sample maintained at various temperatures. A collection of bands near 1550 cm^{-1}, characteristic of carboxylate anions, dominate the spectrum. We previously assigned the band near 1568 cm^{-1} to multiplets while the band near 1547 cm^{-1} was assigned to isolated sodium-carboxylate ion pairs. In part, this assignment was based on the apparent change in the relative intensities of the 1568/1547 cm^{-1} modes with thermal history. However, a closer examination of the spectra, together with the insight gained from the spectra of other group 1 ionomers that we will consider shortly, suggests an alternative explanation. The 1568/1547 cm^{-1} bands could be a doublet that is the result of some local arrangement of carboxylate anions. The change in the relative intensities of these bands is a result of the appearance of an overlapping mode near 1550 cm^{-1}. This latter band is clearly present in the spectrum of the sample held at 130°C and dominates the spectrum of the sample held at 150°C.

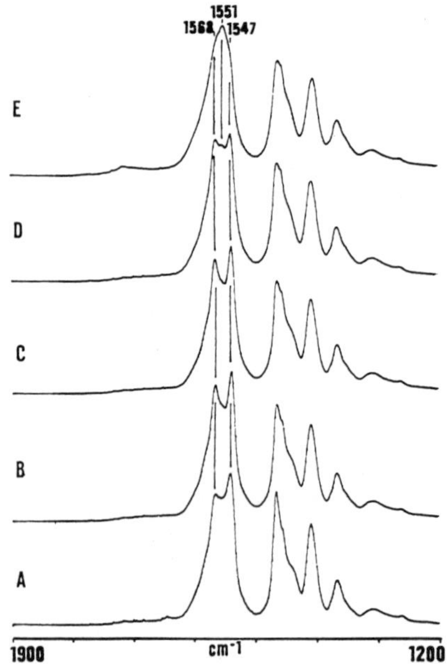

Figure 1. FT-IR spectra in the range 1900-1200 cm^{-1} of a completely ionized sodium ionomer at various temperatures.

Eisenberg [1] has argued that the formation of clusters depends upon the balance between the elastic forces of the polymer chain and the attractions between small ionic domains termed multiplets. As the temperature of the sample is raised, the balance between the elastic forces of the chain and the electrostatic attractions between ionic groups is perturbed, thus influencing the distribution of associated states. The band near 1550 cm^{-1}, which increases in intensity with increasing temperature, could therefore be due to simple ion pairs, while the doublet at 1568/1547 cm^{-1} is more probably associated with some sort of local structure found in multiplets (and clusters).

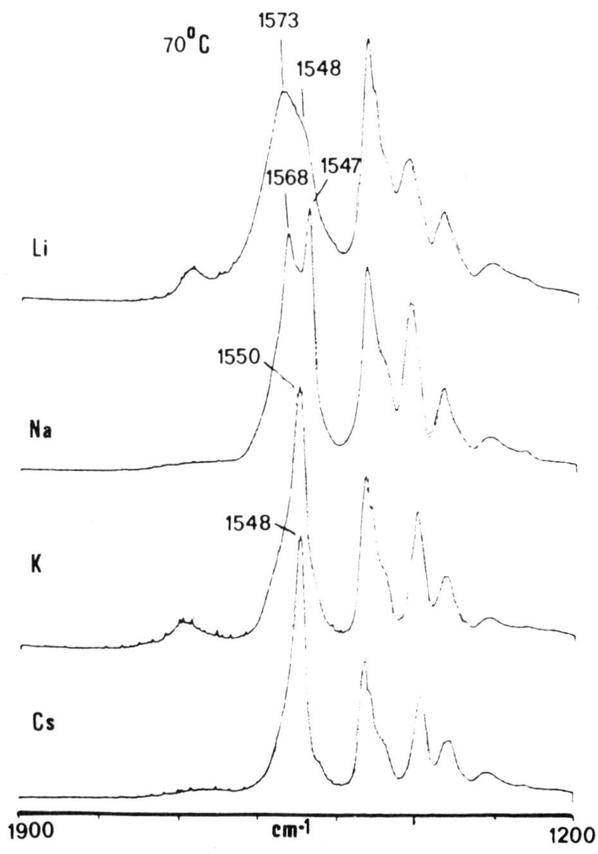

Figure 2. FT-IR spectra in the range 1900-1200 cm^{-1} of the completely ionized alkali metal ionomers at 70°C.

Similar temperature studies were performed on all of the ionomers used in this study. The spectra of samples held in the temperature range 70-90°C were always the sharpest and best defined, probably because such temperatures are high enough to minimize the presence of trace amounts of water (which would hydrate and disrupt ionic structures), but not so high as to favor simple ion pairs over more organized structures. Accordingly, the remaining spectra of groups 1 and group 2 ionomers presented here are those obtained from samples held at 70°C.

The spectra of the lithium, sodium, potassium, and cesium salts are compared in Figure 2. Qualitatively, the spectrum of the lithium salt is similar to that of the sodium ionomer, with a pair of bands near 1573 and 1548 cm^{-1} clearly present. However, the bands are broader and there are additional weak modes near 1590 and 1687 cm^{-1}. In contrast, the spectra of the potassium and cesium salts are of a completely different character. The spectra of both are dominated by a surprisingly sharp single band near 1550 cm^{-1}. Modes near 1675 and 1590 cm^{-1} can again be observed, but are relatively weak. These latter bands can be assigned to the presence of acid salts [5] and will not be considered further.

The spectra of the salts of group II elements are compared in Figure 3. The spectra of all these samples are characterized by a doublet, at 1539 and 1513 cm^{-1} in the spectrum of the barium salt, but at progressively higher frequencies as the size of the cation decreases.

Originally, we considered that these doublets were a result of a splitting due to interaction between pairs of COO^- ions held in close proximity by a divalent counterion [3]. However, this leaves the bands in the spectra of the sodium and lithium ionomers to be rationalized in some fashion. Our doubts concerning such an interpretation were deepened by a consideration of the spectrum of the zinc ionomer, presented in Figure 4. Surprisingly, this material is characterized by a single band at 1585 cm^{-1}. There is little change in the spectrum of the zinc ionomer during the temperature studies. In fact, this observation was the final straw that forced us to abandon our original interpretation. We will show that an analysis based on the coordinating tendencies of the cations provides a more satisfactory and complete explanation for the experimental results.

These observations, taken as a whole, raise some important questions. The relatively sharp singlet and doublets observed in the asymmetric carboxylate stretching region of the spectra of the various ionomers suggest the presence of locally ordered ionic structures (multiplets?); in this context, there is one key question that must be addressed: why do certain ionomers (e.g.

the K, Cs and Zn salts) exhibit sharp singlets while others (e.g. the Na, Ca, Sr, and Ba salts) exhibit distinct doublets?

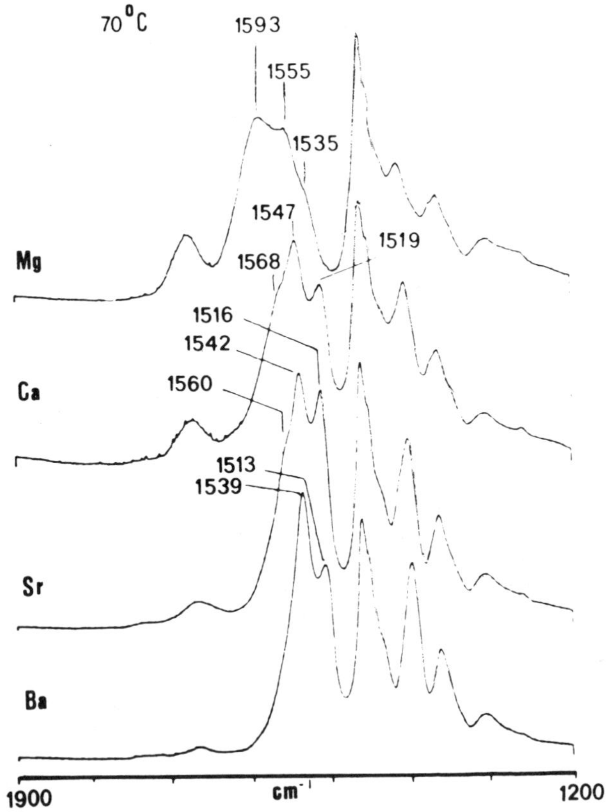

Figure 3. FT-IR spectra in the range 1900-1200 cm^{-1} of the completely ionized alkaline earth ionomers at 70°C.

The symmetric carboxylate stretching vibration is essentially uncoupled from backbone modes. Consequently, in an ordered structure containing a specific number of such interacting groups, we would expect the appearance of infrared bands to be markedly influenced by local symmetry. Thus, it is reasonable to perform a symmetry analysis of the "unit cell" of structure that incorporate the most probable arrangements of carboxylate groups and determine the number of active symmetric and asymmetric carboxylate stretching modes.

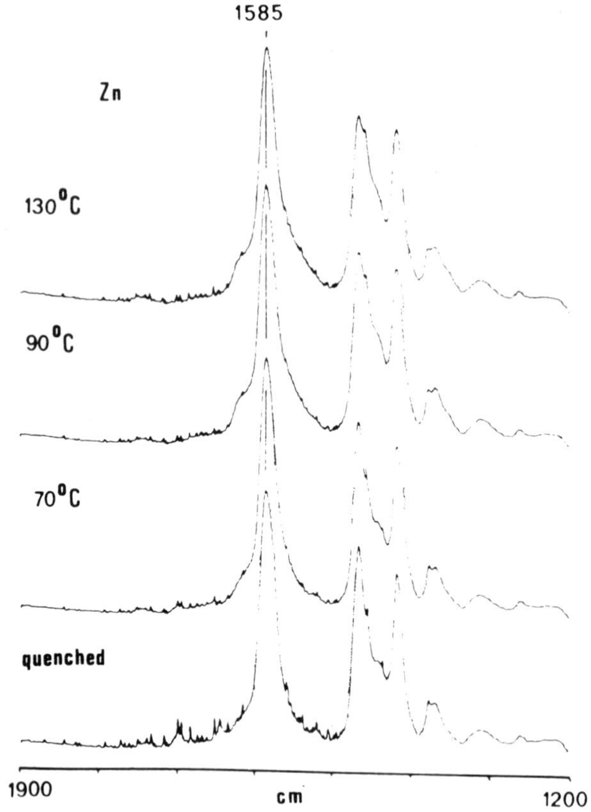

Figure 4. FT-IR spectra in the range 1900-1200 cm^{-1} of a completely ionized zinc ionomer at various temperatures.

The simplest arrangements of carboxylate groups in ionomers of divalent cations would be as a pair, as we originally postulated for the calcium salt [3]. Figure 5 illustrates two probable structures, a planar arrangement (denoted A) and a configuration where the two carboxylate anions are at 90° to one another (B). The former has D_{2h} symmetry while the latter belongs to the point group S_4. Using internal coordinates as a basis, one can determine the symmetry species and hence the vibrational activity of the asymmetric modes [7]. For the planar structure, the two asymmetric modes belong to the B_{1g} and B_{2u} symmetry species respectively, but only the B_{2u} mode is infrared active. The two asymmetric stretching modes in the S_4 structure (Figure 5B) are doubly degenerate (i.e. have the same frequency), belong to the E irreducible representation, and are both Raman and infrared active. In any event, only one discernible infrared band is predicted. Other possible arrangements of the carboxyl groups were analyzed (e.g. structures similar to the planar form where the carboxylate anions are vertically offset above and below the plane), but the result remains the same. For structures involving a single pair of carboxylate anions, a symmetry analysis predicts only one infrared-active asymmetric stretching mode.

On the basis of this somewhat surprising result we decided to extend our symmetry analysis to other possible structures, specifically those that can be postulated on the basis of packing requirements. In a number of previous studies, for example, the seminal theoretical work of Eisenberg [3], it has been pointed out that the coordinating tendency of the cation could be critical in forming multiplets. The coordination number for simple inorganic structures can be predicted to some extent on the basis of the ratio of the radii of the ions involved, as indicated in Table 1. As a first approximation we used the ionic radius of the O^{2-} ion in our calculations together with values for cationic radii listed in standard inorganic texts [8] and obtained the predicted coordination numbers that are also listed in Table 1. The experimentally determined coordination numbers of simple metal oxides are also listed in this table [8].

In the context of this analysis, it is significant that for the alkali metal cations we predict different coordination numbers for the smaller cations (4 or 6 for lithium and sodium) compared to the larger cations (8 for potassium and cesium). In the case of the alkaline earth cations a similar trend is determined, i.e. a coordination number of 4 or 6 for magnesium and calcium and 8 for strontium and barium. However, it is important to realize that the above calculations are crude and even in simple metal oxides the predicted coordinations numbers do not always correspond to those actually found. For example, a coordination number of 6 is observed for all the alkaline earth oxides (see Table 1). For the zinc cation, structures with a coordination number of 4 or 6 are predicted, while a tetrahedral structure with coordination number of 4 is found in zinc oxide.

Having set the stage, we will now turn our attention to a symmetry analysis of 6- and 8-coordinated structures. On the basis of packing characteristics of simple ions the structures illustrated in Figure 5 can be considered reasonable analogues for multiplets occuring in certain ionomers. The octahedral structure, which has a coordination number of 6 (i.e. three carboxylate anions) belongs to the symmetry group D_3. The three asymmetric stretching modes are distributed among the A_2 and E species. Both of these modes are infrared active. On the other hand, the body-centered cubic structure, which has a coordination number of 8 (i.e. four carboxylate anions) belongs to the D_{4h} symmetry group.

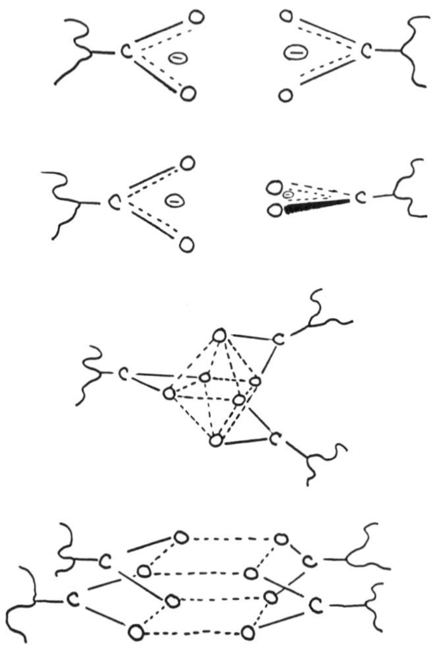

Figure 5. Possible structures for multiplets with coordination numbers of: (A) 4, (B) 4, (C) 6, and (D) 8.

Four asymmetric stretching modes are distributed among the A_{2u}, B_{1u} and E_g symmetry species. The B_{1u} mode is inactive and the E_g mode is only Raman active, leaving solely the A_{2u} mode as an infrared-active vibration. Surprisingly, of those considered, we predict the presence of more than a single infrared-active asymmetric stretching vibration in only one structure, that with six coordinated oxygen atoms (three carboxylate groups).

Let us now return to an examination of our experimental result, using the fresh light of this symmetry analysis. We will first consider the alkali metal ionomers. On the basis of the ratio of ionic radii and known coordinating tendencies, lithium and sodium salts would prefer a coordination number of 6 (although a coordination number of 4 with the oxygen atoms in a square-planar arrangement is also a possibility). In the infrared spectra of the lithium and sodium ionomers (Figures 1 and 2) distinct doublets are observed. This would suggest that the multiplets in the lithium and sodium ionomers form octahedral structures. In contrast, the predicted coordination number for the potassium and cesium salts is 8. Only one infrared-active asymmetric carboxylate stretching vibration is predicted for such structures and that is precisely what is observed in the experimental spectra (Figure 2).

TABLE I

RADIUS RATIO	COORDINATION NUMBER	STRUCTURE
.155	3	triangonal
.225	4	tetrahedral
.414	4	square planar
.414	6	octahedral
.732	8	body centered cubic

CATION	RADIUS RATIO (M/O)	PREDICTED COORDINATION NUMBER	OBSERVED COORDINATION NUMBER
Li	.49	4 or 6	
Na	.68	4 or 6	
K	.95	8	
Cs	1.21	8	
Mg	.46	4 or 6	6
Ca	.71	4 or 6	6
Sr	.81	8	6
Ba	.96	8	6
Zn	.53	4 or 6	4

In the experimental specta of all the alkaline earth ionomers (Mg, Ca, Sr, and Ba; see Figure 3) we observed distinct doublets in the asymmetric carboxylate stretching region, implying a coordination number of 6. Our rough calculations indicate that the magnesium and calcium salts would prefer coordination number of 8.

However, the experimetally observed coordination numbers of the alkaline earth oxides are identical and have a value of 6.

In contrast to the other ionomers of divalent cations considered in this study, the carboxylate stretching region of the spectrum of the zinc salt is dominated by a single band. On the basis of ionic radii a coordination number of 4 or 6 is predicted, but in zinc oxide a coordination number of 4, with the oxygen atoms arranged tetrahedrally, is observed. A symmetry analysis of this structure predicts the observed singlet.

We realize that this analysis is imperfect for a number of reasons. The radius of the oxygen atoms in the carboxylate anions is probably somewhat smaller than that of the O^{2-} ion; the steric packing factors could differ from those in simple oxides; furthermore, additional cations would have to be packed around or between such structures to achieve electrostatic balance. Nevertheless, these are a number of factors that lead us to believe that this analysis has some validity. For example, the structures found in copper carboxylate salts [9] and salts of various stearates [10-12] correspond to those that form the basis of this analysis. In addition, studies of transition-metal ionomers using ESR (as well as infrared and electron microscopy) indicate the formation of complexes involving more than one cation [13-16]. Ultimately, however, our arguments come to rest on one major consideration. Does the analysis successfully predict the experimental observations? We believe it does.

Finally, we wish to clarify the origin of the single broad band that appears upon annealing these samples for extended periods of time. We originally assigned this band to the presence of clusters, on the basis that in a large, generally disordered, aggregate the effect of a distribution of interactions would be to produce a single broad band. Infrared spectroscopy, however, is principally sensitive to short range order, so that an aggregate of locally ordered clusters interspersed with hydrocarbon chains should still largely reflect the spectroscopic features of isolated clusters. This we have confirmed by water immersion studies [6]. For example, upon immersion of a film of the strontium ionomer in water, the well resolved collection of bands between 1600 and 1500 cm^{-1} merge into a broad, asymmetrically skewed band centered at 1535 cm^{-1}, as shown in Figure 6. In addition, there is an enhenced absorbance at 1680 cm^{-1}. This would imply that upon water immersion, acid-salts are formed at the expense of the well ordered multiplet structures. Similar results were obtained for other ionomers.

Although in our initial work samples were annealed at room temperature in a dessicator [3,4], the results of water immersion studies cast considerable doubts upon an assignment of the observed broad band to cluster formation. An interpretation in terms of slow absorption of water is also possible. Accordingly, we conducted two critical experiments. Quenched films of the ionomers were placed in small vacuum desiccators over fresh P_2O_5 which were then evacuated. (In our previous annealing study we had placed the samples in desiccators over anhydrous calcium sulphate.) Figure 7, compares representative spectra of the sodium and strontium ionomers quenched from the melt and after annealing for extended periods of time at room temperature under vacuum over fresh P_2O_5. There are no significant differences between the quenched and annealed samples. This strongly suggests that the broadening of the asymmetric carboxylate stretching vibration upon annealing, that we originally interpreted as evidence for cluster formation, was in fact due to the absorption of a small amount of water.

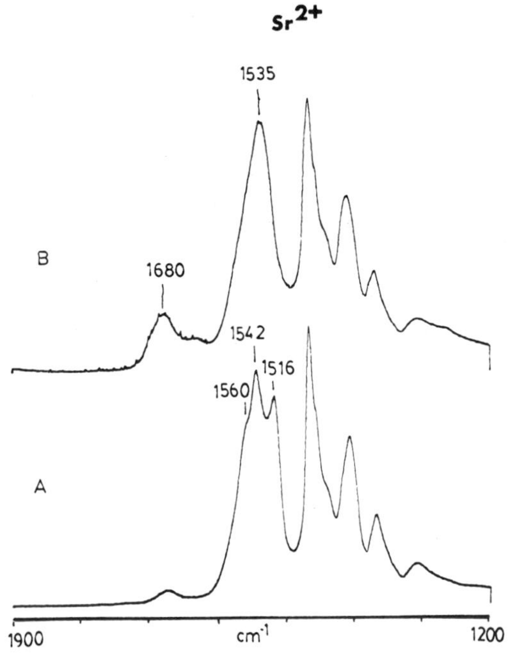

Figure 6. FT-IR spectra in the range 1900-1200 cm^{-1} of (A) strontium ionomer quenched into liquid nitrogen from the melt and (B) strontium ionomer after 21 days in water.

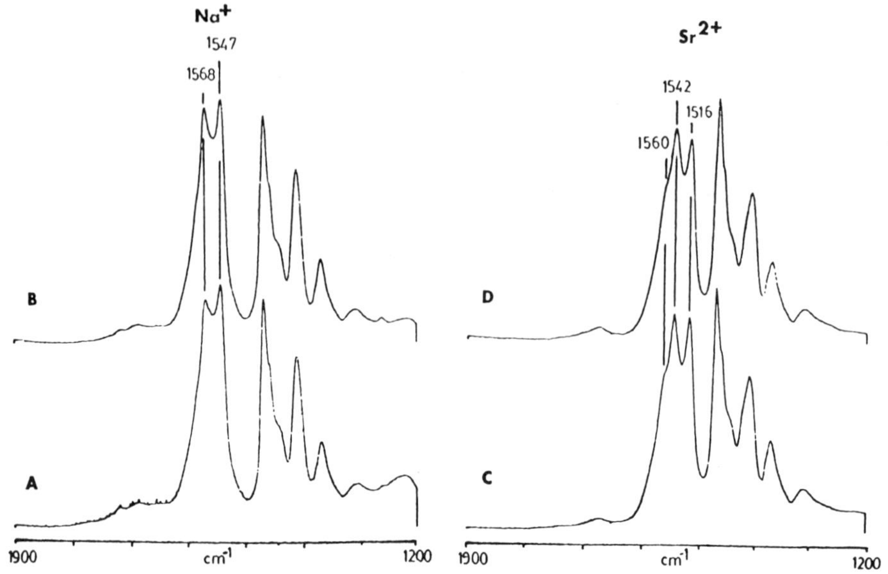

Figure 7 FT-IR spectra in the range 1900-1200 cm^{-1} of (A) the sodium ionomer quenched into liquid nitrogen from the melt (B) strontium ionomer after 89 days under vacuum over P_2O_5 (C) the strontium ionomer quenched from the melt (D) the strontium ionomer after 38 days under vacuum over P_2O_5.

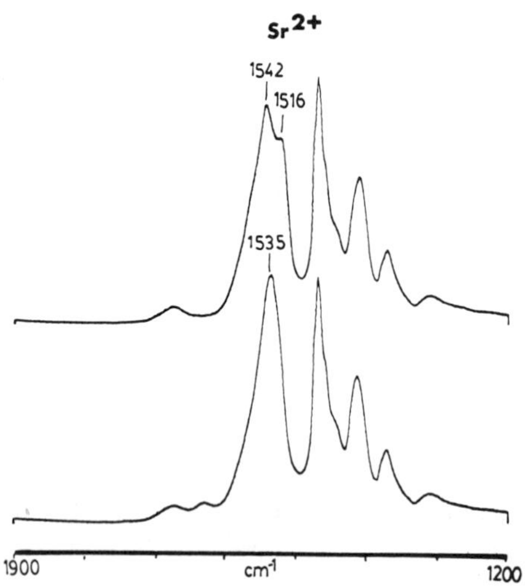

Figure 8 FT-IR spectra in the range 1900-1200 cm^{-1} of the strontium ionomer (A) quenched from the melt and immersed in water for 1 day (B) after 1 week under vacuum over P_2O_5.

The final "nail in the coffin" is introduced when we consider the results of an experiment where a sample of the strontium ionomer was immersed in water for one day (Figure 8A) and subsequently placed over fresh P_2O_5 in a vacuum desiccator for one week (Figure 8B). Upon removing water from the film the characteristic doublet at 1542/1516 cm^{-1} associated with the localized multiplet structure re-emerges.

SUMMARY AND CONCLUSIONS

There are several important conclusions that arise from these studies:

1. In the infrared spectra of ionomers we observe a pattern of bands in the asymmetric carboxylate stretching region of the spectrum that varies with the nature of the cation. On the basis of the coordinating tendency of each cation, various local arrangements of the carboxylate groups can be postulated. A symmetry analysis of these local structures successfully predicts the experimentally observed results.

2. There is little evidence to suggest that infrared spectroscopic studies of the carboxylate stretching region can be used to differentiate between multiplets and clusters. Our previous hypothesis [3,4], which invoked band broadening of the asymmetric carboxylate stretching modes as multiplets aggregate to form clusters, appears moot. It now appears that the elevated temperature results may be explained in terms of a reduction of the concentration of water in the films upon raising temperature.

3. Eisenberg [2] has previously observed that studies on ethylene/methacrylic acid ionomers indicate that clusters are the only ionic species present and that single multiplet structures do not occur even at very low ion concentrations. If this observation applies to our essentially fully ionized samples, then we must conclude that multiplets retain their structural integrity within clusters.

4. Water absorption obviously plays an important role in the overall structure on ionomers. However, different salts are affected to different degrees. Not only does hydration of the ionic domains occur, but water also is instrumetal in the formation of acid-salts. Hydration and acid-salt formation leads to a broadening of the asymmetric carboxylate stretching vibration.

ACKNOWLEDGEMENT

The authors wish to acknowledge the financial support of the National Science Foundation, Grant DMR-8206932, Polymers Program.

REFERENCES

1. A.Eisenberg, Macromolecules, 3, 147 (1970).

2. A.Eisenberg, J. Polym. Sci. Symp., 45, 91 (1974).

3. P.C.Painter, B.A.Brozoski and M.M.Coleman, J. Polym. Sci. Polym. Phys. Ed., 20, 1069 (1982).

4. B.A.Brozoski, M.M.Coleman and P.C.Painter, J. Polym. Sci Polym. Phys. Ed., 21, 301 (1983).

5. B.A.Brozoski, M.M.Coleman and P.C.Painter, Macromolecules, 17, 230 (1984).

6. B.A.Brozoski, P.C.Painter and M.M.Coleman, Macromolecules (in press).

7. P.C.Painter, M.M.Coleman and J.L.Koenig, "The Theory of Vibrational Spectroscopy and its Application to Polymeric Materials", John Wiley and Sons, New York (1982).

8. F.A.Cotton and G.Wilkinson,"Advanced Inorganic Chemistry", Interscience, New York (1982).

9. B.J.Edmindson, A.B.P.Lever, Inorg. Chem., 4, 1608 (1965).

10. T.R.Lomer, Acta Crystallogr., 5, 14 (1952).

11. S.S.Tauale, L.M.Pant and A.B.Biswacs, Acta Crystallogr., 17, 215 (1964).

12. J.H.Dumbleton and T.R.Lomer, Acta Crystallogr., 19, 301 (1965).

13. S.Yano, Y.Fujiwara, K.Aoki and J.Yamauchi, J. Colloid Interface Sci., 61, 258 (1980).

14. J.Yamauchi and S.Yeno, Makromol. Chem., 179, 2799 (1978).

15. M.Pineri, C.T.Meyer, A.Bourret, J. Polym. Sci. Polym. Phys. Ed., 13, 1881 (1975).

16. J.Yamauchi and S.Yano, Macromolecules, 15, 210 (1982).

FOURIER TRANSFORM INFRARED PHOTOACOUSTIC SPECTROSCOPY OF FILMS

N. Teramae and S. Tanaka

Department of Industrial Chemistry
Faculty of Engineering
University of Tokyo
Hongo, Bunkyo-ku, Tokyo 113
Japan

ABSTRACT

Fourier transform infrared photoacoustic spectroscopy (FT-IR PAS) is applied to the nondestructive detection of subsurface layers in a bi-layered film. In the course of the study, the heat generated from the rear surface of a film sample is found to be a main cause giving undesirable photoacoustic (PA) spectral features. This phenomenon is discussed theoretically and experimentally. The heat from the rear surface is found to make the PA spectra of the film samples structureless if the sample was simply placed in a PA cell. Careful positioning of the film sample in a PA cell is required. Taking the above results into consideration, spectral separation of subsurface layer of films has been carried out by applying the subtraction technique to the PA amplitude spectra of bi-layered films with various values of the top layer thickness. It has been found that the structure of substrate layer can be detected to the depth corresponding to the thermal diffusion length.

INTRODUCTION

Nondestructive characterization of materials is of paramount importance in examining the relationship between the functions of materials and the specific chemical structures. It is also important in developing new materials. The chemical characterization of thin polymer films on bulk substrates is one of the neglected aspects of nondestructive evaluation. A large number of techniques, such as Auger electron spectroscopy, x-ray photoelectron spectroscopy and eletron energy loss spectroscopy have been used for the surface characterization [1]. However, the majority of the techniques are not well suited for the analysis of polymer films since irradiation of electron or ion beam may decompose the polymer film. Placing the sample under ultra high vacuum may result in structural alterations. On the other hand, infrared spectroscopy has advantages compared with electron spectroscopy; it does not require the ultra high vacuum system and the instrumental resolution is high. Sample preparation is relatively easy in infrared spectroscopy in contrast with other instrumental analysis techniques. Thus, infrared spectrosopy is a powerful analytical method for the structural elucidation of organic materials. However, there are several kinds of materials which are difficult to examine by conventional infrared measurement techniques.

In recent years, there has been a growing interest in photoacoustic spectroscopy (PAS) as a new analytical tool for the study of solid samples [2,3] and PAS has been successfully extended into the mid-infrared region using Fourier transform infrared spectroscopy (FT-IR) [4-6]. Although the measurements of the PA spectra in the mid-infrared region using a conventional dispersive spectrometer have been reported [7,8], it is convenient to use FT-IR spectrometer for measuring PA spectra. Since the PA signal intensity is proportional to the power of the exciting incident source FT-IR spectrometers have the advantage of higher energy throughput as compared to conventional dispersive spectrometers [9]. The throughput (Jacquinot´s) advantage is particularly suitable for the PA measurements since the high power sources tunable over a wide spectral range do not presently exist in the mid-infrared region. In addition to the Jacquinot´s advantage, FT-IR spectrometers have the multiplex (Fellgett´s) and the frequency accuracy (Conne´s) advantages [9]. The former advntage is useful for increasing the signal-to-noise (S/N) ratio of the spectrum and the latter is particularly useful for spectral addition and for using a spectral subtraction technique.

The first attempt of FT-IR PAS was reported using methanol vapor as a sample [10]. Succesively, FT-IR PAS has been demonstrated as a usuful technique for obtaining mid-infrared spectra of solid samples [11-13] and has been applied to solve problems in various chemical fields. Qualitative analysis of organic [14-16] and inorganic [17,18] powders, and of conducting polymers [19-22] has been reported. The usefulness of FT-IR PAS to the quantitative analysis of multicomponent solid samples [23,24] and of the samples on the TLC plate [25] has also been demonstrated. FT-IR PAS is a very simple method for obtaining absorption spectrum of a solid sample with minimal preparation. This method is particularly useful for measuring the spectra of compounds which suffer from the structural alteration [26,27] arising from the sample preparation and for eliminating the Christiansen effect [28]. It is of interest to characterize the surface of materials with FT-IR PAS. Information obtained from the PA spectra has been compared with that obtained from diffuse reflectance or attenuated total reflectance (ATR) spectra [29,30]. FT-IR PAS has also been utilized for the analysis of catalytic surface [31-33], electrode surface [34], coal surface [35], and surface species changed photochemically [36,37] or thermally [38].

One advantage of PAS over the conventional absorption spectroscopy is that the PA technique can give information about the optical characteristics of a sample as a function of depth beneath its surface [2]. If we consider thermally thick samples which can be applied to most of the organic materials studied by FT-IR PAS, the PA signal intensity is proportional to the thermal diffusion length and has a $f^{-3/2}$ dependence, where f is the modulation frequency. Accordingly, depth profiling information can be obtained by varying the modulation frequency. The following characteristics of the interferometry must be taken into account. In the conventional measurement system for PAS a fixed modulation frequency is used for obtaining PA spectra. On the other hand, in the case of rapid scanning FT-IR spectrometers the frequency of the radiation being emitted from the source corresponds to the modulation frequency through the relationship [9], $f=2V\nu$, where f is the modulation frequency in Hz, V is the velocity of the moving mirror of the interferometer in cm/sec, and ν is the wavenumber in cm^{-1}. This multi-frequency modulation leads to two major complications. One is that the FT-IR PAS spectra tend to be affected by the Helmholtz resonance effect [39,40]. The PA signal intensity can be enhanced at the Helmholtz resonance frequency. However, it should be noted that the phase changes significantly around the resonance frequency [41,42], so that careful phase correction must be taken into account when using a resonant PA cell. It should also be noted that the response of a resonant PA cell varies with the velocity of the moving mirror. The other complication is that the sampling depth varies with frequency and care must be taken for normalizing FT-IR PAS spectra [43,44].

Although FT-IR PAS has several complications arising from the multi-frequency modulation, this technique undoubtedly has a potential as a new method for surface analysis in the infrared region. In this paper, we describe an application of FT-IR PAS to the characterization of thin polymer films on bulk substrate and make some experimental and theoretical discussions about the characteristic features of film samples. We also describe the application of FT-IR PAS to the spectral separation of subsurface layer in a multi-layered sample.

EXPERIMENTAL

The measurement system for FT-IR PAS is similar to the one we used in earlier experiments [37-39]. It consists of a Digilab

FTS-15 FT-IR spectrometer and the PAS attachment described below. Figure 1 shows a schematic diagram of the system which was primarily designed so as to examine in-situ the photochemical changes of a photosensitive polymer [37]. The infrared radiation from a nichrome wire collimated by mirrors is modulated with a Michelson interferometer in which a movable mirror was scanned at a constant velocity of 0.16 cm/sec. Accordingly, the wavenumber range 4000-400 cm^{-1} corresponds to the acoustic frequency range 1280-128 Hz. The modulated infrared beam is then reflected by a plane mirror placed in the sample compartment of the spectrometer to the right angle off-axis toroidal mirror which causes a 6:1 reduction of the beam. The toroidal mirror and the PA cell were placed inside an acoustically insulated compartment which was set up outside the spectrometer. The PA signals from the microphone attached to the PA cell were amplified with a Brookdeal 9454 A.C. Amplifier and were fed to the HgCdTe detector amplifier section of the FT-IR spectrometer.

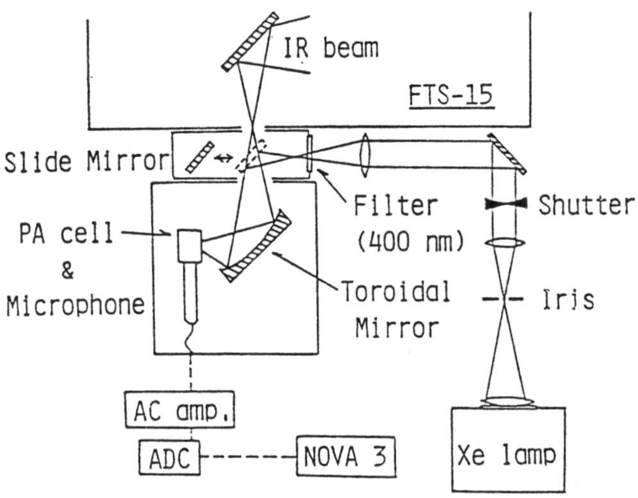

Figure 1. Schematic diagram of the PA attachment.

In order to irradiate the sample with visible light, a 300 W xenon lamp (Varian VIX-300F) was used. The frequency of the excitation light was 400 nm, which was selected with an interference filter. The mirror placed between the FT-IR spectrometer and the compartment which contains a PA cell and a toroidal mirror was used so that the visible light for photochemical reactions or the infrared light for spectral measurement could irradiate the sample alternatively by translating a slide stage. The irradiation was carried out under ambient conditions. The period of the irradiation was controlled by a mechanical shutter. By using this optical system, we could elucidate in-situ the structural changes of solid samples which might alter their chemical structure photochemically.

The samples studied were polymer films. In this case, a PA cell with a large sample cavity is not necessary. According to the results on the optimum dimensions of a PA cell, the maximum PA signal occurs at l_g/μ_g=1.4 [45] or 1.8 [46], where l_g and μ_g are the length of the gas column in a PA cell and the thermal diffusion length of the gas, respectively. Taking these results into account, a PA cell was designed so as to minimize the dead volume and to enhance the sensitivity.

A schematic drawing of the PA cell used in this study is shown in Figure 2. The cell was made of brass and its internal volume was appoximately 0.04 cm^3. A KBr plate (2 mm thick) was used as a cell window. A condenser microphone, 1/2 inch Bruel Kjaer (B & K) Model 4165 with a B & K Model 2619 preamplifier and B & K Model 2807 power supply were used for detection. The sample can be placed in a PA cell from either the top or the bottom. Most PA cells will have their own acoustic resonant frequencies and the response of the PA cells will be affected by the volume of a sample placed in the cell cavity. In the case of our cells, the effect of the volume of a film sample to the response of the PA cell can be neglected since the volume of the cell cavity can be kept almost constant when placing a sample on the sample holder from the bottom. The cell can be used for powder samples and rubbers as well.

Characteristics of the PA cells were examined by using a He-Ne laser (Uniphase, Model 1105P, 5 mW) as an excitation source. Light modulation was carried out using a light chopper (NF Circuit Design Block, Model CH-353) and the modulated PA signal was processed with an autophase lock-in amplifier (NF Circuit Design Block, LI-574A). Figure 3 shows the dependence of the PA cells on the modulation frequency. The open and closed circles in Figure 3 denote the signal intensity obtained with the PA cell depicted in Figure 2 and that obtained with a PA cell which has two valves for changing the coupling gas, respectively. The coupling gas used was ambient air. As can be seen from Figure 3, the PA signal is inversely proportional to the modulation frequency. Since the modulation frequency used in this study is in the range of 1280-128 Hz, our cells are evaluated as non-resonant PA cells and have a flat response in the spectral region of interest. The output signal level from the microphone was 130 mV (p-p at 51.1 Hz wih 5 mW He-Ne laser and a carbon black (benzene soot) sample. At this modulation frequency, output signal level from the lock-in amplifier was ca. 40 mV which is three times greater than that of commercial PA cell (4.5 mV rms with a 2 mV He-Ne laser at 50 Hz).

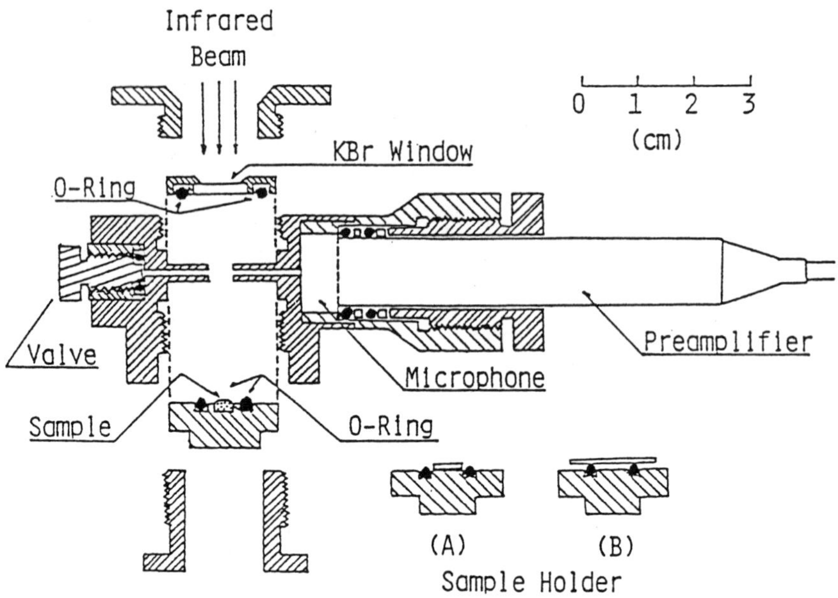

Figure 2. Schematic drawing of the PA cell.

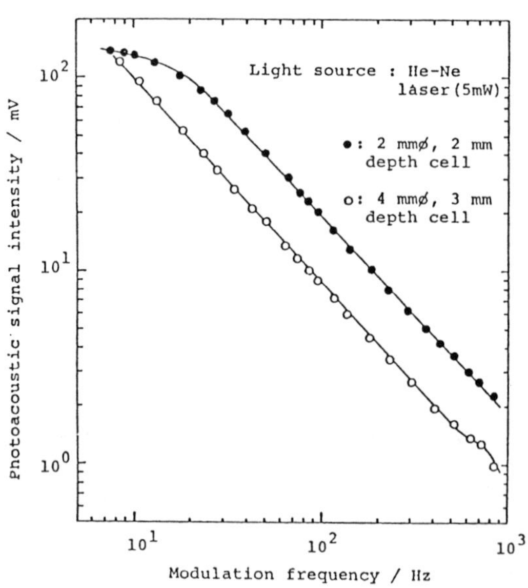

Figure 3. Dependence of PA signal intensity of carbon black on the modulation frequency.

In the measurement of FT-IR PAS, the output signal level (centerburst in the interferogram) from the microphone was ca. 10 mV (peak-to-peak) at 0.16 cm/sec mirror velocity with a nichrome source and a carbon black sample. The noise level was ca. 20 μV (peak-to-peak). Thus, a maximum S/N ratio of 500 could be obtained from a single scan. The signal intensity can be enhanced using helium as a coupling gas. Typically, FT-IR PAS spectra presented in this study were measured at 8 cm^{-1} resolution with 900 scans.

RESULTS AND DISCUSSION

A. FT-IR Photoacoustic Spectra of Films

It has been well recognized that the PA signal intensity of film-like samples is weaker than that of powder samples [47,48] since the former has the lower surface area than the latter. As an example of the performance of our instruments, PA spectra of 36 μm-thick poly(ethylene terephthalate) (PET) film are shown in Figure 4. In this figure, the number of repeated scans (NSS) used for measuring the spectra of A, B, C and D was 9, 25, 400, and 1600 respectively. It took about 10 seconds to measure the spetrum of A at 8 cm^{-1} resolution. As shown in Figure 4, it took only a few minutes to obtain high quality PA spectra.

One of the advantages of PAS, which is often mentioned, is that spectra corresponding to the absorption spectra can be obtained with minimal sample preparations; by simply putting a sample into a PA cell. However, the PA spectra of multi-layered films as shown in Figure 2 A were found to show distorted spectral features around the strong absorption bands when the sample was simply placed on the sample holder. This phenomenon is shown in Figure 5 A which represents the PA spectrum of a bi-layered film of polyethylene (PE)/ PET. Top layer is PE of about 40 μm-thick and the second layer is PET of about 10 μm-thick. In the spectrum A, it is noticeable that the PA signals around the strong absorption bands at 1725, 1270, and 1120 cm^{-1} do not correspond to the features observed in the absorption spectrum. The spectrum shows a pattern similar to self-absorption.

Taking into account the velocity of the moving mirror and the thermal properties of PET film with thermal diffusivity of about 10^{-3} (cm^2/sec) [2], the thermal diffusion length in spectral range of 4000 to 400 cm^{-1} can be estimated as 5 to 16 μm. In this case, since the top layer is about 40 μm-thick, one might have expected that a PA spectrum corresponding to the absorption spectrum of the top layer, PE, might be observed judging from both the one-dimensional PA (Rosencwaig-Gersho) theory [49] for homogeneous materials and/or several extended PA theories for layered materials [50-53]. However, the observed PA spectrum shown in Figure 5 A does not fulfil this expectation. The contribution from the subsurface layer, PET, to the PA spectrum can be easily recognized.

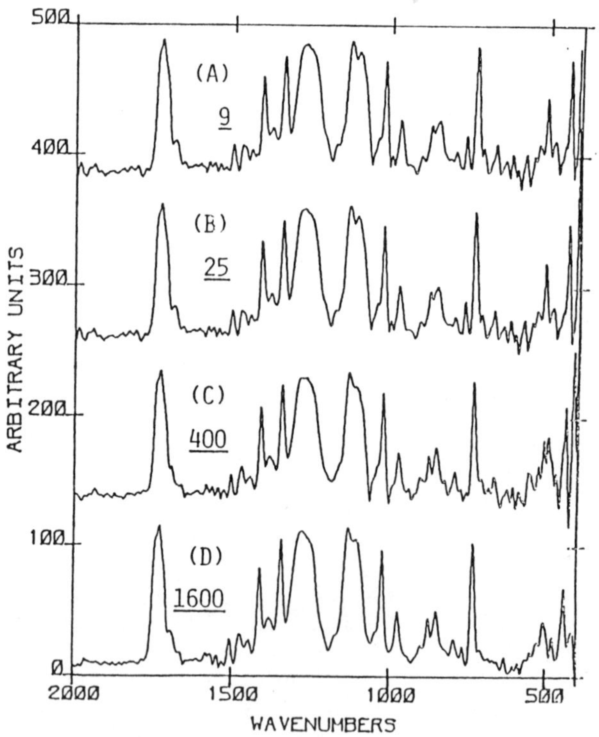

Figure 4. Effect of number of repeated scans on FTIR-PA spectrum of 36 μm-thick PET film. NSS9 (A), 25 (B), 400 (C), 1600 (D). Mirror velocity=0.16 cm/sec, and resolution=8 cm^{-1}.

In order to explain the above phenomenon, it might be relevant to take the following characteristics of FT-IR into account: (i) there is insufficient phase correction in the FT-IR spectrometer used here, though the PA spectrum shown in Figure 5 A is presented as a power spectrum, and (ii) the effect of an air layer under the sample is not negligible, since the modulation frequency of the FT-IR spectrometer is higher than the conventional PA spectrometer. When an additional film was attached to the rear surface of the sample with an adhesive tape, its PA spectrum was found to correspond to the absorption spectrum of the top layer as shown in Figure 5 B, and 5 C. This observation suggests that the heat generated at the rear surface of the sample gives a significant contribution to the PA signal of films. In addition, the air layer and the higher modulation frequency play also an important role in FT-IR PAS spectra of the film samples.

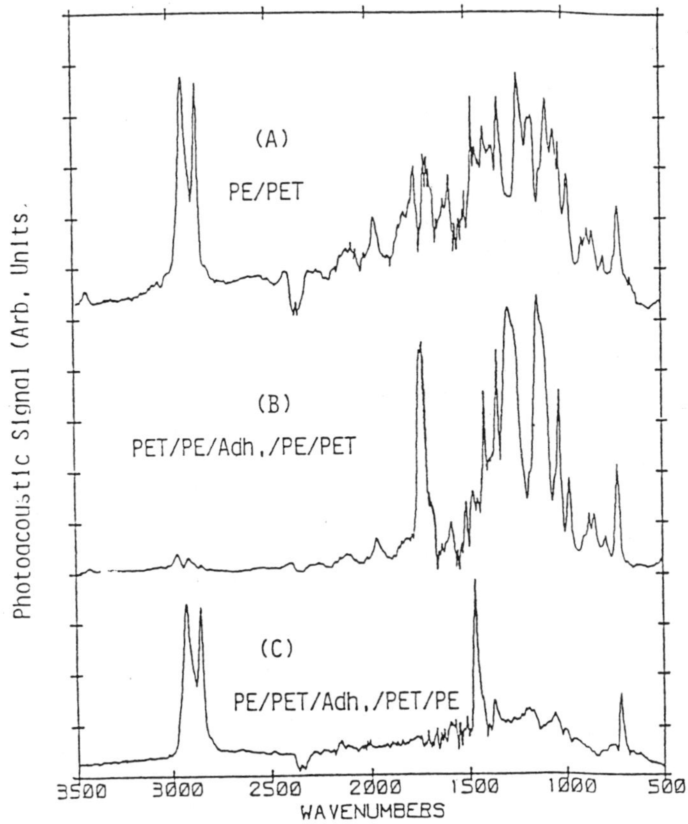

Figure 5. FTIR-PA spectra of layered films. PE, PET, and Adh. represent polyethylene, polyethyleneterephthalate, and adhesive tape, respectively. Top layer of (A), (B), and (C) are PE, PET, and PE, respectively.

In order to confirm the above suggestion, dependence of the PA spectra on the position of the sample in a PA cell was examined. Figure 6 shows a schematic drawing of the PA cell and the position of the film sample. At the position A in Figure 6, only the heat generated from the rear surface of the sample can contribute to the PA signal. When the sample is placed at the position B, the heat generated from both the top and the rear surface of the sample can contribute to the PA signal. Only the heat from the top surface of the sample can contribute to the PA signal at the position C. The positioning at C is also shown in Figure 2 B. As shown in Figure 6, a KBr window was used as a backing material so as to minimize the effect of the reflecting light from the backing material on the PA signal.

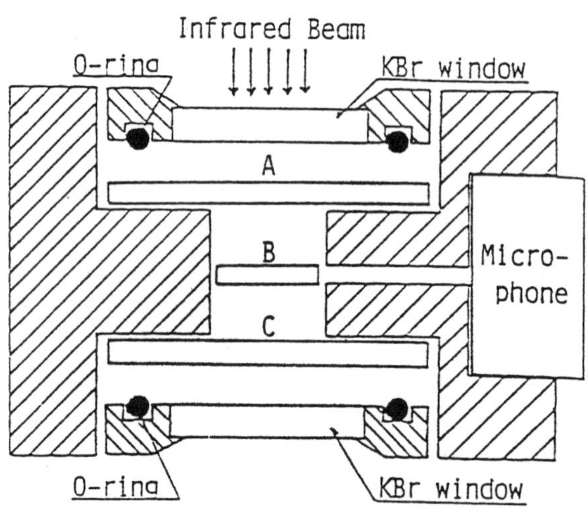

Figure 6. Schematic drawing of a PA cell and the placing position for a film sample. (A):at the entrance window, (B):in the cell cavity, (C):at the exit window.

Figure 7 shows the dependence of the PA spectra on the position in which the sample is placed. The spectra at the top, middle, and bottom were obtained by placing the sample in positions of A, B, and C as depicted in Figure 6. In the top spectrum, A, a spectral pattern similar to self-absorption can be recognized in the vicinity of strong absorption band. The middle spectrum is quite similar to the one shown in Figure 5 A, which indicates that the spectral features similar to self-absorption in Figure 5 A could well have been caused by the heat generated at the rear surface of the sample. The bottom spectrum in Figure 7, in which the contribution to the PA signal from the heat at rear surface can be discounted, corresponds to the absorption spectrum of the top layer, PE. From these results it is clear that when measuring the PA spectrum of a film-like sample, simply putting the sample into a PA cell cavity is not a desirable method for obtaining a FT-IR PA spectrum that resembles the absorption spectrum. Thus, in order to obtain a reasonable FT-IR PA spectrum, careful positioning of the sample is required. This leads to elimination of the heat generated at the rear surface of the sample as shown in Figure 6 C. Using a suitable material at the rear surface may eliminate this undesirable signal. This is shown in Figure 5 B and 5 C.

B. Theoretical Consideration for the Heat from Rear Surface

In order to evaluate the heat generated from the rear surface of a sample the PA signals of films were examined theoretically and experimentally using a mono-layered film as a model sample.

Figure 8 depicts a one-dimensional gas piston model in which modulated light shines a sample on a backing material. Generated heat corresponding to the amount of the absorbed light is detected as pressure variations of a coupling gas. The temperature generated from the front surface of the sample, θF, is given by Rosencwaig and Gersho [49], and can be expressed as follows:

$$\theta_F = \frac{(r-1)(b+1)\theta^{\sigma\ell} - (r+1)(b-1)\theta^{-\sigma\ell} + 2(b-r)\theta^{-\beta\ell}}{(g+1)(b+1)\theta^{\sigma\ell} - (g-1)(b-1)\theta^{-\sigma\ell}} \cdot E \qquad (1)$$

where

where $E=\beta I_0/2k_s(\beta^2-\sigma_s^2)$, $b=k_b a_b/k_s a_s$

$g=k_g a_g/k_s a_s$, $r=(1-j)\beta/2a_s$, $\sigma=(1+j)a_s$

Parameters, I_o, β, l, a_i, and k denote the intensity of light,

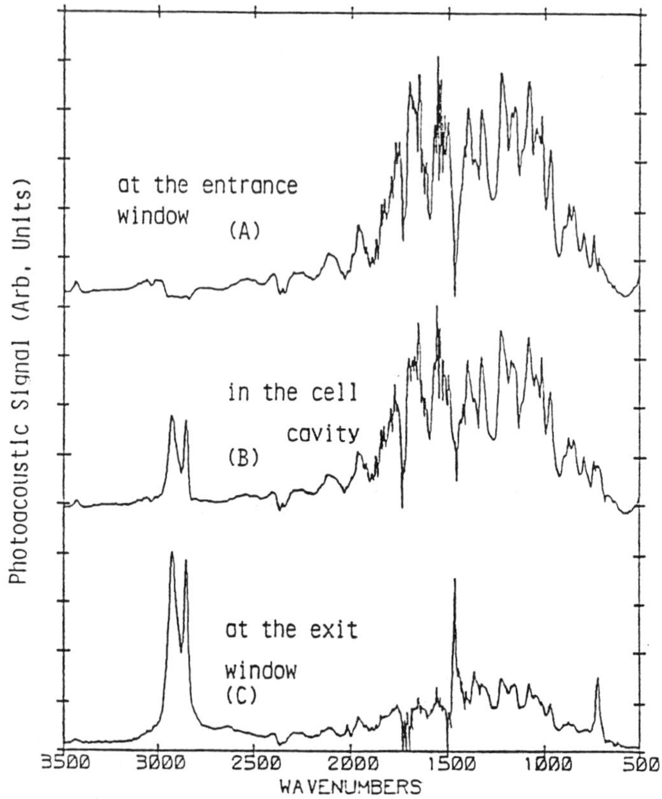

Figure 7 Dependence of FTIR-PA spectra of bi-layered film on the placing position in a PA cell. The top layer is PE.

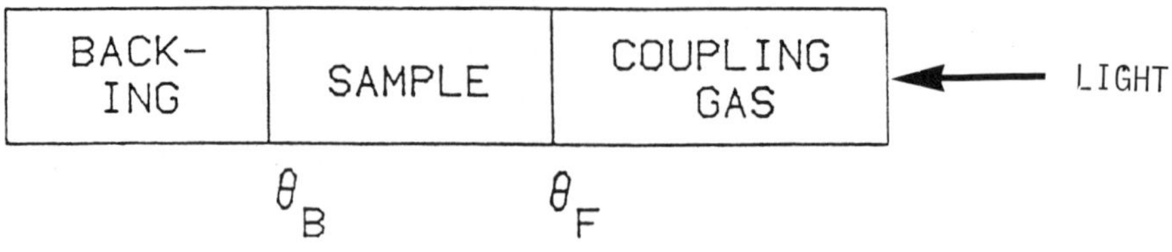

Figure 8 One-dimensional gas piston model.

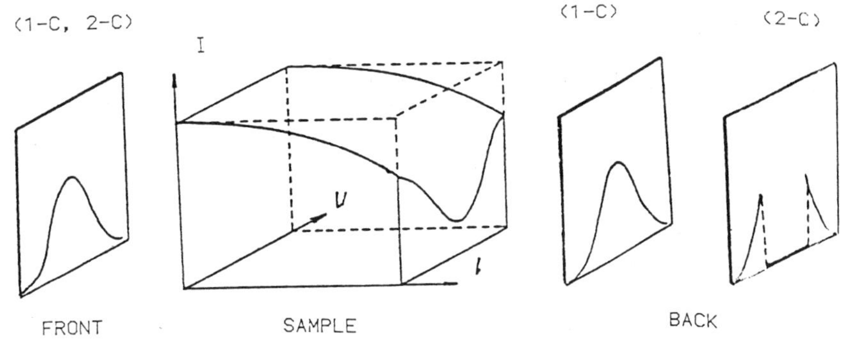

Figure 9 Schematic drawing for the photoacoustic signals arising from the front and rear surfaces of a sample.

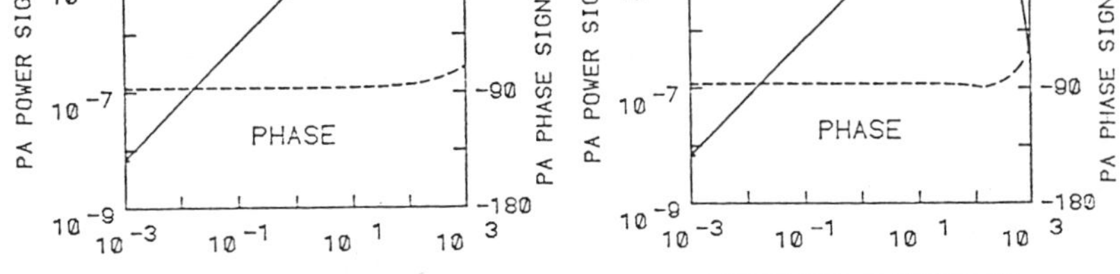

Figure 10 Photoacoustic power (solid line) and phase (broken line) signals from the front (left) and rear (right) surfaces of a sample. (sample thickness=0.01 cm, thermal diffusion length=0.0008 cm).

optical absorption coefficient, sample thickness, thermal diffusion coefficient of material i (g:gas, s:sample, b:backing), and thermal conductivity, respectively.

We can obtain an expression for the heat generated at the rear surface, θ_B, by solving the thermal diffusion equation by the same approach as in the Rosencwaig-Gersho theory [49]. The solution is given by

$$\theta_B = \frac{2(g+r)-(r+1)(g+1)\theta^{\sigma\ell}\theta^{-\beta 1}+(r-1)(g-1)\theta^{-\sigma\ell}\theta^{-\beta 1}}{(g+1)(b+1)\theta^{\sigma\ell}-(g-1)(b-1)\theta^{-01}}\cdot E \qquad (2)$$

Since the expressions given by Eqs. (1) and (2) are somewhat difficult to interpret intuitively, special cases are examined by using the same approximation as in the Rosencwaig-Gersho theory. In this study we consider the PA signal of thermally thick sample ($1>1/a_s$). Assuming the existence of air layer between the rear surface of a sample and a backing material, the following two cases are examined: (i) Optically transparent and thermally thick (Case 1-C), (ii)Optically opaque and thermally thick (Case 2-C). Resultant complex PA magnitude signals generated from the front and the rear surfaces of a sample are summarized in Table 1.

As can be seen in Table 1, in the case of 1-C, the same PA signal including phase arises from both surfaces and this signal is proportional to the optical absorption coefficient. On the other hand, in the case of 2-C, no PA signal arises from the rear surface, though the same signal as in the case of 1-C can be generated from the front surface.

	FRONT SURFACE	BACK SURFACE
(CASE 1-C) $\frac{1}{\beta} > \ell > \frac{1}{a_s}$	$\frac{-j}{2a_g}\left(\frac{\mu_s}{k_s}\right)\beta\mu_s$	$\frac{-j}{2a_g}\left(\frac{\mu_s}{k_s}\right)\beta\mu_s$
(CASE 2-C) $\ell > \frac{1}{\beta} > \frac{1}{a_s}$	$\frac{-j}{2a_g}\left(\frac{\mu_s}{k_s}\right)\beta\mu_s$	$\fallingdotseq 0$

If we consider a Gaussian absorption band, it is expected from the expressions in Table 1 that a Gaussian spectral feature is always obtained from the front surface, and that a self-absorption-like PA spectral feature will be observed for the rear surface in the case of 2-C. This expectation is shown as a schematic drawing in Figure 9, where I, l, and ν, represent the light intensity, sample thickness, and wavenumbers, respectively.

In Figure 10, we show the computer-generated plots derived directly from Eqs. (1) and (2) for the magnitude Q and phase Φ of the PA signal as a function of optical absorption coefficient. In this figure, we consider the case in which the sample thickness is 100 μm and the thermal diffusion length is 8 μm, and the magnitude (solid line) is shown as log-log plots and the phase (dotted line) is shown as semi-log plots. Upon increasing the absorption coefficient up to the value that the optical absorption length ($\ell_\beta = 1/\beta$) is coincident with the sample thickness, the PA amplitude signals from both surfaces increase linearly and the phase signals keep a constant value of -90 degrees for both surfaces. If the absorption coefficient increases further, the phase signal varies gradually from -90 to -45 degrees for front and rear surfaces, but the amplitude signals of the rear surface decrease drastically in contrast to the case of front surface. These calculated results are in good agreement with the expectation from the results shown in Table 1.

Considering the Gaussian absorption profile, PA signal amplitude is calculated as shown in Figure 11. From the front surface, the PA signals keep the Gaussian absorption profile. On the other hand, the PA signals generated from the rear surface show self-absorption-like profile whose central part has approximately no intensity. Considering the fact that the phase does not depend on the variation of the absorption coefficient keeping almost the same value, observed PA signals can be considered as a simple sum of the signals from front and rear surfaces. It is, therefore, expected that the diffuse PA spectral features would be observed for the closely neighboring strong absorption bands if the existence of air layer cannot be negligible and the heat effect from rear surface contributes to the PA signal greatly. In this case, observed FT-IR PA spectrum would be structureless if many sharp and strong absorption bands are closely present.

In order to confirm the above expectation, FT-IR PA spectra of PET film was measured at the several placing positions of the sample. The results are shown in Figure 12. In this figure, top spectrum A was obtained without considering heat effect from rear surface and the sample was placed as shown in Figure 2 B. Spectra B, C, and D were obtained by placing the sample at the position A, B, and C in Figure 6, respectively. As shown in Figure 12 B, the

PA signal around the strong absorption bands appeared as having self-absorption-like features.

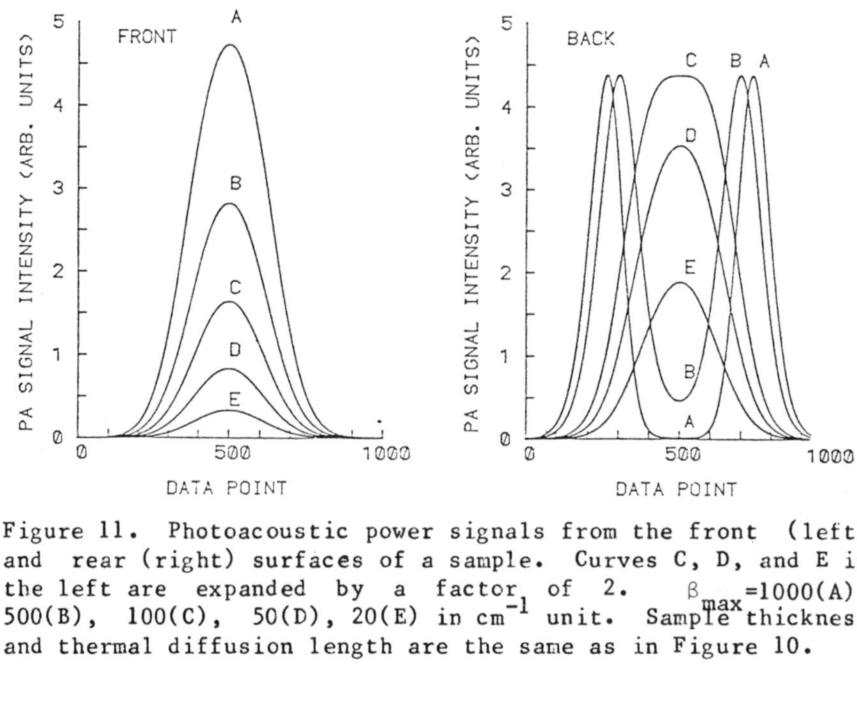

Figure 11. Photoacoustic power signals from the front (left) and rear (right) surfaces of a sample. Curves C, D, and E in the left are expanded by a factor of 2. β_{max}=1000(A), 500(B), 100(C), 50(D), 20(E) in cm^{-1} unit. Sample thickness and thermal diffusion length are the same as in Figure 10.

It is also recognized that the spectrum A and C resembles with each other and that the spectrum D shows the most distinct spectral features. Comparing the spectrum D with A or C, it is concluded that the heat from rear surface is the main cause of making the PA spectrum of film samples structureless.

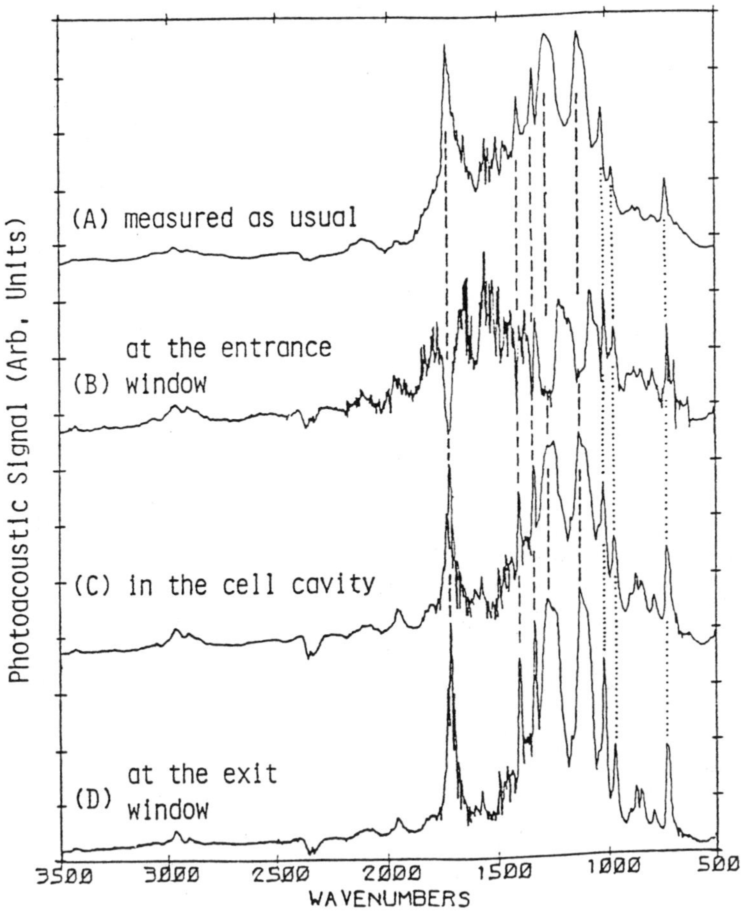

Figure 12. Dependence of FTIR-PA spectrum of PET film on the placing position in a PA cell.

It may be required for applying the above results obtained for the mono-layered film to the bi-layered film that phase lag should be taken into account. Nevertheless, the film sample should be placed at the position where PA-signal is not affected by the heat generated from the rear surface of a sample.

Detection of Subsurface Layer by Subtraction Technique

In this section we describe the application of FT-IR PAS to the non-destructive detection of a subsurface layer of a multi-layered film using a subtraction technique. Unfortunately, it was not possible to change the velocity of the movable mirror of our FT-IR spectrometer and, therefore, we examined the possibility of depth profiling by preparing several bi-layered film samples with top layers of different thicknesses. The top layer was epoxy resin (EPO) and the second layer was 23 μm-thick PET or 23 μm-thick polypropylene (PP). The EPO layers were formed on the PET or PP substrate by the photopolymerization of 3,4-epoxycyclohexyl-methyl-3′,4′-epoxycyclohexane carboxylate.

Figure 13 shows the PA spectra of bi-layered EPO/PP films in which the top layers were 23, 9, 5, and 3 μm-thick. In this figure, open circles marked at about 3400, 1730, 1255, 1190 and 1090 cm^{-1} denote the charateristic absorption bands of the top layer, EPO, and closed circles at around 2800, 1380, 990, and 840 cm^{-1} denote the characteristic bands of the subsurface layer, PP. Figure 14 shows the PA spectra of bi-layered EPO/PET films in which the top layers were 25, 9, 5, and 3 μm-thick. In this figure, the open circles marked at about 1190 cm^{-1} denote the characteristic bands of the top layer, EPO, and the closed circles marked at about 1410, 1345, 1020, 980 and 730 cm^{-1} denote the characteristic bands of the subsurface layer, PET.

In order to obtain the spectrum of sursurface layer, spectral subtraction was attempted using the spectra shown in Figures 13 and 14. It was assumed that the top spectrum A of Figures 13 and 14 corresponds to the absorption spectrum of the top layer, and that the PA signals of the bilayered films measured in this study could be regarded as the sum of signals for the top and the second layers. The spectra for the subsurface layers were obtained by subtracting the spectrum A from the spectra B, C, and D in Figures 13 and 14, taking the intensity of the 1190 cm^{-1} band as zero. The difference spectra are shown in Figures 15 and 16 and are found to be quite similar to the spectra of the subsurface layers.

Therefore, using FT-IR PA amplitude spectra, it is possible to detect and elucidate the subsurface layer down to a depth corresponding to half of the thermal diffusion length. FT-IR ATR spectra of the same samples were also measured at the incident angle of 45 degrees but no distinct spectral differences were found. Thus, the PA technique can give information about the structure of a sample to a deeper level than the ATR.

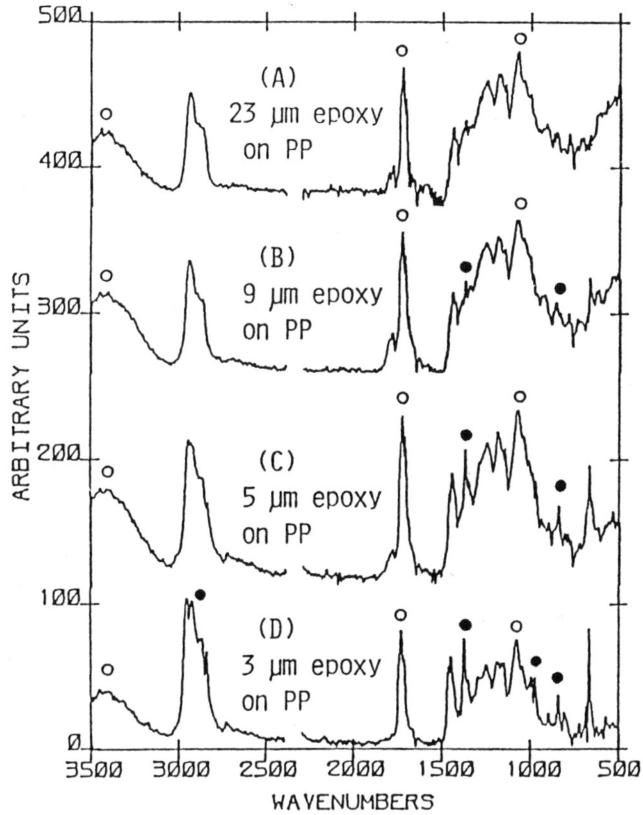

Figure 13. FTIR-PA spectra of bi-layered films. (Epoxy resin on polypropylene film). ○ denotes the characteristic peak of the top layer, and ● denotes that of the subsurface layer.

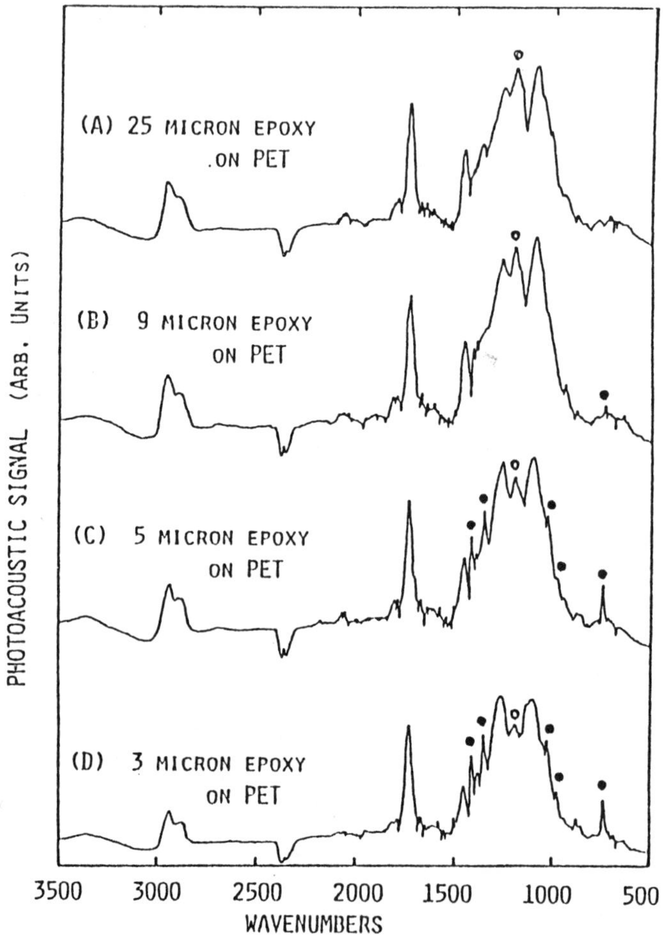

Figure 14 FTIR-PA spectra of bi-layered films. (Epoxy resin on polythyleneterephthalate film) ○ and ● are same as in Figure 13.

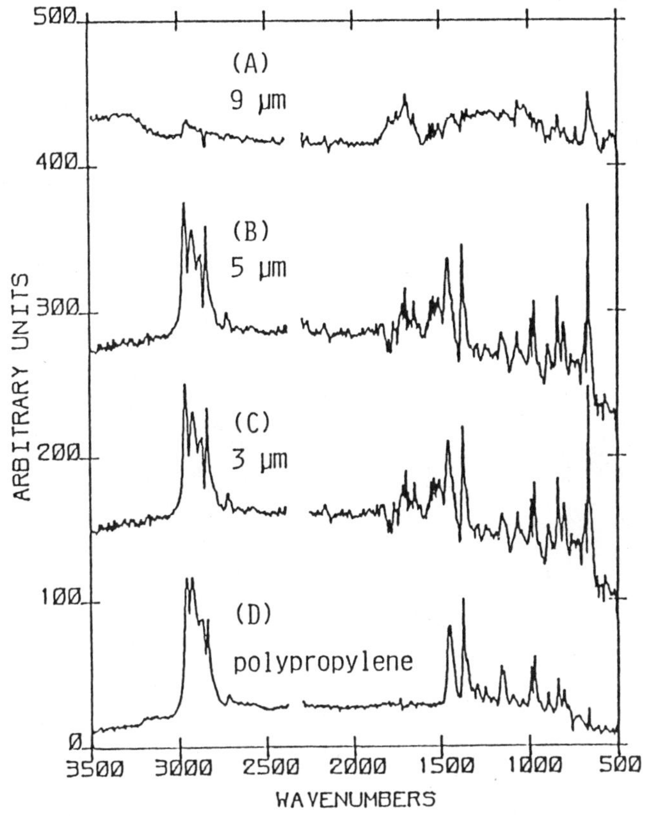

Figure 15 Subtracted FTIR-PA spectra of sub-surface layer in EPO/PP bi-layered film. (D) is a FTIR-PA spectrum of um thick PP.

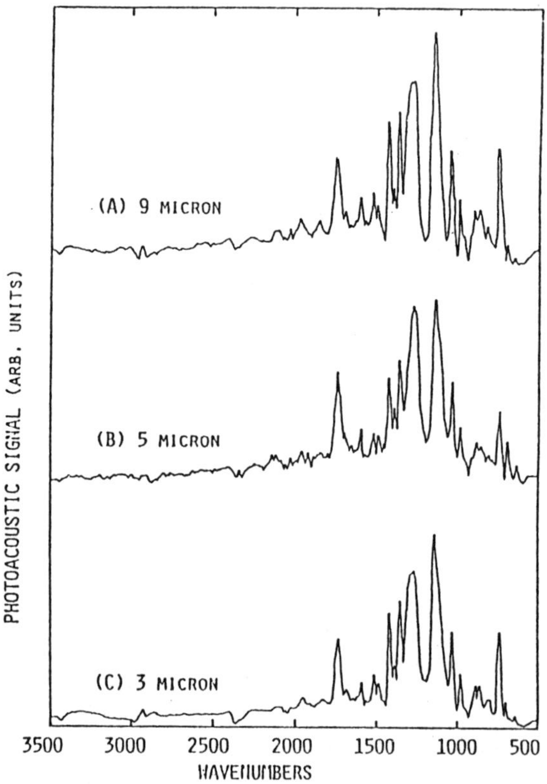

Figure 16 Subtracted FTIR-PA spectra of sub-surface layer in EPO/PET bi-layered film.

CONCLUSIONS

The FT-IR PA spectra of films have been measured and examined. It was shown that the heat generated from the rear surface of a film-like samples gives rise to spurious PA signals and this effect can be eliminated by placing the sample in a suitable position in the PA cell. The possibility of detecting a subsurface layer of a multi-layered film is also demonstrated. This is achieved by the subtraction technique using FT-IR PA amplitude spectra.

REFERENCES

1. G.A.Somorjai and F.Zeaera, J. Phys. Chem., 86, 3070 (1982).

2. A.Rosencwaig,"Photoacoustics and Photoacoustic Spectroscopy", John Wiley & Sons, New York (1980).

3. G.A.West, J.J.Barrett, D.Siebert and K.V.Reddy, Rev. Sci. Instrum., 54, 797 (1983).

4. D.W.Vidrine, in "Fourier Transform Infrared Spectroscopy", Vol.3, J.R.Ferraro and L.J.Basile, Eds., Academic, New York (1982) pp .125-148.

5. J.F.Mclelland, Anal. Chem. 55, 89A (1983).

6. J.A.Gardella, Jr., D.-Z.Jiang, W.P.McKenna and E.M.Eyring, Appl. Surf. Sci., 15, 36 (1983).

7. M.J.D.Low and G.A.Parodi, Appl. Spectrosc., 34, 76 (1980).

8. Ibid., J. Photoacoustics, 1, 131 (1982).

9. P.R.Griffiths,"Chemical Infrared Fourier Transform Spectroscopy", John Wiley & Sons, New York (1975).

10. G.Busse and B.Bullemer, Infrared Phys., 18, 255 (1978); and Ibid., 18, 631 (1978).

11. M.G.Rokley, Chem. Phys. Lett., 68, 455 (1979).

12. M.G.Rokley, D.M.Davis and H.H.Richardson, Science, 210, 918 (1980).

13. D.W.Vidrine, Appl. Spectrosc., 34, 314 (1980).

14. M.G.Rockley, Appl. Spectrosc., 34, 405 (1980).

15. Ibid., Chem. Phys. Lett., 75, 370 (1980).

16. M.G.Rockley, H.H.Richardson and D.M.Davis, J. Photoacoustics, 1, 145 (1982).

17. J.A.Gardella, Jr., E.M.Eyring, J.C.Klein and M.B.Carvalho, Appl. Spectrosc., 36, 574 (1982).

18. J.H.Nelson, J.J.Macdougall, F.G.Baglin, D.W.Freeman, M.Nadler and J.L.Hendrix, Appl. Spectrosc., 36, 574 (1982).

19. S.M.Riseman, S.I.Yaniger, E.M.Eyring, D.Macinnes, A.G.MacDiarmid and A.J.Heeger, Appl. Spectrosc., 35, 557 (1981).

20. S.I.Yaniger, S.M.Riseman, T.Frigo and E.M.Eyring, J. Chem. Phys., 76, 4298 (1982).

21. F.G.Will, R.S.McDonald, R.D.Gleim and M.R.Winkle, J. Chem. Phys., 78, 5847 (1983).

22. S.I.Yaniger, D.J.Rose, W.P.McKenna and E.M.Eyring, Appl. Spectrosc., 38, 7 (1984).

23. M.G.Rockey, D.M.Davis and H.H.Richardson, Appl. Spectrosc., 35, 185 (1981).

24. M.G.Rokley, M.Woodard, H.H.Richardson, D.M.Davis, N.Purdie and J.M.Bowen, Anal. Chem., 55, 32 (1983).

25. L.B.Lloyd, R.C.Yeates and E.M.Eyring, Anal. Chem., 54, 549 (1982).

26. S.R.Lowry, D.G.Mead and D.W.Vidrine, Anal. Chem., 54, 546 (1982).

27. V.Renugopalakrishnan and R.S.Bhatnagar, J. Am. Chem. Soc., 106, 2217 (1984).

28. G.Laufer, J.T.Huneke, B.S.H.Royce and Y.C.Teng, Appl. Phys. Lett., 37, 517 (1980).

29. K.Krishnan, Appl. Spectrosc., 35, 549 (1981).

30. K.Krishnan, S.Hill, J.P.Hobbs and C.S.P.Sung, Appl. Spectrosc., 36, 257 (1982).

31. S.M.Riseman, F.E.Massoth, G.M.Dhar and E.M.Eyring, J. Phys. Chem., 86, 1760 (1982).

32. J.B.Kinney and R.H.Staley, J. Phys. Chem., 87, 3735 (1983).

33. J.A.Gardella, Jr., D.-Z.Jiang and E.M.Eyring, Appl. Spectrosc., 37, 131 (1983).

34. M.D.Porter, D.H.Karweik, T.Kuwana, W.B.Theis, G.B.Norris and T.O.Tiernan, Appl. Spectros., 38, 11 (1984).

35. M.G.Rockley and J.P.Devlin, Appl. Spectrosc., 34, 407 (1980).

36. J.B.Kinney, R.H.Staley, C.L.Reichel and M.S.Wrighton, J. Am. Chem. Soc., 103, 4273 (1981).

37. N.Teramae, T.Yamanoto, M.Hiroguchi, T.Matsui and S.Tanaka, Chem. Lett., 37 (1982).

38. N.Teramae, M.Hiroguchi and S.Tanaka, Bull. Chem. Soc. Jpn., 55, 2097 (1982).

39. N.Teramae, M.Hiroguchi and S.Tanaka, Chem. Lett., 1091 (1981).

40. J.B.Kinney and R.H.Staley, Anal. Chem., 55, 343 (1983).

41. N.C.Fernelius, Appl. Opt., 18, 1784 (1979).

42. W.A.McClenny, C.A.Bennett, Jr., G.M.Russwurm and R.Richmond, Appl.Opt., 20, 650 (1981).

43. J.M.Chalmers, B.J.Stay, G.F.Kirkbright, D.E.M.Spillane and B.Beadle, Analyst, 106, 1179 (1981).

44. Y.C.Teng and B.S.H.Royce, Appl. Opt., 21, 77 (1982).

45. L.C.Aamodt, J.C.Murphy and J.G.Parker, J. Appl. Phys., 48, 927 (1977).

46. A.C.Tam, Y.H.Wong, Appl. Phys. Lett., 36, 471 (1980).

47. M.J.Adams, A.A.King and G.F.Kirkbright, Analyst, 101, 73 (1976).

48. J.J.Freeman, R.M.Friedman and H.S.Reichard, J. Phys. Chem., 84, 315 (1980).

49. A.Rosencwaig and A.Gersho, J. Appl. Phys., 47, 64 (1976).

50. N.C.Fernelius, J. Appl. Phys., 51, 650 (1980).

51. P.Helander, I.Lundstrom and M.McQueen, J. Appl. Phys., 52, 1146 (1981).

52. Y.Fujii, A.Moritani and J.Nakai, J. Appl. Phys. Jpn., 20, 361 (1981) and Errata, 20, 1005 (1981).

53. M.Morita, J. Appl. Phys. Jpn., 20, 835 (1981).

FT-IR AS A TOOL FOR THE CHARACTERIZATION OF INDUSTRIAL MATERIALS

A. Ishitani

Material Science Laboratories
Toray Research Center, Inc.
1-1 Sonoyama 1 Chome, Otsu-shi, Shiga 520
Japan

INTRODUCTION

The characterization of industrially used materials has presented many difficulties which are not experienced in academic scientific work. A wide range of materials such as metals, semiconductors, ceramics and polymers have to be covered. These materials are brought in frequently as alloys, copolymers and mixtures. Multilayered structures and composites have become very common in advanced technology. The small amounts of additives and impurities are also important. There are also considerable limitations due to variety in sample size and shape, the scarcity in quantities and from the lack of proper information. In addition, quickness, clarity of conclusions, detection of small differences and quantitative comparisons are commonly required under the above difficult situations.

Infrared spectroscopy has many intrinsic advantages such as wide applicability, nondestructiveness and the capability of giving detailed structural information in industrial application. Furthermore, the introduction of FT-IR has brought on such additional advantages as high sensitivity, high precision, quickness in measurement and extensive data processing capability, which have turned infrared spectroscopy from merely a method of compound identification to a comprehensive system of material characterization. The technical development of FT-IR in various measurement modes carried out at the Toray Research Center, Inc., is summerized in this paper.

BULK ANALYSIS

The transmission mode coupled with the digital difference spectrum technique is frequently used. Low transmission samples like carbon fibers [1] or bilayer membranes [2] dispersed in water are major applications of this technique. Thermal or photochemical reactions taking place within a polymer matrix can also be successfully analyzed. Quality control problems concerning slight fluctuations in composition and structure in different lots of products are also effectively dealt with by the same technique.

Work on a synthetic bilayer membrane [2] is an example of bulk analysis. Dialkylammonium salts developed by Kunitake [3] form a stable bilayer structure vesicles in water like phospholipids in a biomembrane. Selectively functioning membranes formed from these compounds will have interesting applications as drug-carriers or matrices for very specific organic reactions. Figure 1 depicts the spectrum of dialkylammonium salt in low concentration (10 mM) obtained by the digital difference spectrum technique. A spectrum of the CH stretch region is beautifully obtained. Figure 2 shows the temperature variation of ν^{S}_{CH} band of the vesicles solution between 20 and 55°C. The shift to higher wavenumbers and the broadening of the band is clearly seen in the figure. The temperature is controlled with a precision of ±0.1°C in the experiment.

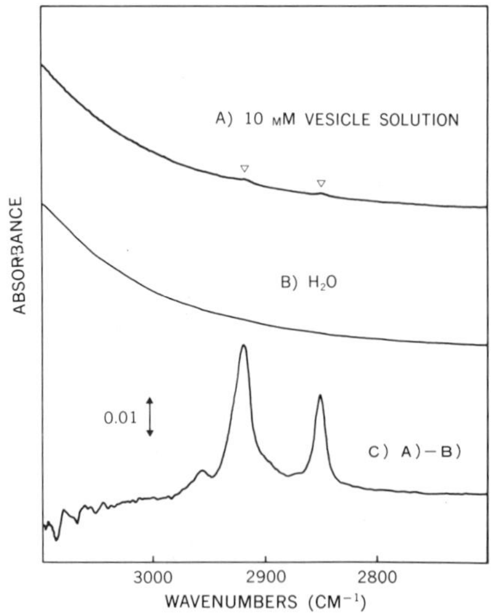

Figure 1. A digital difference spectrum of dialkylammonium salt vesicles dispersed in water. Concentration: 10mM, window: CaF_2, spacer: 50 μm, resolution: 2 cm^{-1}, accumulation: 1000 scans.

The frequency shifts observed at various temperatures are plotted in Figure 3. A sudden change of the frequency is seen at 42°C. This is due to the gel-liquid crystal transition which is often observed in phospholipids vesicles. There is another small pre-phase transition around 25°C. A polypeptide, gramicidin D, is embedded in the vesicles as a protein model. Gramicidin D is a biologically active polypeptide with 15 amino acid residues and is known to form ion channels in membranes. By putting gramicidin D in the vesicle, the above phase transition temperature is lowered to around 40°C, and the sharpness of the transition is lost due to the increase in randomness of the vesicle structure by inclusion of the polypeptide.

An advantage of the IR probe in the membrane study is that various components of the membrane system such as the hydrophobic and hydrophilic parts of the vesicle molecule and the embedded peptides or proteins can be observed independently but simultaneously. It works as an effective multiprobe.

Figure 4 shows the temperature variation of the amide I band of gramicidin D embedded in the vesicles. There is no shift of the band seen in this case, but the increase in intensity together with the narrowing of the band shape is observed with the elevation of the temperature. Plotting the intensity variation in Figure 5 reveals a similar transition at exactly the same temperature in the peptide conformation change as in the matrix phase transition.

The ability to obtain a detailed analysis of a low transmittance system such as a membrane dispersed in water exemplifies the potential of FT-IR in the bulk analysis.

SURFACE ANALYSIS

Surface analysis has become immensely important in characterization of industrial materials in the last decade. Four major techniques, electron probe microanalysis(EPMA), x-ray photoelectron spectroscopy (XPS), Auger electron spectroscopy (AES) and secondary ion mass spectrometry (SIMS), are extensively used for this purpose. The measurement of vibrational spectra of surfaces by FT-IR utilizing various kinds of attachments has proven to be very useful when coupled with the above conventional surface techniques [4,5]. Advantages over other surface techniques include 1) the measurement under ambient atmosphere, 2) nondestructiveness without charge up problems for insulators, and 3) the capability to give detailed structural information. The disadvantage of FT-IR surface analysis is the relatively poor surface sensitivity; however, it has reached a level of 0.01 μm for wide range of samples.

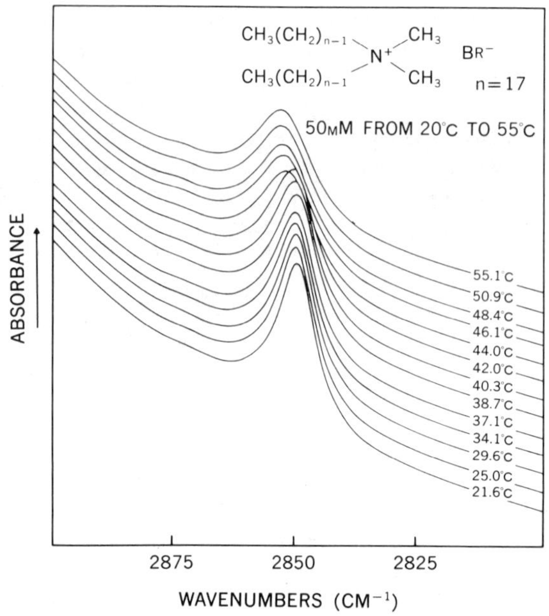

Figure 2. Variation of ν_{CH}^{s} band of the vesicles solution with temperature variation between 20 and 55°C.

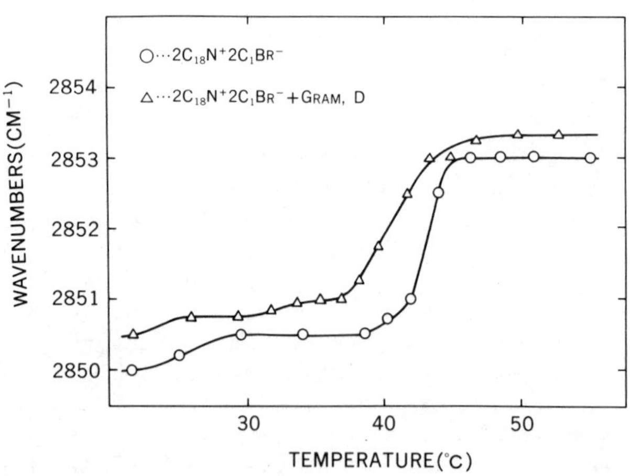

Figure 3. The wave number shift of a ν_{CH}^{s} band of the solutions due to phase transitions for vesicles with and without the polypeptide gramicidin D.

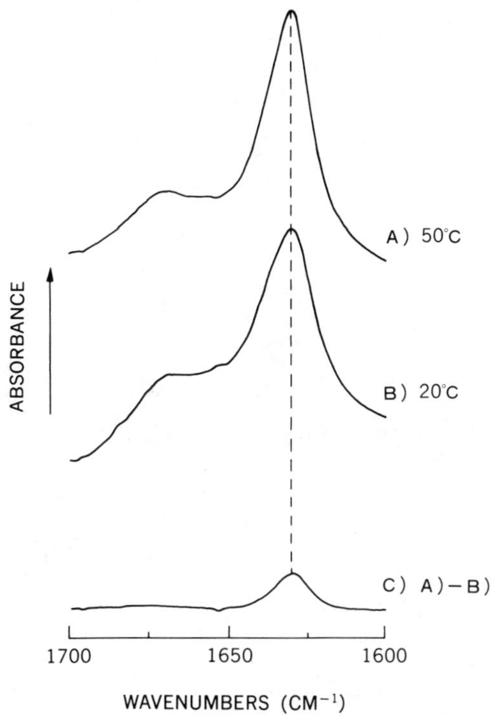

Figure 4. Variation of intensity and shape of amide I band of gramicidin D in the vesicles with temperature variation.

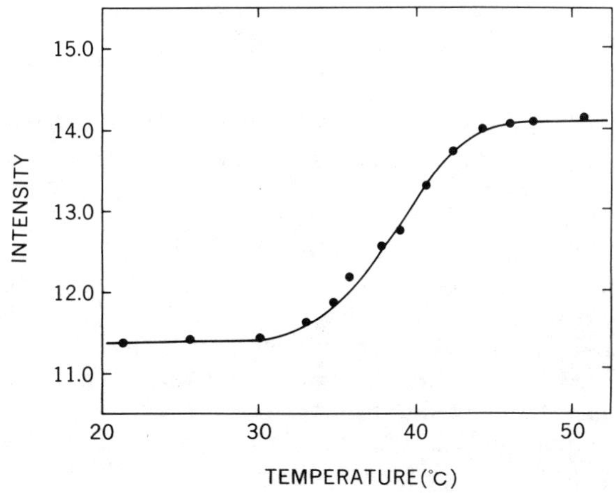

Figure 5. Intensity variation of amide I band of gramicidin D with the phase transition of the matrix.

A. Attenuated Total Reflection Spectroscopy (ATR)

ATR is the most commonly used technique. Enhancement of the surface sensitivity is attained by subtracting the substrate signal. High surface sensitivity up to 0.001 μm can be attained when an organic thin film is on a substrate which has broad and simple spectrum. Figure 6 shows the spectra of a single and a nine molecular-layer Langmuir-Blodgett film of Cd arachidate on glass [4]. The monomolecular layer of film has a thickness of 2.8 nm. The two carbonyl stretch bands are split because of the interaction of the carboxylate groups with the surface of the glass.

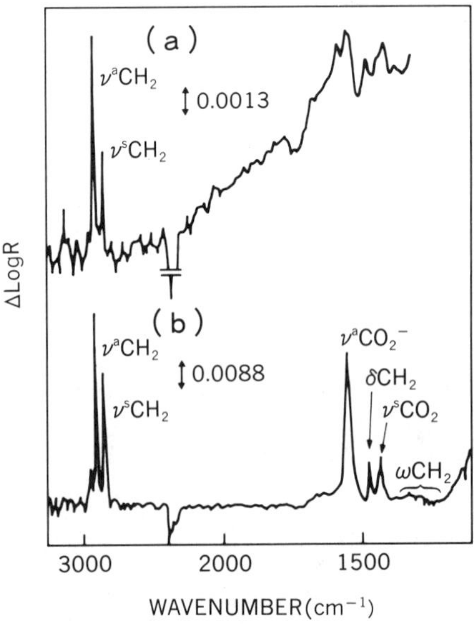

Figure 6. FT-IR ATR difference spectra for a) monolayer and (b) nine-layer Langmuir-Blodgett films of Cd arachidate on glass.

When a substrate has an intense, narrow shaped and complicated spectrum, like that from a thin polymer film on another polymer substrate, subtraction becomes difficult. Sampling conditions like the material and shape of the internal reflection element (IRE), incident angle, and wavelength region being studied are important factors in attaining good surface sensitivity. The quality of the obtained difference spectrum also depends on the combination of polymers. Good contact with the IRE is essential. Fluctuations in the signals, which are due to the differences in orientation and crystallinity as well as to the content of water, additives, and oligomers, affect much of the results. The balance of band intensities between the overlayer and the substrate is also important.

Figure 7 depicts the measurement of 9.0 nm poly(methyl methacrylate) (PMMA) overlayer on Nylon 6. ATR with spectral subtraction reveals most of the PMMA bands on a reasonably flat background except in the intense amide I and amide II band regions.

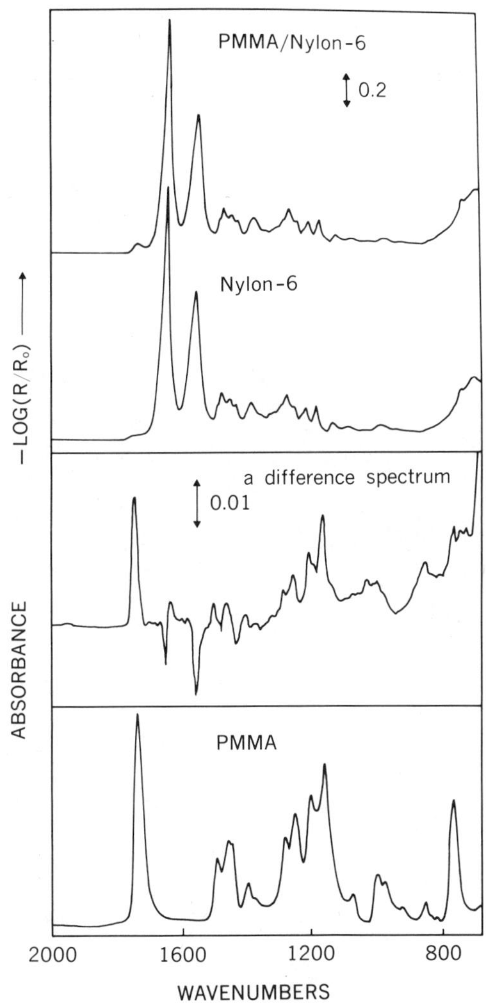

Figure 7. A digital difference spectrum of PMMA thin fils (90 Å) on Nylon 6 substrate. IRE: Ge, angle of incidence 45 degrees, 12 reflection, resolution 4 cm^{-1}, and accumulation 400 scans.

The observation of an oxide layer on a silicon wafer and its interaction with polymers like photoresist is being planned with a special set-up of ATR. Here a silicon single crystal is used for the IRE by cutting it in a 52x20x2 mm block with 45° angled edges. The surface of the IRE, the silicon (100) plane, is oxidized under various conditions and compared. Figure 8 shows an example of silicon oxidized under wet conditions at three different temperatures. The OH stretching mode at 3670 cm^{-1} due to the free Si-OH groups decreases with elevating temperature because of dehydration reaction.

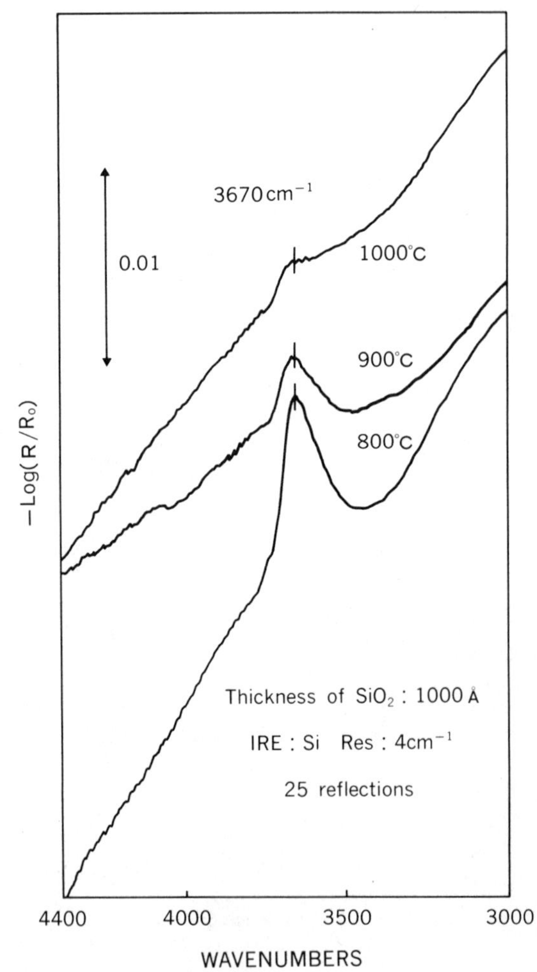

Figure 8. FT-IR ATR measurement of 100 Å-thick SiO_2 layers formed on silicon wafer surface under different temperature by special set up of using the wafer as IRE.

An interesting application of ATR in industrial problems in the nondestructive depth profiling of surface area. The penetration depth of the evanescent waves is controlled by the incident angle variation, the choice of IRE materials and by the appropriate wavelength region. Other applications include a quantitative analysis with ATR [6] and a study of molecular orientation on surfaces utilizing a special double edged IRE plate [7]. However, these detailed and quantitative analyses can be carried out only for relatively thick layers of submicron order.

B. Reflection-Absorption Spectroscopy (RAS) and Emission Spectroscopy (EMS)

Due to the high refractive indices of metals, ATR is inadequate in measuring thin films on metals; therefore, thin films on metals are measured either by RAS or EMS mode. Flat, smooth and highly reflecting surfaces are well characterized by RAS. Oxide layers on surfaces and polymer-metal interfaces are frequently analyzed. The high sensitivity of FT-IR enables the measurement of several square millimeter area samples.

Figure 9 compares a RAS spectrum of a copper plate heated in air at 150°C which has 15.0 nm cuprous oxide layer with that of the sample heated at 200°C with 32.0 nm cuprous and cupric oxides double layer. These measurements support the result from XPS and ellipsometry and give semi-quantitative compositions of copper plates heated at different temperature as indicated in the figure [5].

EMS is an alternate of RAS in metal surface analysis. It has a reasonable surface sensitivity up to around 10 nm for most polymers when a sample is heated between 100 and 150°C. The spectrum of poly(co-acryronitrile styrene) films on aluminum are shown in Figure 10 [8]. In Figure 11 emissivity is plotted against film thickness. Good linearity holds up to 100 nm, above which saturation begins due to selfabsorption.

The advantage of EMS over RAS is seen on non-flat, rough surfaced metal. Figure 12 compares RAS and EMS spectra of a polymer film coated on copper wire [8]. Better baselines and signal-to-noise ratios are attained by EMS.

C. Diffuse Reflectance Spectroscopy (DRS) and Photoacoustic Spectroscopy (PAS)

DRS and PAS are not necessarily surface techniques. DRS is useful for fine powders. Figure 13 shows the successful measurement on a toner powder which is used for xerographic printing. Despite the difficulty of low transmission due to the presence of

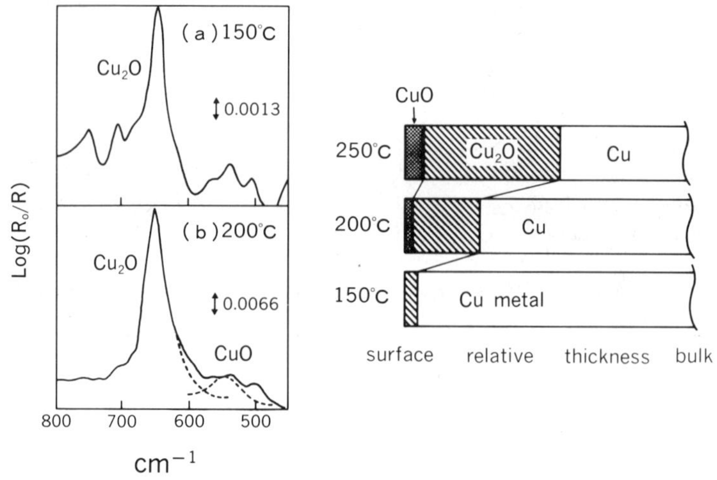

Figure 9 FT-IR-RAS spectra of surface oxidized copper plates and surface composition calculated from them.

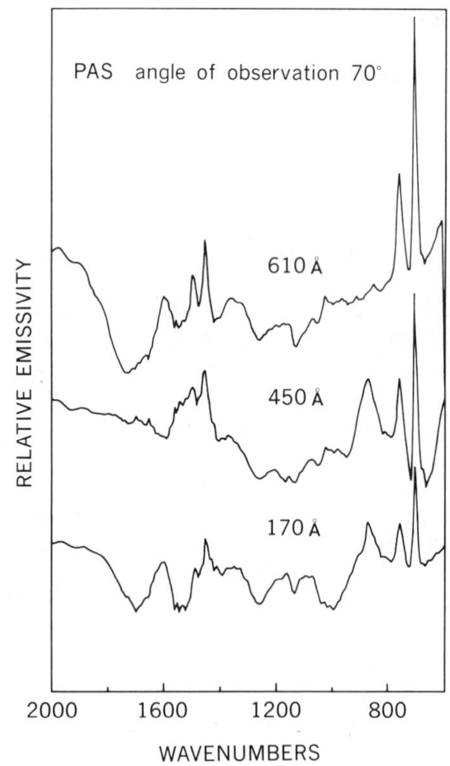

Figure 10 Emission spectra of thin films of PAS.

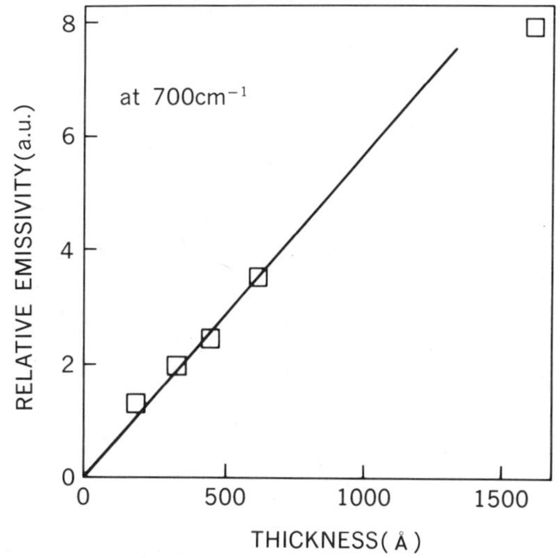

Figure 11 Dependence of emission intensity on thickness of the thin polymer films on aluminum.

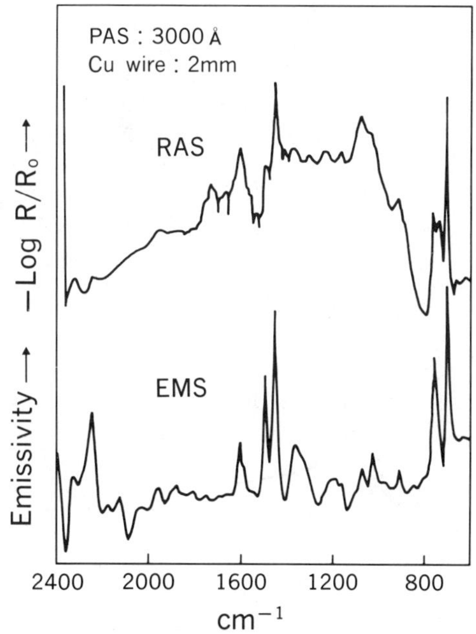

Figure 12 Comparison between FT-IR-EMS and FT-IR-RAS for a thin polymer film coated on copper wires.

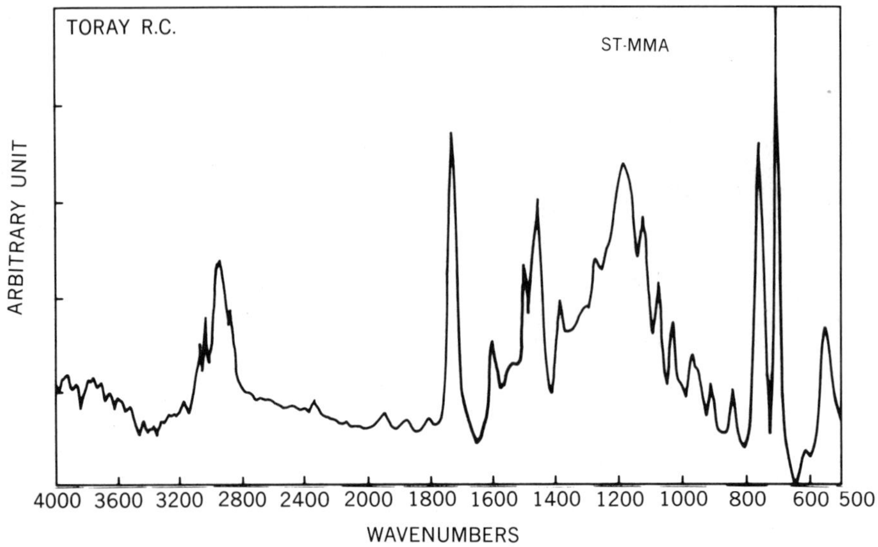

Figure 13 FT-IR-DRS spectrum of a binder polymer in toner.

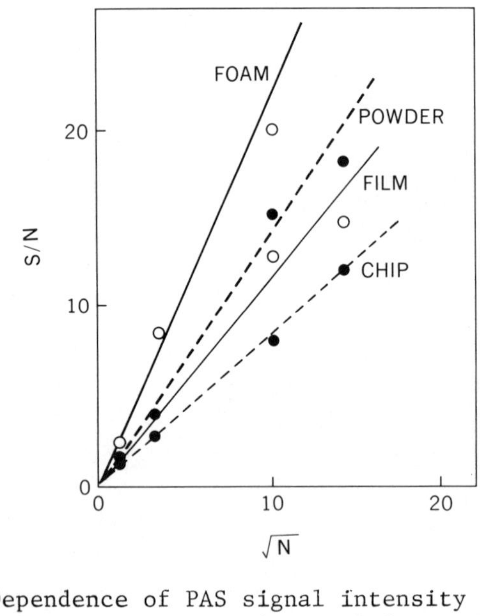

Figure 14 Dependence of PAS signal intensity (1460 cm^{-1}) of polystyrene on sample morphology.

carbon black, the binder polymer is identified as poly(co-styrene methyl methacrylate) from the spectrum.

PAS is useful for irregularly shaped samples. PAS does not require any pretreatment of the sample. Figure 14 compares the quality of spectrum obtained from polystyrene (PS) samples with various morphologies. A sample with larger surface area like a form is most favorable for PAS.

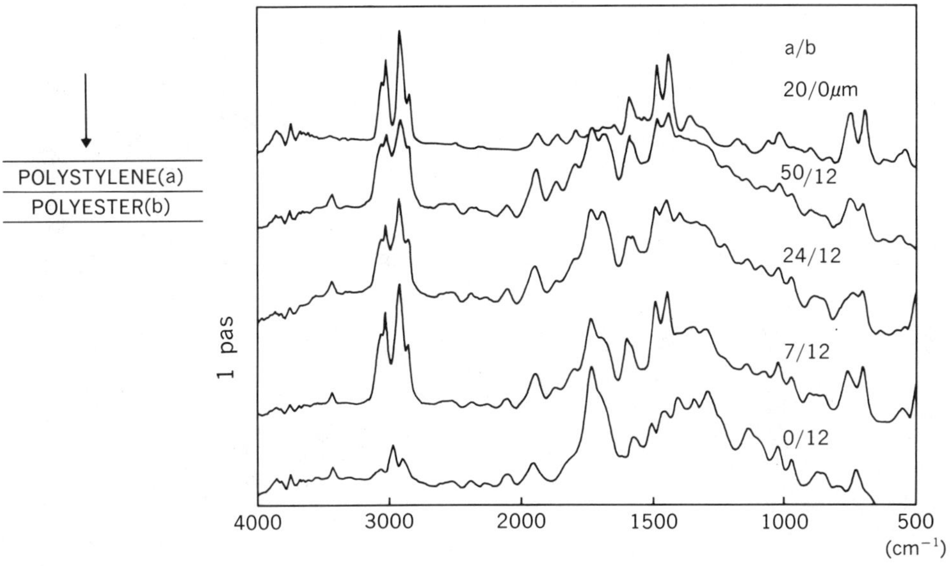

Figure 15. PAS spectra of laminated films of polystyrene and polyester.

Another interesting aspect of PAS is that it can measure beneath the surface. Figure 15 shows PAS measurements of laminated polystyrene film overlayers of different thicknesses on a 12 μm poly(ethylene terephthalate) (PET) film substrate. The PAS signal of the underlying PET seems to come through even the 50μm-thick PS film in this system. Figure 16 shows the utilization of this deep sampling capability combined with digital spectral subtraction to elucidate layer compositions.

D. Microanalysis

Microanalysis is particularly important in the characterization of industrial materials. Heterogeneous compositions are becoming more common in composite materials and multilayer laminated films. Important problems include the identification of in-

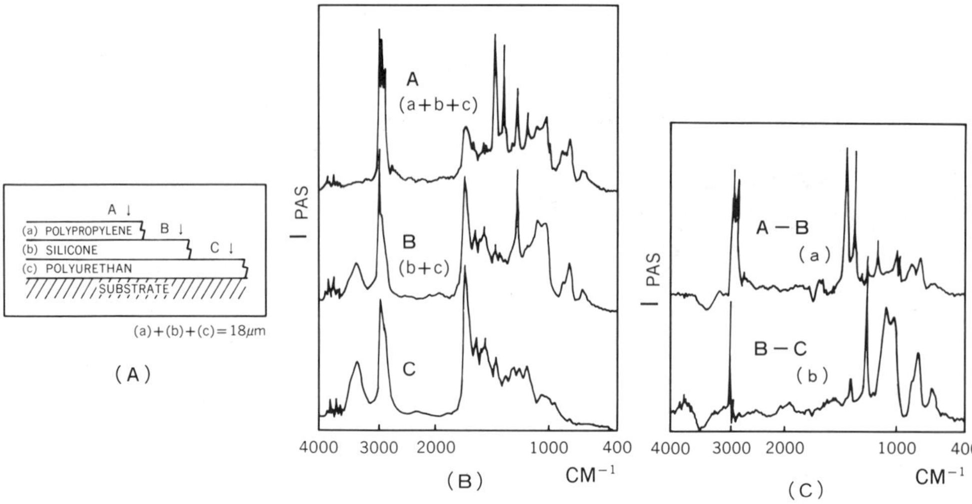

Figure 16 PAS difference spectra obtained from spectra of three layers composition of polypropylene, silicone and polyurethane on aluminum substrate.

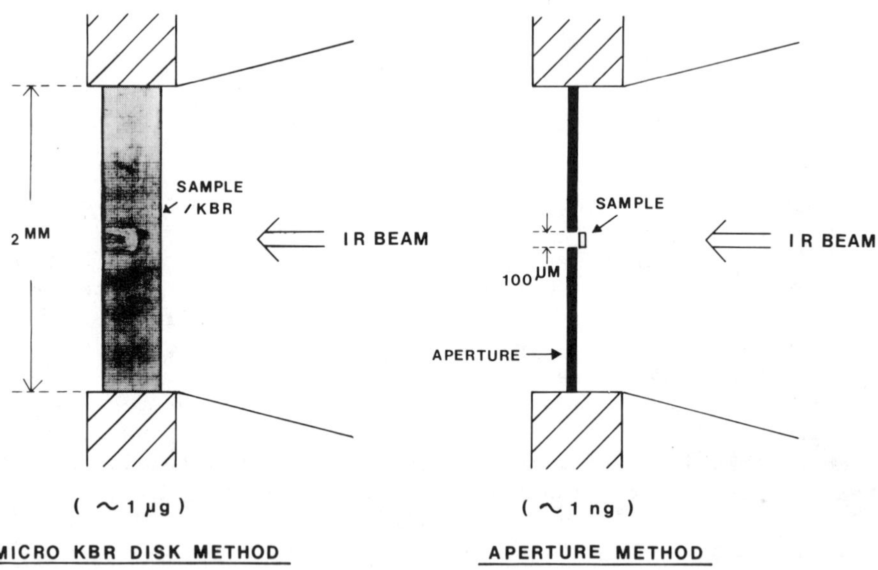

Figure 17 Two methods to measure small samples by FT-IR.

clusion particles, the lateral and depth distribution of additives in polymer matrix, and the depth variation of crystallinity. Conventional microanalysis techniques like EPMA, scanning Auger microscopy (SAM) and SIMS give information only on the elements. Raman microprobe and x-ray microdiffractometry have many limitations in applicable samples. IR microanalysis is expected to become the most general method although it is still in a premature stage.

Samples as small as 100 μm in diameter and around 1 μg in weight can be measured easily by preparing a small KBr disk. Soluble samples are measured at high sensitivity by DRS to the limit of nanogram order after putting them on fine KCl powder. However, microanalysis of a sample of micron order with nanogram level weight can only be attained with the aperture technique: by using an IR microscope with a movable X-Y stage or by placing a sample particle on the aperture by hand and focusing the beam on it with a condensor. The two techniques are illustrated in Figure 17. We have been practicing the microsample handling technique using tantalum apertures of a transmission electron microscope.

The limitation of the aperture diameter is due to the decrease in beam intensity which is caused by diffraction. Figures 18 and 19 show the dependence of the single beam intensity and the signal-to-noise ratio on aperture diameter. For both apertures, a linear relationship holds for apertures greater than 20 μm but an abrupt departure from the linearity is seen at 10 μm. The quality of spectra of a 1.5 μm-thick PET film put on 10, 100 and 1000 μm diameter apertures are compared in Figure 20. the spectra were measured at 8 cm^{-1} resolution with the accumulation of 1000 scans. Even in the smallest aperture of 20 μm diameter, where the weight of the measured PET is 660 pg, the spectral features hold well. There is no shift or deformation of the bands; however, there is a slight degradation of relative intensities between the peaks.

Sample morphology is very important in FT-IR microanalysis. A thin, flat sample like a film is desirable. However, fibers or particles are often encountered. Figure 21 shows scanning electron micrographs (SEM) of variously shaped samples set on apertures. Figure 22 compares the spectra taken from the samples shown in Figure 21. Although the measured weight of the fiber is three times larger than the film, the fiber gives a much poorer spectrum than the film. Flattening the fiber by pressing improves the spectrum greatly. Such small know-hows are very important for obtaining good spectra in microanalysis.

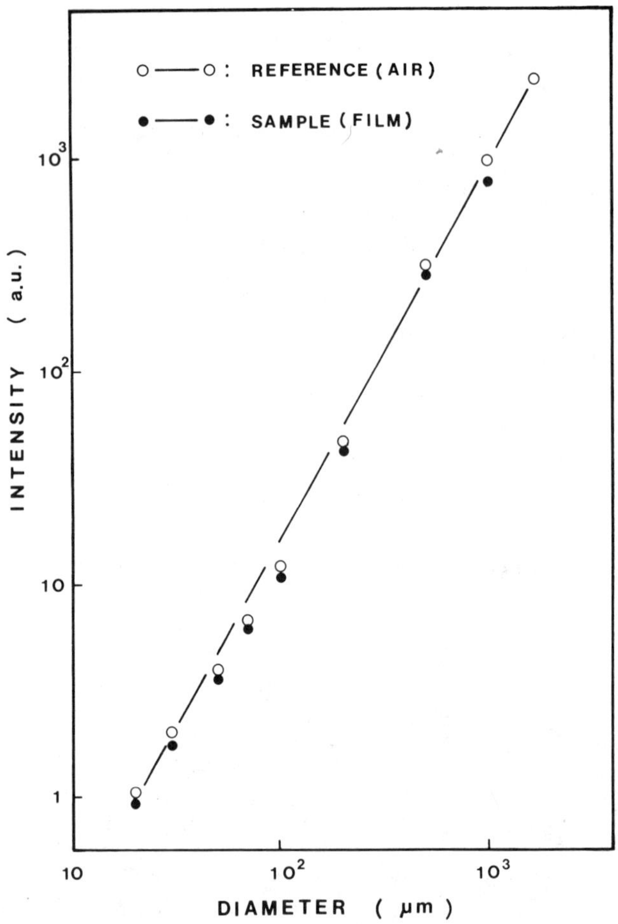

Figure 18. Dependence of single beam intensity on the aperture diameter at 2000 cm^{-1}.

Another example is shown in Figure 23 where an inclusion on a composite fiber is located, washed, and fixed on an aperture with glue and then measured. All the procedures are carried out under an optical microscope. The aperture has 100 μm diameter. From the spectrum obtained by the accumulation of 2000 scans with a resolution of 4 cm^{-1}, the inclusion was identified as a PET gel.

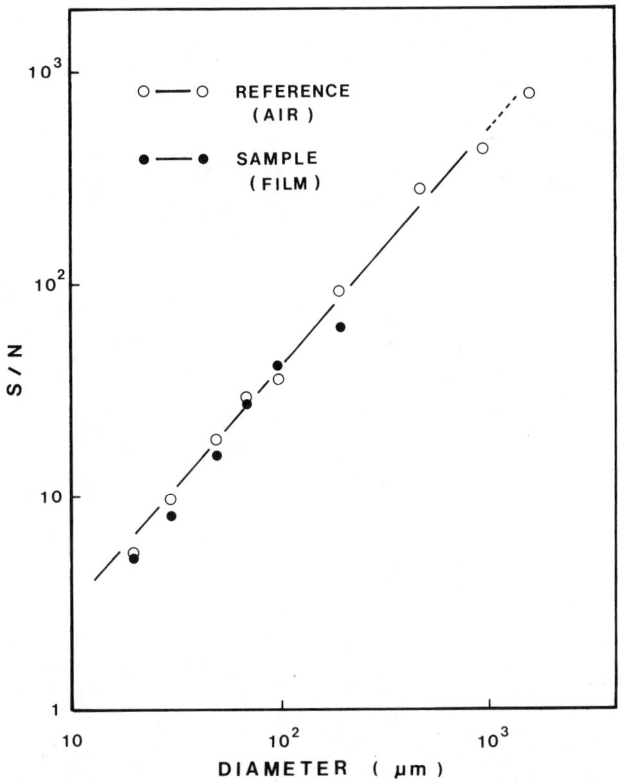

Figure 19. Dependence of signal-to-noise ratio on the aperture diameter at 2000 cm^{-1}.

E. Combination with Separation Techniques

GC-IR is in the most advanced stage; however, finding a real need for this technique is difficult in our research. Knowledge and experience of GC is indispensable for its effective use; as the sensitivity can be improved drastically by the proper choice of GC conditions [10,11]. LC-IR is more important than GC-IR due to insufficient development of LC-MS. The batch sample preparation procedure combined with DRS measurement is the most practical and is capable of nanogram order analysis. The <u>in-situ</u> DRS measurement on extended spots of thin layer chromatography (TLC) is successful for sub-microgram order samples [12]; however, higher sensitivity can be obtained by DRS measurements of an extracted sample which is dispersed on fine KCl powder.

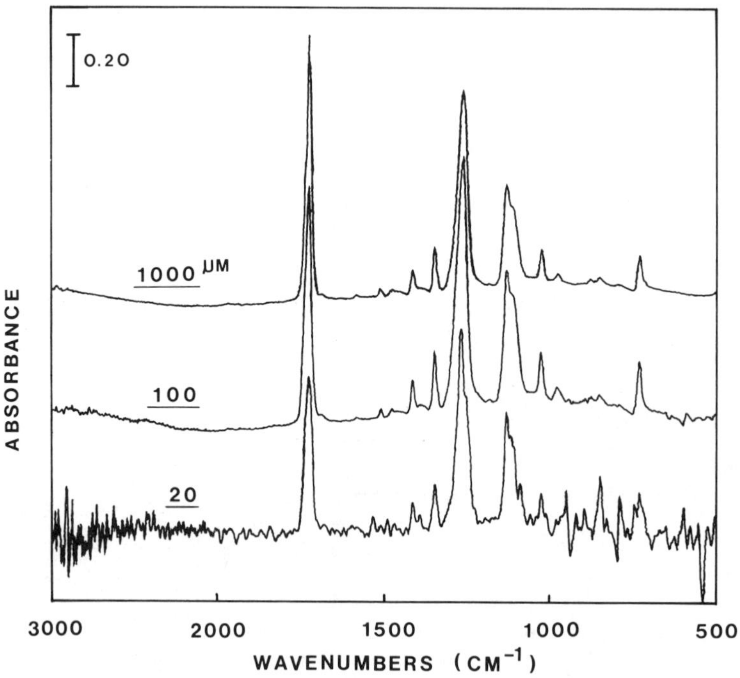

Figure 20. Comparison of quality of spectra of 1.5 μm-thick PET film on apertures with different diamters.

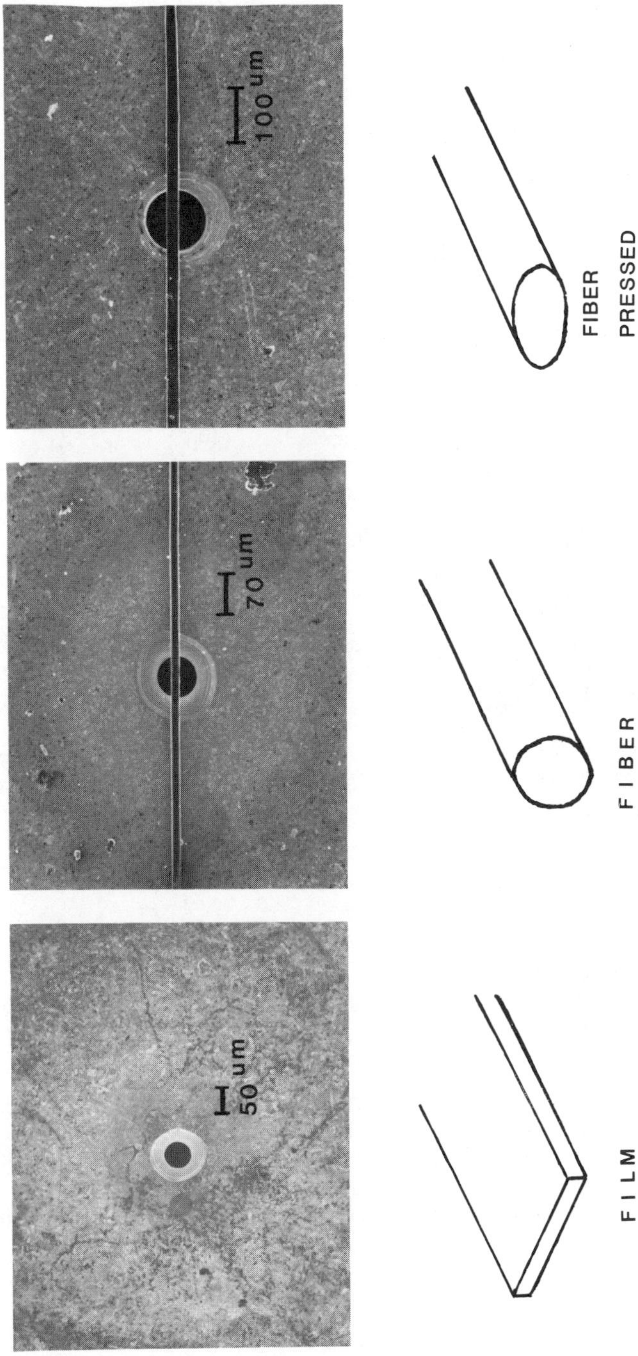

Figure 21 SEM micrographs of various shaped samples set on apertures.

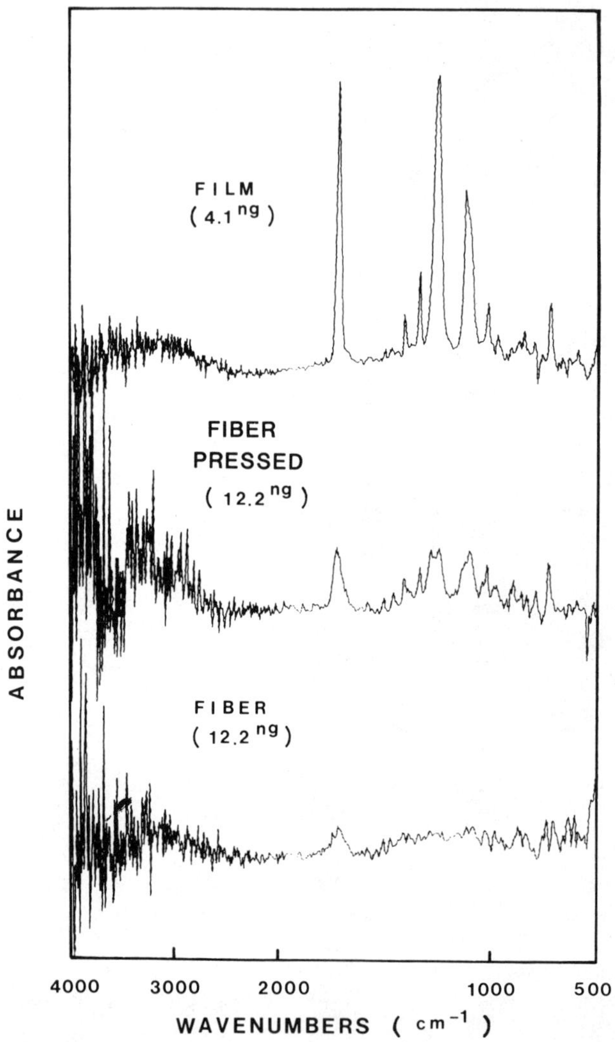

Figure 22 Effect of sample morphology on quality of spectra (50 μmΦ aperture.

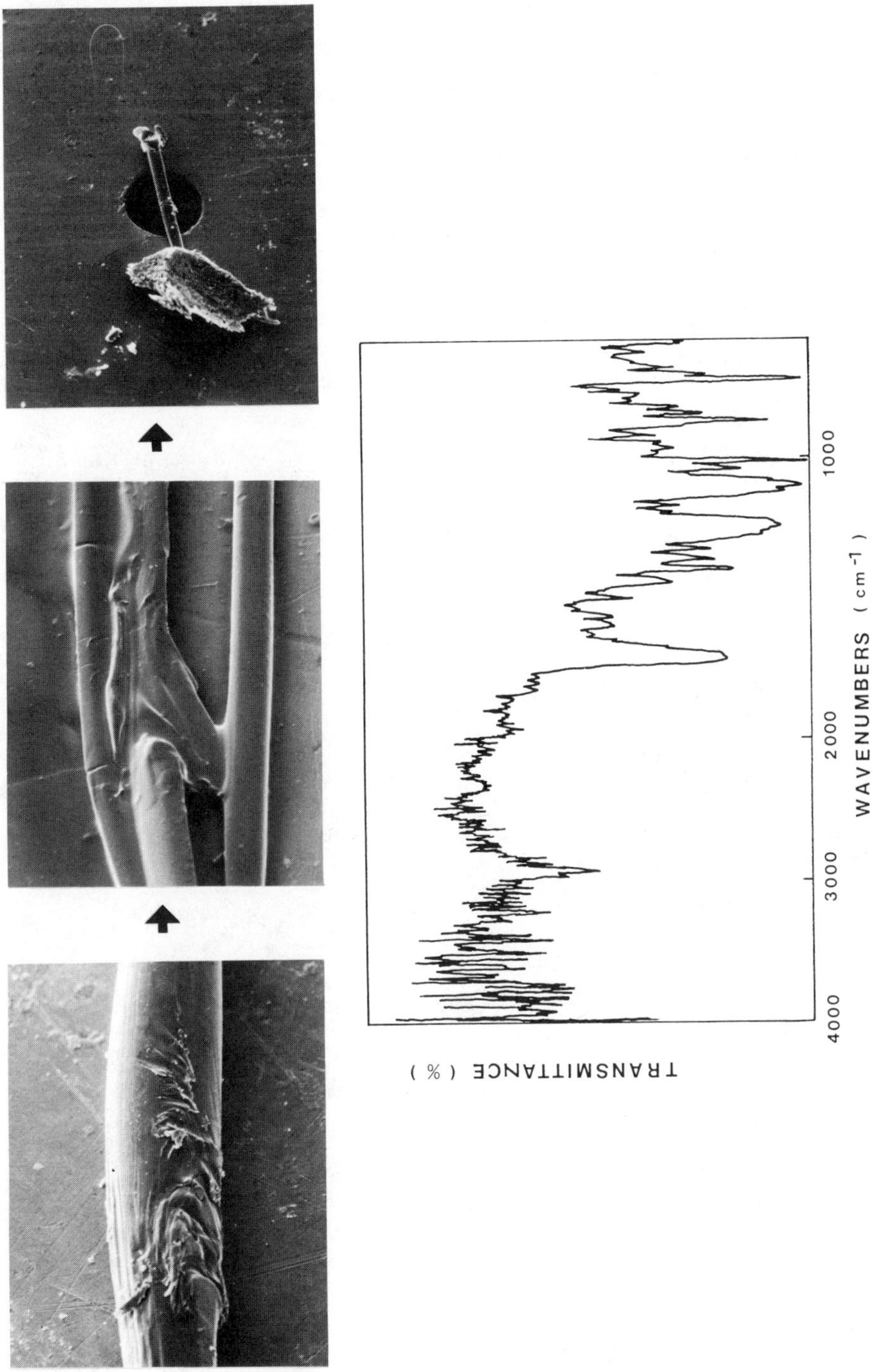

Figure 23 An example of analysis of an inclusion embedded in a composite fiber by the microsampling technique.

REFERENCES

1. I.Simada, T.Takahagi, M.Fukuhara, K.Morita and A.Ishitani, submitted to J. Polym. Sci.

2. I.Shimada, H.Ishida and A.Ishitani, Report of Research Project on Bilayer Membrane by Science and Technology Agency, Japan, 1983.

3. T.Kunitake and Y.Okahata, J. Am. Chem. Soc., 99, 3860 (1977).

4. T.Ohnishi, A.Ishitani, H.Ishida, N.Yamamoto and H.Tsubomura, J. Chem. Phys., 82, 1989 (1978).

5. A.Ishitani, H.Ishida, F.Soeda and Y.Nagasawa, Anal. Chem. 54, 682 (1982).

6. F.M.Mirabella, J. Polym. Sci. Poylm. Phys. Ed., 20, 2309 (1982).

7. C.S.P.Sung, Macromolecules, 1981, 591 (1981).

8. Y.Nagasawa and A.Ishitani, Appl. Spectrosc., 38, 168 (1984).

9. K.Shoda, H.Ishida and A.Ishitani, Annual Meeting of Anal. Chem. Soc., Japan (1982).

10. R.Kamoto, H.Ishida and A.Ishitani, 19th Meeting of Applied Spectroscopy, Tokyo (1982).

11. S.Onnagawa, A.Ishitani and K.Nakayama, 18th Meeting of Applied Spectroscopy, Tokyo (1982).

12. S.Onnagawa and A.Ishitani, Annual Meeting of Anal. Chem. Soc., Japan (1983).

13. S.Onnagawa, H.Ishida and A.Ishitani, 42th Meeting of Anal. Chem., Japan (1981).

FT-IR OF THE POLYMER-REINFORCEMENT INTERPHASE IN COMPOSITE MATERIALS

Andrew Garton

Division of Chemistry
National Research Council of Canada
Ottawa, Ontario
Canada K1A OR6

ABSTRACT

A modified internal reflection spectroscopy (IRS) technique is described, where the IRS element is coated with a thin (<100 nm) layer of a material intended to simulate the surface of a reinforcement. Examples of coatings include silica, poly(p-phenyleneterephthalamide), carbonized poly(acrylonitrile) and an aminosilane coupling agent. The way in which these surfaces affect the crosslinking kinetics and final crosslinked state of the first 200-400 nm of an amine- or anhydride-cured epoxy resin was then determined by infrared (IR) spectroscopy. Chemical differences detected by spectroscopy are shown to correlate with physical property differences of reinforced epoxy composites.

INTRODUCTION

The attractiveness of composite materials lies in their ability to exploit the high specific strength and stiffness of reinforcements such as glass, carbon or aramids, while retaining the processibility associated with their thermoplastic or thermosetting polymer matrices. It has long been recognized that the interface between the reinforcement and the matrix material is crucial to the performance and durability of the composite [1-3], and this recognition has prompted extensive research to characterize the interfacial region [1-6]. A semi-quantitative picture which is evolving is one of a region of finite thickness (the interphase) over which the composition and properties change from those of the bulk of the reinforcement to those of the bulk of the matrix (Figure 1).

The work described here is an attempt to characterize the interphase by infrared (IR) spectroscopy. The principal advantages of the chosen IR technique are its specificity in examining only the interphase, and its ability to examine the interphase *in situ*, i.e. with the reinforcement surface still in intimate contact with the matrix material. The principal disadvantage of the technique is the necessity to model the surface of the reinforcement as a film, not in its original fibrous state. No claim is made to the universal preferability of the chosen technique over alternative techniques, such as ESCA or Auger spectroscopies. Only by application of a range of techniques will a comprehensive description of the interphase evolve.

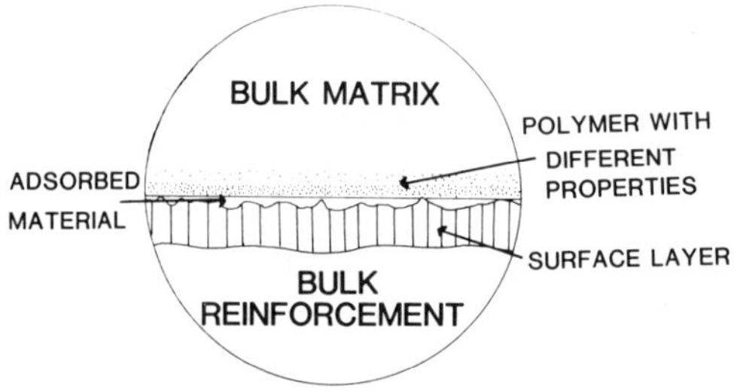

Figure 1. A pictorial representation of the composite interphase (after L.T.Drzal [6]).

Description of Spectroscopic Technique

Internal reflection spectroscopy (IRS) is a well-known technique which allows IR spectroscopic analysis of thin layers or surfaces [7]. The beam from the spectrometer passes down the IRS element by a series of internal reflections. At each reflection, a standing wave is set up, extending out from the element suface into the sample, which is pressed against the element (Figure 2). The "depth of penetration" (dp) of the standing wave depends on the angle of incidence, the wavelength, and the refractive indices of the sample and of the element. Because a high angle of incidence (usually 60°) and a high refractive index element (usually germanium) are used here, dp is typically only about 300 nm at 1800 cm^{-1} (a useful IR frequency).

The refinement of the IRS technique, which is described here, is to coat the element with a thin layer (<100 nm) of a second material which is chosen to simulate the surface of a reinforcement [8,9]. The coatings described here include silica (to simulate glass), poly(p-phenyleneterephthalamide) (to simulate aramid fibers), carbonized poly(acrylonitrile) (to simulate carbon fibers) and a silane (to simulate a coupling agent). The coating thickness is less than d_p, so the standing wave penetrates the surface coating and samples the first 200-400 nm (at mid-IR frequencies) of matrix material which is laid over the surface coating. It is possible, therefore, to observe directly the effect of a reinforcement surface on the crosslinking kinetics and final crosslinked state of a thermosetting matrix immediately adjacent to that surface. Correspondingly, the chemical effects of a reinforcement surface on the thermoplastic matrix can be observed, although such studies are not reported here.

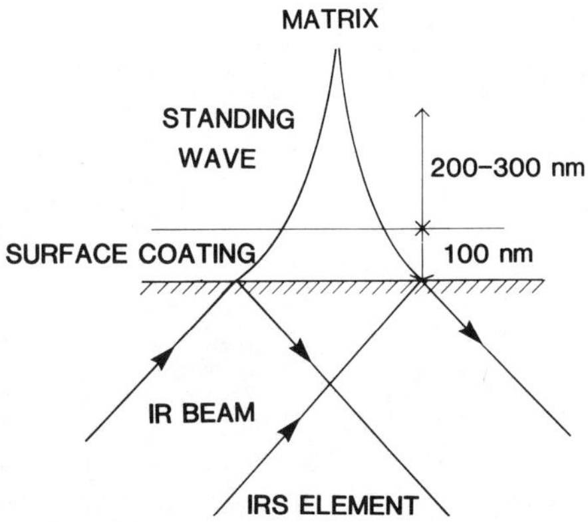

Figure 2. A pictorial representation of the IRS technique.

IRS spectra were obtained at 4 cm^{-1} resolution using a Nicolet 7199 FT-IR and a Wilks Model 10 IRS attachment. After each experiment the element was cleaned by burning off the matrix material in a furnace and the element was then repolished with alumina. Where necessary, deconvolution of overlapping absorptions was carried out using conventional least squares curve fitting techniques.

Materials

The matrix material chosen for this study was an epoxy resin crosslinked with either an anhydride or an aromatic amine curing agent. The epoxy resin was a commercial diglycidyl-type resin (Epon 828, Shell, largely the diglycidylether of bisphenol A). The curing agents were Nadic methyl anhydride (NMA, Fisher, vacuum distilled, 93 parts per hundred of resin) with benzyldimethylamine catalyst (BDMA, Ciba Geigy, vacuum distilled, 1 part) or diaminodiphenylmethane (DDM, Ciba-Geigy, 30 parts, uncatalyzed). The resin, curing agent, and catalyst were mixed in the conventional fashion [8-10] and degassed, before being placed on the IRS element as a ca. 100 μm-thick layer and covered with a clean Pyrex microscope slide. Except when stated differently, the cure cycle with NMA was 3 h at 100°C + 3 h at 150°C and with DDM was 1 h at 100°C + 1 h at 150°C.

The control surface against which other surface coatings were compared was a freshly polished, vacuum dried, germanium IRS element. The control surface was then modified by humid aging (2 h at 100% R.H. and 26°C) or oxidation (10 min at 440°C in air). A thin silane coating was applied by spraying with a dilute aqueous solution of γ-aminopropyltriethyoxysilane (APS, Fisher). An aramid coating was applied by dipping the element into a 0.5% solution of poly(p-phenyleneterephthalamide) (Kevlar 49, DuPont) in concentrated sulphuric acid. The coated element was then immersed in water to remove the acid, and the coagulated aramid coating was removed, by wiping, from all but the desired sampling area (40 mm x 20 mm on one side of the IRS element) and dried at >150°C in vacuum. Some initial attempts were also made to simulate carbon fiber surfaces by applying a thin coating of poly(acrylonitrile) (PAN) from dimethyl sulphoxide solution. The PAN coating was then partially carbonized by heating in air at 175°C for 48 h and in vacuum at 400°C for 100 h. A silica surface was produced by heating a silicon IRS element in air to 400°C.

Unidirectional aramid/epoxy composites were made by winding an aramid fiber (Kevlar 49, DuPont, 1140 denier) on to a metal former and subjecting the samples to various drying histories (see later). The fiber was then impregnated with the resin/curing agent/catalyst mixture in a specially modified vacuum oven [9]. Because the impregnation step was carried out in vacuum, the air content of the final composite was low. The fiber content of the composites was in the range 50-60 vol%.

Analysis of Spectra

For this modified IRS technique to be useful, there must be "windows" in the spectrum of the surface coating, i.e., spectral regions where the coating absorbs only weakly, which coincide with informative absorptions in the spectrum of the matrix material. Weak, relatively broad, absorptions due to the surface coating may be removed by conventional spectral subtraction techniques, but a complete subtraction of the coating spectrum is usually not possible. A spectrum obtained of a thin surface coating alone on an IRS element differs from a spectrum of the same coating obtained with an overlayer of a second material [7].

The IRS spectra of the DGEBA-NMA system on a clean dry germanium element, before and after curing, are shown in Figure 3.

The consumption of anhydride curing agent and build-up of ester crosslinks may be followed by their absorptions at 1780 cm^{-1} and 1740 cm^{-1}, respectively [8,9]. To obtain concentration data, it was necessary to ratio these absorptions to a reference absorp-

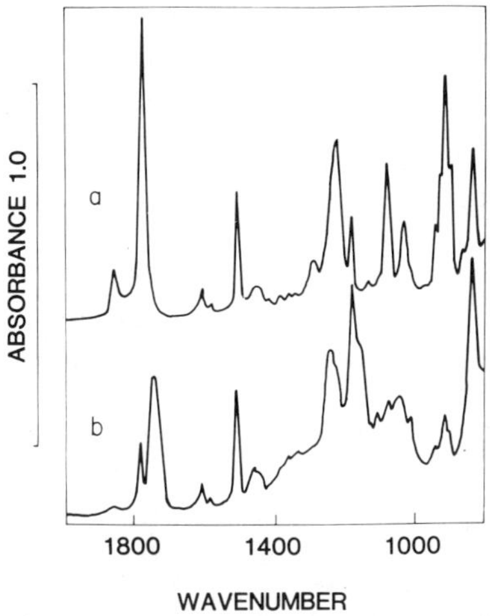

Figure 3 IRS spectra (germanium absorptions subtracted) of DGEBA/NMA/BDMA on a freshly polished, dry germanium element.
(a) Initial (b) After 3 h at 100°C + 3 h at 150°C.

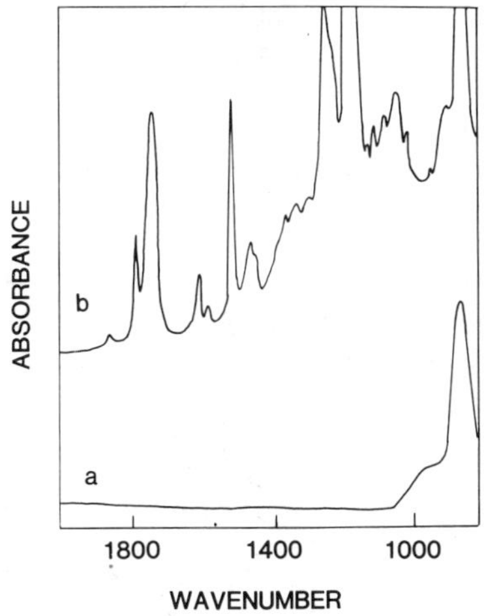

Figure 4 IRS spectra (germanium absorptions subtracted) obtained using a germanium element with a thin germanium oxide coating.

tion. For the DGEBA-NMA system, the initial 1780 cm^{-1} anhydride absorption in each experiment was chosen as a reference, while for the DGEBA-DDM system, the initial 915 cm^{-1} epoxide absorption was used. The reason for using an external referencing procedure of this sort rather than an internal reference (i.e., ratioing to an invariant absorption in each spectrum) was that for some surface coatings, the overlap between the spectrum of the coating and the spectrum of the matrix was such that no internal reference absorption, characteristic only of the matrix, was available. The use of an external reference necessitates that the optical contact between the element and the matrix (equivalent to sample thickness in transmission measurements) remained constant during each crosslinking experiment. The constancy of optical contact was checked by monitoring an absorption (e.g., phenyl at 1509 cm^{-1}) which is likely to be changed little by the crosslinking process.

Another complication to the use of absorption ratios as concentration indices in IRS is the wavelength dependence of dp [7]. The nature of the wavelength dependence is also affected by the presence of a surface coating of variable thickness, if that thickness becomes significant relative to d_p. For this reason an absorption is ratioed only to its own initial value or to another absorption close in frequency to itself (e.g., 1740 : 1780 cm^{-1}), when surface coatings of appreciable thickness (100 nm) are used.

The simplest case for spectral analysis with a modified IRS element is when the effect of humid-aged surfaces or thin oxide layers are to be examined, because there is little overlap between the spectrum of the matrix and the spectrum of the coating material (Figure 4). The absence of overlap also allows limited use of the spectral subtraction ability of FT-IR. Figure 5 shows the difference spectrum between DGEBA-NMA cured on humid-aged germanium and the same matrix cured on dry germanium. The scaling factor for the subtraction was chosen such that the phenyl absorption at 1509 cm^{-1} was minimized (i.e. the IR spectra were normalized to constant optical contact). The negative absorptions at 1740 cm^{-1} and 1780 cm^{-1} (relative to a reasonable pseudo-baseline) indicate that the humid-aged surface reduced the yield of esters and reduced the residual anhydride level. The positive absorption at 1705 cm^{-1} indicates that acid groups may have been produced in the interphase.

The aminosilane coating has a more complex IR spectrum than the oxides, but still has sufficient spectral "windows" to allow examination of the matrix material (Figure 6). Spectral subt-

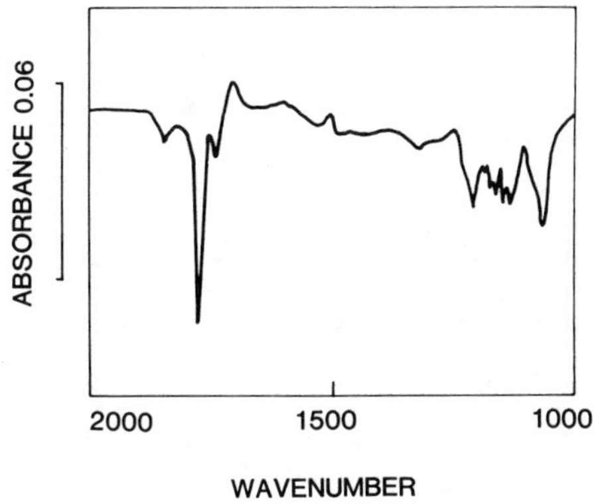

Figure 5. The difference spectrum obtained by subtraction of the spectrum of DGEBA/NMA/BDMA crosslinked on dry germanium from the spectrum of the same resin crosslinked on humid-aged germanium. Spectra scaled so as to minimize the phenyl absorption at 1509 cm^{-1} in the difference spectrum.

raction is again useful at least in a semi-quantitative fashion to demonstrate that the aminosilane coating reduced the yield of esters (1740 cm^{-1}), reduced the residual anhydride level (1780 cm^{-1}) and was responsible for the production of amide groups (1650 cm^{-1} and 1550 cm^{-1}) in the interphase (Figure 7).

The spectra of the aramid coating and the carbonized PAN coating (Figure 8) were sufficiently complex and intense that spectral subtraction was inappropriate over most of the IR spectral region. However, the ester and anhydride concentration indices were extractable, as is discussed below. For the DGEBA-DDM system, the degree of spectral overlap with the aramid coating was such that only the residual epoxide level (915 cm^{-1}) was accessible with any confidence.

The Effect of Surfaces on Crosslinking Processes

Figure 9 shows the crosslinking kinetics of the DGEBA-NMA system on clean, dry, germanium, on humid-aged germanium and on a silica surface, all obtained with a d_p of about 400 nm (at 1800 cm^{-1}). The humid-aged and oxide surfaces reduced the level of residual anhydride curing agent. The silica surface also significantly reduced the yield of ester products.

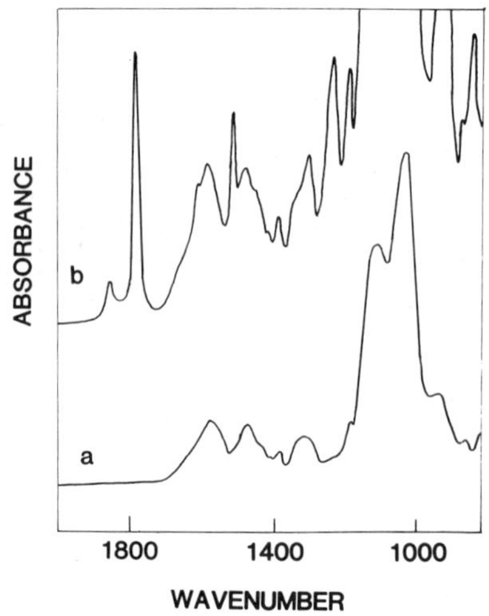

Figure 6 IRS spectra (germanium absorptions subtracted) obtained using a germanium element with a thin APS coating.
(a) APS coating
(b) With an overlayer of DGEBA/NMA/BDMA (uncrosslinked).

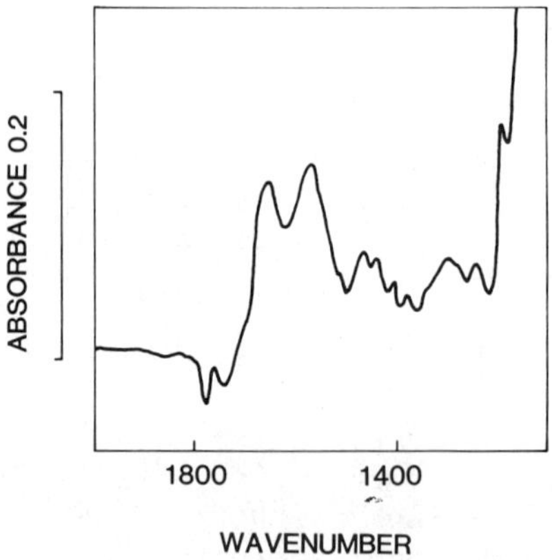

Figure 7 The difference spectrum obtained by subtraction of the spectrum of DEGBA/NMA/BDMA crosslinked on dry germanium from the spectrum of the same resin crosslinked on APS-treated germanium. The spectra were scaled to minimize the phenyl absorption at 150° cm^{-1} in the difference spectrum.

As was indicated in Figure 7, the APS coating reduced the yield of esters and reduced the residual anhydride level (Figure 10). The magnitude of the chemical effect varied with the thickness of the silane layer. This observation indicates that appreciable interpenetration of the silane and epoxy layers must occur, thus allowing the amino functionalities in the interior of the silane layers to interfere with the crosslinking chemistry of the matrix [8].

The aramid coating also reduced the yield of ester products compared to the germanium control and, as reported elsewhere, the magnitude of the chemical effect depended on the humidity history of the coating [9]. The data in Figure 11 refer to a d_p of 200-300 nm (at 1800 cm^{-1}) because a 60° incidence IRS element used. When stringent precautions were taken to exclude moisture, the aramid coating had only a slight effect on the crosslinking kinetics. Even brief exposure of the aramid coating to a humid environment, however, resulted in an appreciable reduction in the final ester concentration over the sampling depth. For the DGEBA-DDM system, the aramid coating had little effect on the residual epoxy level (Table 1), at least under the relatively mild cure conditions described here [9].

The data for the carbonized PAN coatings are only at preliminary stage, but significant differences have been observed in the first 200-300 nm of matrix material adjacent to the carbonized PAN surface, compared to the control germanium surface.

Chemical Structure-Physical Property Correlations

A detailed analysis of the dynamic mechanical and thermal properties of aramid/epoxy and carbon fiber/epoxy unidirectional composites are presented elesewhere [9,11]. One example is included here to demonstrate that the chemical differences, detected by spectroscopy, correlates with physical property differences in composites.

Figure 12 shows the energy dissipation in forced torsion at 1 Hz for a series of aramid/epoxy composites. The torsion direction was perpendicular to the fiber direction and so the response of the composite was largely controlled by the energy dissipation in the matrix and in the interphase. The unreinforced DGEBA-NMA system exhibits in peak in tan δ at 148°C (after a cure cycle of 8 h at 100°C + 8h at 160°C). The term "conditioned" refers to aramid fiber which has been stored at 65% relative humidity and 23°C for several months before use. The term "vacuum dry" refers to a sample which was vacuum dried at 100°C overnight and impregnated after only a brief exposure (1 min) to the laboratory atmosphere. The composite made with "vacuum dried" aramid showed

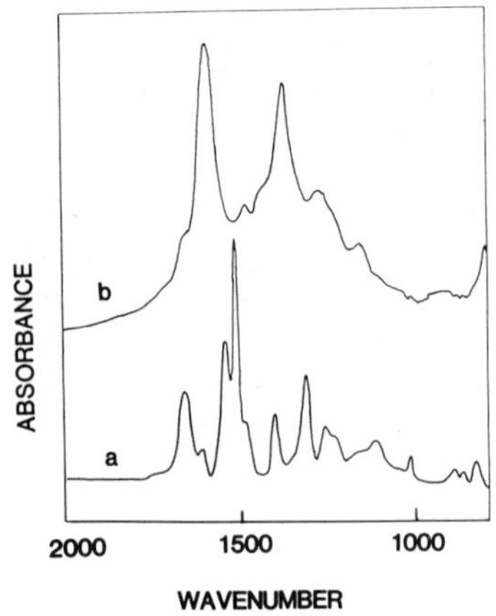

Figure 8 IRS spectra (germanium absorptions subtracted) of thin coatings of:
(a) poly(p-phenyleneterephthalamide)
(b) carbonized poly(acrylonitrile).

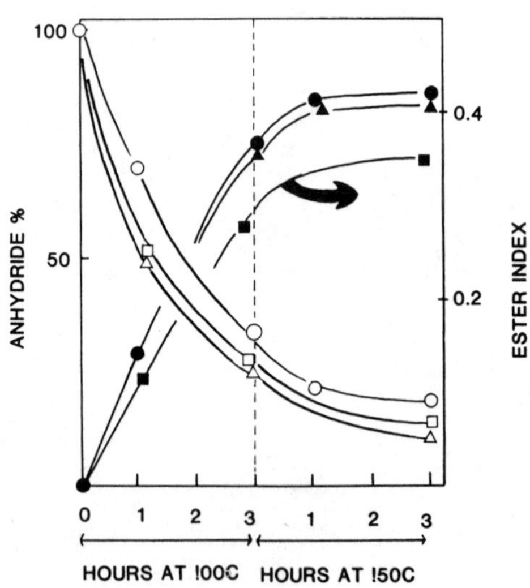

Figure 9 A comparison of reaction kinetics for the DGEBA/NMA/BDMA system.
○● dry germanium, 45° incidence IRS
△▲ humid-aged germanium, 45° incidence IRS
□■ SiO_2 coated silicon, 45° incidence IRS.

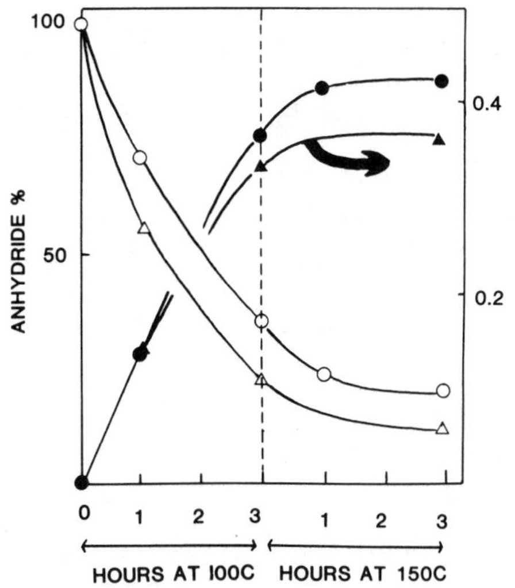

Figure 10 A comparison of reaction kinetics for the DGEBA/NMA/BDMA system.
○● dry germanium, 45° incidence IRS
□■ APS treated germanium, 45° incidence IRS.

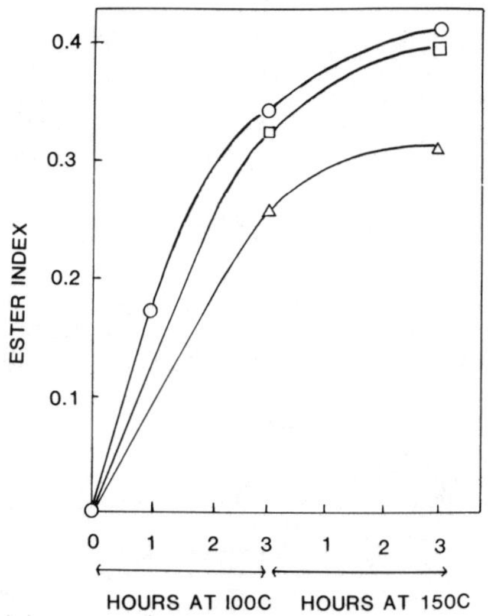

Figure 11 A comparison of reaction kinetics for the DGEBA/NMA/BDMA system.
○ Dry germanium, 60° incidence IRS
□ aramid coated germanium (vacuum dried) 60° incidence IRS
△ aramid coated germanium (conditioned) 60° incidence.

a tan δ maximum at a temperature close to that of the unreinforced plastic. The composite made with "conditioned" aramid fiber exhibits a tan δ maximum about $50^{o}C$ lower than the unreinforced plastic, as might be expected from Figure 11, which shows that a conditioned aramid surface significantly modifies the crosslinking chemistry. Samples made using the DGEBA-DDM epoxy matrix showed little effect in dynamic mechanical response with changing the humidity history of the aramid fiber [9], as might be expected from the insensitivity of the crosslinking chemistry to the aramid coating (Table 1).

TABLE I

Residual epoxide in DDM-cured epoxy resin after 1 h at 100°C + 1 h at 150°C

Sample	Residual Epoxide
Control	10% (Average of 3)
Aramid Coated	9% (Average of 3)

DISCUSSION

A range of surfaces are shown to modify the crosslinking kinetics and final crosslinked state of the first few hundred nanometers of epoxy resin adjacent to those surfaces. These chemical differences produce changes in the physical properties of the composites.

For the anhydride-cured system, the chemical differences in the interphase can be rationalized through a series of straightforward chemical reactions, as is discussed more fully elesewhere [8,9,11]. Adsorbed or absorbed water can potentially react with the anhydride to form acid groups. Acid groups so formed, or acidic functionalities on the surfaces, may further change the course of the crosslinking process, with the introduction of side reactions such as etherification [10]. The amino functionality of the silane can react either with the anhydride forming an amide group, or with the DGEBA by opening of the epoxide ring.

Extensions and refinements of this work fall in two major categories. Firstly, it is necessary to continue improving the accuracy with which the thin film coatings simulate the surfaces

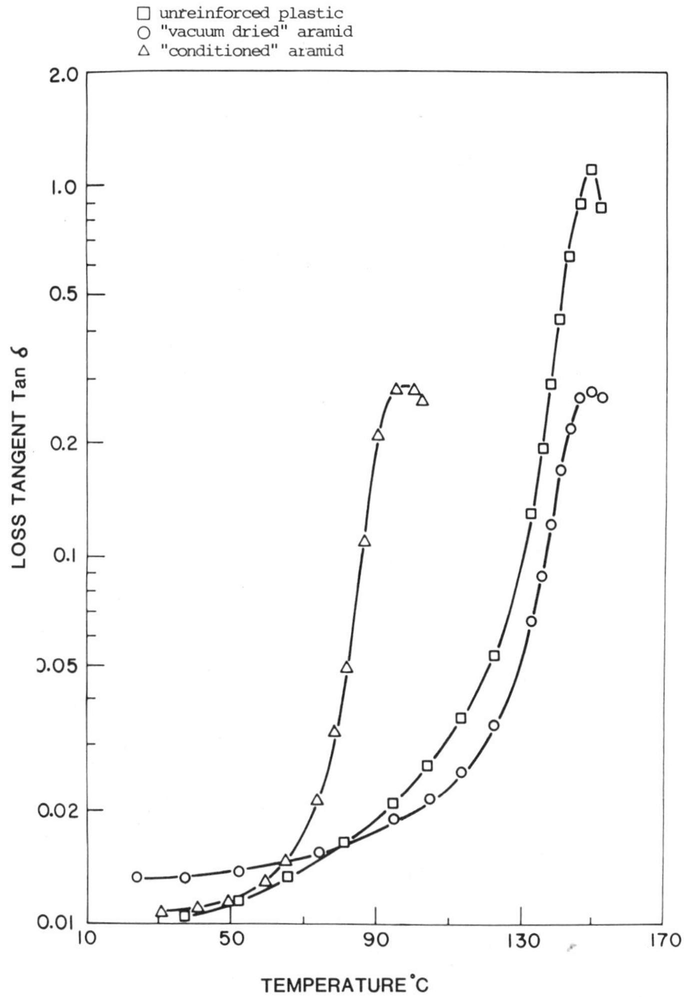

Figure 12 Dynamic Mechanical Analysis of Aramid/NMA cured epoxy composites (at 1 Hz, cure cycle 4 h at 100°C + 8 h at 160°C.)
□ unreinforced plastic
○ "vacuum dried" aramid
△ "conditioned" aramid.

of fibrous reinforcements. In particular, the carbonization temperature of the PAN film is being increased to a more realistic level. Secondly, we hope to reduce the depth of penetration significantly by using higher angles of incidence and shorter wavelength radiation. Raman IRS would clearly be an improvement in this regard [12] but difficulties exist in obtaining elements of suitable geometry and with the low fluorescence necessary for Raman spectroscopy.

REFERENCES

1. P.W.Erickson, Proc. 25th Ann. Tech. Conf., Reinforced Plastic/Composites Div., SPI, Section 13-A (1970).

2. E.P.Plueddemann in "Interfaces in Polymer Matrix Composites", E.P.Plueddemann, Ed., Academic, New York (1974).

3. A.J.Kinloch, J. Mater. Sci., 15, 2141 (1980).

4. H.Ishida, Polymer Composites, 5, 101 (1984).

5. W.L.Baun, Appl. Surface Sci., 4, 291 (1980).

6. L.T.Drzal, SAMPE J., 19, 7 (1983).

7. N.J.Harrick, "Internal Reflection Spectroscopy", John Wiley, New York (1967).

8. A.Garton, J. Polym. Sci. Polym. Chem. Ed., 22, 1495 (1984).

9. A.Garton and J.Daly, J. Polym. Sci. Polym. Chem. Ed., 21, 1031 (1985).

10. H.Lee and K.Neville, "Handbook of Epoxy Resins", McGraw-Hill, New York (1974).

11. A.Garton and J.Daly, Polym. Composites, 6, 195 (1985).

12. R.I.Iwamoto, M.Miya, K.Ohta and S.Mima, J. Chem. Phys., 74, 351 (1979).

FOURIER TRANSFORM DIFFUSE REFLECTANCE INFRARED STUDY OF FIBERS, POLYMER FILMS, AND COATINGS

Martin T. McKenzie*, Scott R. Culler+
and Jack L. Koenig**

Department of Macromolecular Science
Case Western Reserve University
Cleveland, Ohio 44106

* Present address: DuPont Co., Marshall Lab.
Philadelphia, PA
\+ Present address: 3M Co., Dental Products
Minnesota, MN
** Author to whom correspondence should be addressed

ABSTRACT

A unique modification of the diffuse reflectance FT-IR (DRIFT) technique has been utilized in the study of surface structure by vibrational spectroscopy. This approach has extended the use of DRIFT to fiber, polymer film, and coating surface characterization. High sensitivity, elimination of orientation effects, and a non-destructive sampling procedure are valuable aspects of the technique. The modification of the standard DRIFT experiment involves the use of a granular overlayer such as KBr whereby the relative contribution of the surface to the total spectrum can be controlled by the amount and particle size of the overlayer. Examples of this approach are described in this paper.

INTRODUCTION

The surface structure of materials is related to a wide variety of important end-use properties. For example, the interfacial structure of surface modified fibers is directly related to their performance in composite applications. Further, the surface structure of films and coatings is related to phenomena such as adhesion and durability. It is not surprising, therefore, that considerable attention to characterization of surface structure is evident in the literature [1,2 for example]. Of the many techniques available for surface characterization, vibrational spectroscopy, and in particular FT-IR, provides a combination of sensitivity and detailed chemical information not available from any single alternate approach.

There are several infrared techniques that have been used for surface studies. The most established of these is attenuated total relectance (ATR). This method has been primarily used for non-granular samples which can make good optical contact with the ATR crystalline element. Another technique for surface analysis is photoacoustic spectroscopy (PAS). PAS can be used to study a wide variety of sample forms but has been primarily used to study granular materials. However, the sensitivity of PAS is not presently equal to other FT-IR techniques and the quantitative application of PAS is not completely understood at this time. DRIFT was originally designed to study powdered samples and is known for high sensitivity and quantitative application [3,4]. Kubelka and Munk [5,6], and later Hecht [7] discussed the theoretical aspects of the diffuse reflectance experiment. In general, the theory and application of DRIFT to powders has been well explored.

This paper describes the extension of DRIFT to surface characterization of fiber and polymer film samples. The scattering of incident infrared radiation is made diffuse through the use of a KBr overlayer. This approach eliminates the orientation effects observed in studies of fiber materials and enables enhancement of the spectra of surface species on films. The method is non-destructive and reproducible. Results for a model system are presented in conjunction with those for E-glass fiber and PMMA film samples.

EXPERIMENTAL

The E-glass fibers used in this study were supplied by Crane Glass and Co. The fiber diameter is 3.8 microns, and the surface area is reported to be 0.5 m^2/g. The fibers were heat treated at 500°C for 24 h to remove sizing and/or lunbricants and stored in a dry atmosphere until used. The poly(ethylene terephthalate) (PET) film was purchased from E.I. DuPont de Nemours and Co. while the poly(vinylidene fluoride) (PVF_2) film was cast from a 5% solution of dimethylacetamide at 100°C on a glass plate. Poly(methyl methacrylate) (PMMA) film was supplied by Dow Chemical Co. The KBr was purchased from Aldrich Chemical Co. at 99+% purity. Consistent KBr particle size distribution was achieved through the use of standard sieves.

The fiber experiment involved treatment with -aminopropyltriethoxysilane (APS) from a 3% by weight aqueous solution. The coupling agent was hydrolyzed for 30 min in distilled water. This was followed by exposure of the glass to the solution for 3 minutes and subsequent drying at ambient conditions for 24 h. Film thicknesses for the PVF_2 and PET experiment were obtained by transmission. PET thickness was calculated from the interference fringes observed in the spectrum using a refractive index of 1.58. The thickness of the PVF_2 film was not sufficient to observe fringes. Therefore, the thickness of a larger sample was measured by the fringe method (n=1.42) and the integrated intensity of the C-H stretch region measured. By comparison of the intensities in this region, the thinner PVF_2 film thickness was determined to be 1.5 microns.

A FTS-20 Digilab FT-IR equipped with a narrow band pass MCT detector was used in conjunction with a Digilab diffuse reflectance attachment (DRA-100) to collect the data. KBr powder of particle size <74 microns was used as the reference material. A minimum of 100 scans were averaged in the data collection. The scans were recorded in the single beam mode at a resolution of 2 cm^{-1}. All of the spectra reported in this paper are plotted in Kubelka-Munk units. Quantitative spectral analysis was accomplished through the use of a linear least-squares program.

RESULTS AND DISCUSSION

The surface characterization of fiber materials by FT-IR is not without complications [1]. Nonetheless, the surface properties of such systems are intimately related to the performance of the fibers in composite materials. An example of the complications in spectroscopic surface analysis is shown in Figure 1. The bottom and middle spectra represent DRIFT scans of E-glass fibers taken with samples as received (following heat treatment). Proper sample size was achieved through the use of a No.6 cork bore such that the circular glass sample fit the sample cup precisely. Figure 1 bottom represents the initial spectrum of the glass fibers. Figure 1 middle is the result of rotation of the sample cup in the DRIFT attachment by 90 degrees. No other perturbation of the sample was performed. Figure 1 top is the difference spectrum of the middle and bottom spectra.

It can be seen from the difference spectrum in Figure 1 that merely rotating the sample cup in the DRIFT attachment causes a shift in the frequencies and intensities of the bands below 1400 cm^{-1}. This dependence on sample orientation limits the utility of diffuse reflectance FT-IR to the 4000-1600 cm^{-1} range for glass fiber samples. Characterization of surface agents on fibers is difficult using this restricted spectral range. Our solution to this problem involves the use of a diffusely

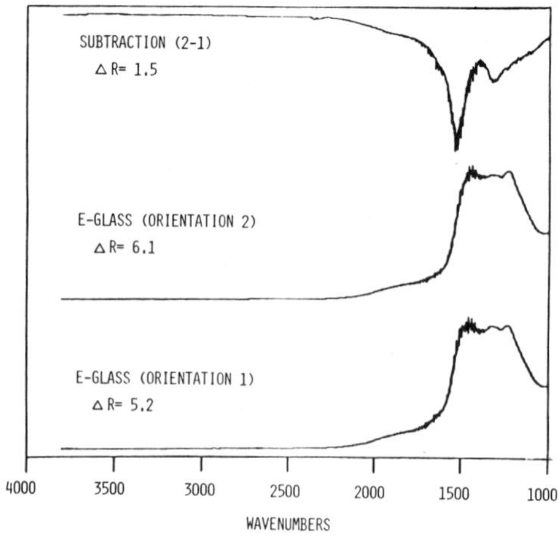

Figure 1 Demonstration of orientation dependence of DRIFT experiment for fibers.

scattering, non-absorbing overlayer above the fiber samples. Typically, 50-75 milligrams of KBr powder are used for these experiments. The quantity of KBr used is a function of the particle size of the salt. The results of the overlayer approach are illustrated in Figure 2. Both the top and middle of the figure represent the DRIFT spectra of E-glass fibers as a function of sample orientation. Two observations are immediately apparent. First, the intensity of the bands due to the glass are reduced

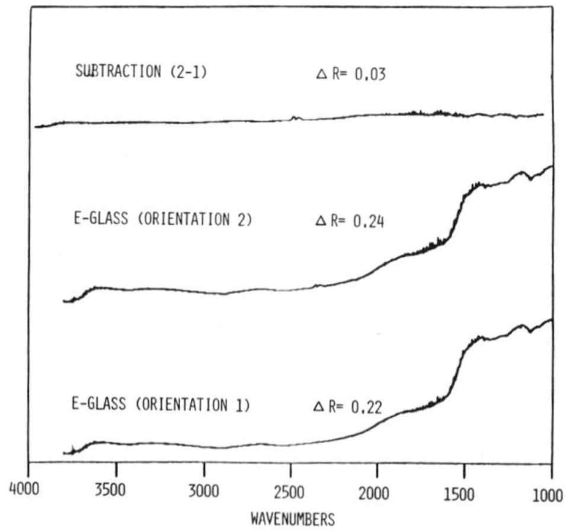

Figure 2 Use of KBr overlayer to remove orientation dependence for E-glass fibers.

relative to Figure 1. Second, and most important, is that the difference spectrum shown at the top does not contain artifacts such as that in Figure 1. In other words, the spectra have been made orientation independent.

The impact of this is shown in Figure 3. In this experiment, E-glass had been treated with a 3% solution of an amino functional silane, -aminopropyltriethoxysilane (APS). In order to characterize the surface modification by FT-IR, the DRIFT spectrum of the untreated fiber was recorded using the overlayer approach. The same amount of KBr overlayer was used to record the spectrum of the treated fiber sample. Elimination of the spectral contributions from the glass substrate was accomplished by spectral difference. The result of this is shown at the top of Figure 3. The elimination of orientation induced artifacts in the difference spectrum by the overlayer approach provides data which can be interpreted over the complete spectral range available from the instrument (4000-700 cm^{-1}). The chemistry of the surface agent which can be deduced from the difference spectrum is in excellent agreement with that reported by Culler et al. [8] in a previous paper. Further, this was accomplished with no destructive sample manipulations which could potentially alter the nature of the surface.

In addition to the elimination of orientation effects, it appeared from the dramatic reduction in the glass band intensities that the overlayer approach might also serve to enhance the contribution of the surface species to the total spectrum while reducing the contribution of the substrate. In order to test for this effect and to further understand the overlayer procedure, a model system was devised which was intended to simulate a coating on a substrate. The system consists of a 29 micron PET film (substrate) and a 1.5 micron PVF_2 film (coating). The spectra of the individual components are shown in Figure 4 bottom and top respectively.

The model system was covered with increasing amounts of KBr of particle size 45-74 microns. With increasing quantities of KBr in the overlayer, the spectra in the C-H stretch region (3200-2700 cm^{-1}) increasingly resembled that of the "coating", PVF_2, rather than the substrate, PET. A quantitative correlation between overlayer weight (at a given particle size) and surface/substrate spectral contribution is shown in Figure 5. A linear least squares program was used to fit the pure material spectra to the composite spectra. The trend in the percent contribution of the substrate (PET) is shown by the circles, while the trend for the contribution of the PVF_2 is shown by the "x" curve. Note that at roughly 60-70 mg coverage, the contribution from the substrate is becoming leveled. This enables the recording of reproducible spectra of the surface material in a convenient fashion.

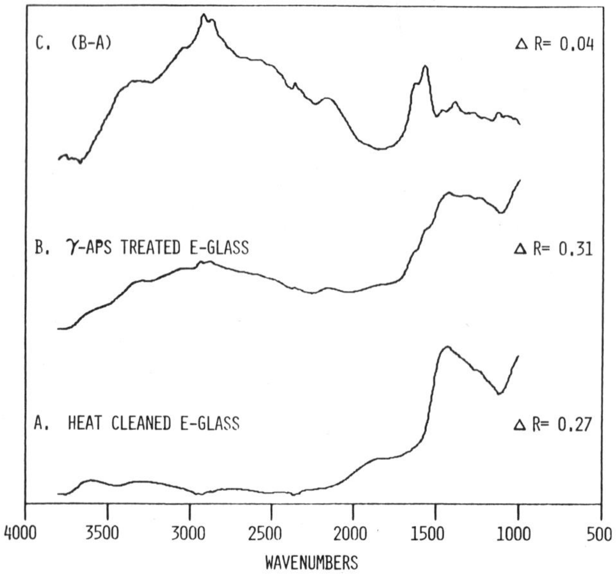

Figure 3 Use of DRIFT/overlayer method to study surface modified glass fibers.

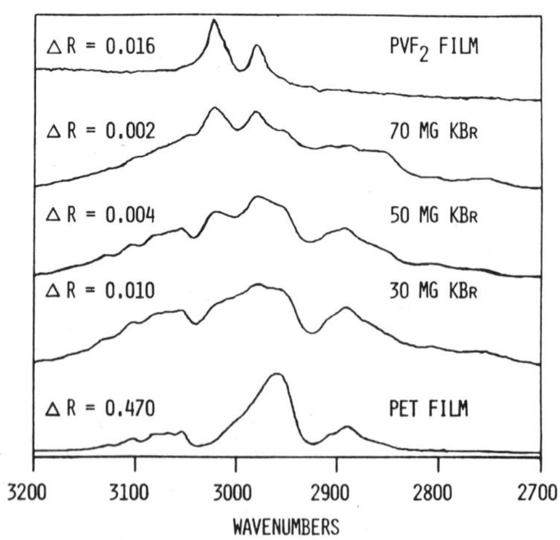

Figure 4 Results of model coating/substrate experiment.

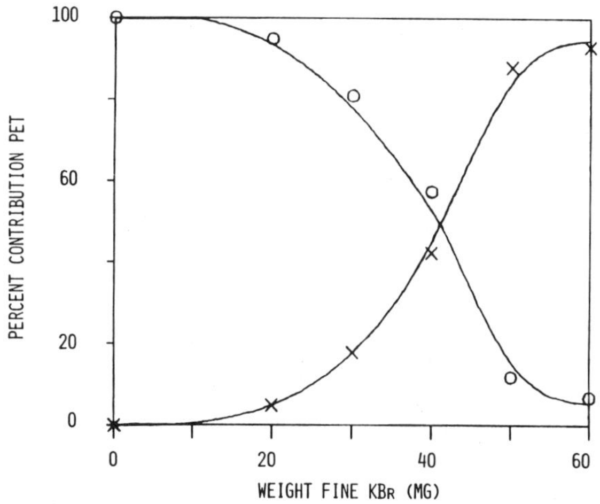

Figure 5 Plot of least squares fit results for model experiment spectra.

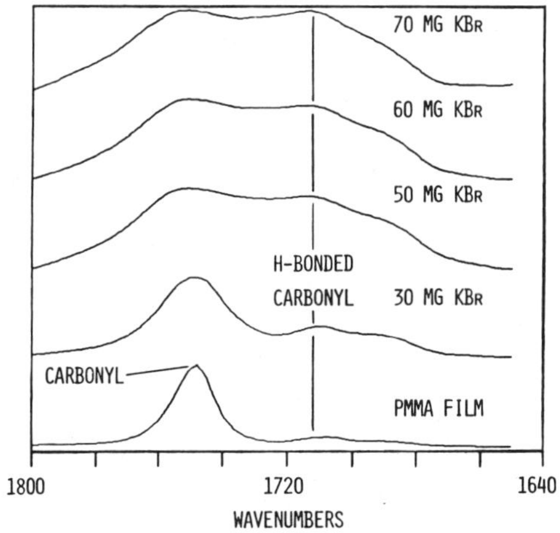

Figure 6 Use of DRIFT/overlayer method to study polymer film surfaces (PMMA).

The utility of the overlayer method has been demonstrated for surface modified fibers and a coating/substrate model system. It was not clear, however, if the method could be extended to studies of polymer film surfaces. An example of the results of this experiment is shown in Figure 6. A poly(methyl methacrylate) (PMMA) film was examined by DRIFT using increasing amounts of the same KBr material mentioned above. With increasing overlayer thickness, and hence increased surface contribution, the intensity of the carbonyl band at 1750 cm^{-1} decreases while the intensities of two lower frequency bands increase. It should be noted that all of these bands were observed in a transmission experiment but at grossly different relative intensities.

The precise assignment of the lower frequency bands (1710 and 1680 cm^{-1}) is not clear from this experiment. One explanation is that hydrogen bonding of the carbonyl species at the surface with adsorbed moisture causes a shift in the fundamental frequency. The magnitude and direction of the shift is consistent with this interpretation. An alternate explanation is that one of the bands is due to residual acid groups in the methacrylate. It is conceivable that such groups would be concentrated at the surface of the film. The other band could be assigned to a hydrogen bonded moiety (acid or ester carbonyl). Regardless, it is clear from this study that characterization of film and fiber surfaces can be conveniently and non-destructively explored by this method.

REFERENCES

1. M.T. McKenzie, S.R. Culler and J.L. Koenig, Appl. Spectrosc., 38, 786 (1984).

2. S.R. Culler, M.T. McKenzie, L.J. Fina, H. Ishida and J.L. Koenig, Appl. Spectrosc., 38, 791 (1984).

3. M.P. Fuller and P.R. Griffiths, Anal. Chem., 50, 1906 (1980).

4. M.P. Fuller and P.R. Griffiths, Appl. Spectrosc., 34, 533 (1980).

5. P. Kubelka and F. Munk, Z. Tech. Phys., 12, 593 (1931).

6. P. Kubelka, J. Opt. Soc. Am., 38, 448 (1948).

7. W.W. Wendlandt and H.G. Hecht, "Reflectance Spectroscopy", Interscience, New York (1966).

8. E.P. Plueddemann, "Silane Coupling Agents", Plenum, New York (1982).

COMPARISON OF FT-IR TRANSMISSION, SPECULAR REFLECTANCE, AND ATTENUATED TOTAL REFLECTANCE SPECTRA OF POLYMERS

R.T. Graf, J.L. Koenig and H. Ishida*

Department of Macromolecular Science
Case Western Reserve University
Cleveland, OH 44106

ABSTRACT

Thin films of poly(methyl methacrylate) were deposited on germanium substrates and analyzed by Fourier transform infrared spectroscopy. Three types of experiments were performed: transmission, specular reflectance, and attenuated total reflectance. The spectra obtained from these experiments differed in the position, shape, and intensity of the bands even though all the experiments were performed on the same sample. These differences were the result of optical distortion effects. To account for these effects the optical constants of the polymer were determined and the expected band profiles for each experiment were calculated. The calculated and experimental spectra agreed well in both position and intensity of the bands.

INTRODUCTION

For the analysis of polymers by infrared spectroscopy various types of experiments are used depending upon the sample. Three of the most common infrared experiments are transmission, specular reflectance, and attenuated total reflectance (ATR). Although transmission is the preferred technique due to its simplicity, one must often use one of the other techniques because of the nature of the sample. The theory, practice, and applications of specular reflectance spectroscopy [1-4] and attenuated total reflectance spectroscopy [5-6] have been reviewed elsewhere. In all cases one would like to obtain an absorption spectrum which is

*: To whom correspondence should be addressed.

characteristic only of the chemical structure of the sample rather than the geometry of the experiment. Furthermore, one would like to directly compare the spectra from different techniques. However, because of the nature of the three different experiments, this is not strictly possible. The band profiles obtained from these three experiments can be markedly different, even for the same sample. The physics of these experiments is important in understanding the reasons for these differences [7-9].

The refractive index n and absorption index k determine the response of a material to incident electromagnetic radiation. Collectively n and k are referred to as the optical constants. At an interface between two phases a propagating electromagnetic wave will be both reflected and transmitted. The amount of the wave that is either reflected or transmitted is determined by the optical constants of the two phases and the angle of incidence. The Fresnel relationships give the proportions of reflected and transmitted radiation at an interface. Thus, while the absorption index k governs the attenuation of radiation within a single medium, the distribution of radiation across several media or phases is governed by both optical constants for each phase. In essence, the measured absorbance is a complex function of both the refractive index and the absorption index. Figure 1 shows schematic diagrams for each of the three experiments.

The quantities n and k are not strictly constants since they vary with frequency. This phenomenon, known as optical dispersion, has an important ramification. An operation known as a Kramers-Kronig transformation relates the dispersion of $k(\nu)$ to the dispersion of $n(\nu)$ [9]. Given an absorption index spectrum, the corresponding refractive index spectrum can be calculated or vice-versa. The following equation gives the exact Kramers-Kronig relationship.

$$n(\nu') = n- = 2/p \int_{0}^{\infty} \frac{k(\nu)\ \nu}{\nu^{2} - \nu'^{2}}\, d\nu \qquad (1)$$

n- = refractive index at infinite frequency

The main source of error with Kramers-Kronig analysis is the limited frequency region over which spectroscopic information is usually available. Jones et. al. [10] have discussed the errors involved in using a finite frequency interval and the effect of neglecting neighboring bands. They concluded that a slightly altered form of equation [1] known as the subtractive Kramers-Kronig procedure produced superior results in error simulation studies. Jezierski et. al. also have analyzed the accuracy of the Kramers-Kronig procedure [11].

Previous authors have noted [12-14] the distorting effect of specular reflectance spectroscopy on peak positions and intensities. Other authors [15-17] have discussed the importance of the optical constants in determining the bandshapes of ATR spectra. Allara et. al. [14] have given the most systematic study of band distortion in specular reflectance spectroscopy. They studied thin films of poly (methyl methacrylate) on gold and silicon substrates. The angle of incidence was varied between 11° and 82°, and the thickness of the polymer was varied between 15 angstroms and 2 microns in their study. These authors calculated the expected profile of the carbonyl band, for each case and compared it with the experimental results. For their calculations they measured the optical constants of the carbonyl band of PMMA by two different techniques. One method involved a Kramers-Kronig analysis of the transmission data, while the other relied on solution of the optical equations for a combination of reflection and transmission data. Both methods gave comparable results. Crawford et. al. [17] discuss the optical distortion of ATR relative to transmission and use it to accurately determine the optical constants. Their results showed that the 670 cm^{-1} band of benzene became distorted as the angle of incidence decreased from 70° to 45° on a KRS-5 hemicylinder. Band inversion, in fact, was observed at 20°. Harrick has discussed the types of optical distortion effects encountered in reflection spectroscopy [15]. In general, for strong bands, thick films, and substrates of high refractive index, optical distortion for both specular reflectance and attenuated total reflectance will be significant.

EXPERIMENTAL

Thin films of poly(methyl methacrylate) were cast from dichloromethane (spectroscopic grade) solution onto germanium substrates in a saturated solvent vapor environment. The substrates were optically polished ATR elements from Wilks Scientific. These ATR elements measured 50x18x2 mm and were trapezoidal in shape. The samples were subsequently placed in a vacuum oven at ambient temperature for 12 hours to remove residual solvent. The oven temperature was raised to 110° C. and held for 12 more hours to minimize any residual orientation.

Spectra were run on a Digilab FTS-20E spectrometer equipped with a mercury-cadmium-telluride detector at a resolution of 4 cm^{-1} throughout the mid-infrared region. For each polymer-ATR element sample, an identical pristine ATR element was run as reference. For the specular reflectance spectra, a Harrick specular reflectance attachment was used, while a Wilks ATR attachment was used for the ATR experiments. All spectra are plotted in $-\log_{10}(I/Io)$ format.

The thickness of the PMMA film on the substrate was determined from the transmission spectrum. Using previously derived optical constants for PMMA, and literature values for the germanium substrate [18], we could obtain the film thickness by least-squares analysis of the film|substrate transmission spectrum. The least-squares routine operated by comparing the observed spectrum with a calculated one for a given thickness value. The thickness values were refined until the best fit was obtained.

RESULTS AND DISCUSSION

Three experiments (transmission, specular reflectance, ATR) are performed on a single sample. The sample system consists of a thin polymer film deposited on an ATR crystal substrate (see figure 1). The expected spectra for each experiment is calculated using the optical constants. The optical constants for the polymer have been previously determined by the Kramers-Kronig method. The spectra from all three experiments are compared with the calculated spectra for each experiment to check agreement between theory and experiment.

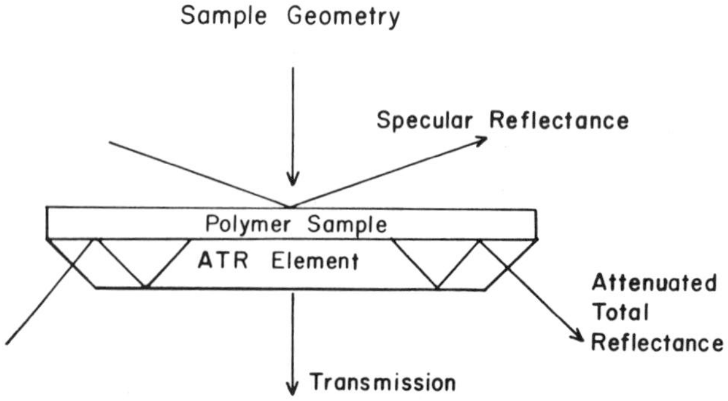

Figure 1. Diagram of the system used to perform all three experiments on the same polymer sample.

A mathematical algorithm described by Heavens [19-20] is used to simulate each of the three experiments. It is capable of handling any number of phases, and gives exact solutions within the limits of classical electromagnetic theory and linear optics. The algorithm assumes that each phase is isotropic and homogeneous, and that all interfaces are plane parallel. The optical constants of each phase must be specified, and the thickness of each layer and the angle of incidence must also be given. Incident radiation is resolved into components parallel and perpendicular to the plane of incidence. It computes the transmittance and reflectance of each phase present and then coherently sums these values to give a total transmittance and reflectance. A schematic diagram of this algorithm is shown in figure 2.

In practice of course, the optical constants must be deduced from the measured spectra. Since, some optical distortion will be present in any measured spectrum, an iterative procedure must be used to extract the optical constants [14,21-22] The following brief summary gives the steps necessary to extract $k(\nu)$ and $n(\nu)$ from the transmission spectrum $A(\nu)$ of a thin film sample by the well-known Kramers-Kronig iterative procedure. The thickness and baseline refractive index are determined from the interference fringes or some other means. An approximate absorption index is then calculated based on equation 2.

$$A = -\log_{10}(I/Io) = 4\ p\ k\ \nu\ d\ /\ 2.303 \qquad (2)$$

The Kramers-Kronig transformation (equation 1) is then used to calculate an approximate refractive index. The optical constant spectra are then used to simulate a thin film transmission experiment. This simulated spectrum is compared with the actual spectrum. The difference between the experimental and calculated spectra is used to modify the approximation of $k(\nu)$. The whole cycle is then repeated until the agreement between the calculated and experimental values is acceptable. If the thickness and baseline refractive index were correct, then this iterative procedure will converge to the correct optical constant spectra.

Once the optical constants have been determined, any of the three types of spectroscopic experiments can be modelled. By comparing the calculated spectra with the actual experimental spectra one can account for band shifts and relative intensity changes which are artifacts of the experiment. This ability makes it possible to quantitatively compare spectra generated by different techniques. The following example shows how different techniques can generate distorted spectra of the same sample.

Figure 3 compares the three experiments for a thin film of PMMA on a Ge substrate. This is the same sample system as shown schematically in figure 1. These spectra show large changes in peak position and shape. The carbonyl band is shifted 12 cm^{-1} to higher frequencies for specular reflectance and 3 cm^{-1} to lower frequencies for ATR. The specular reflectance and ATR spectra show an asymmetric carbonyl band, while the transmission spectrum gives the correct profile. Yet, the same sample was used for all three experiments.

Using optical constants, previously derived for PMMA [23], the expected band profiles for specular reflection and ATR experiments can be calculated. The results of these calculations along with the experimental results from figure 3 are shown in figures 4 and 5. The peak shifts and shape changes evident in figure 3 are completely accounted for by the calculations. Note

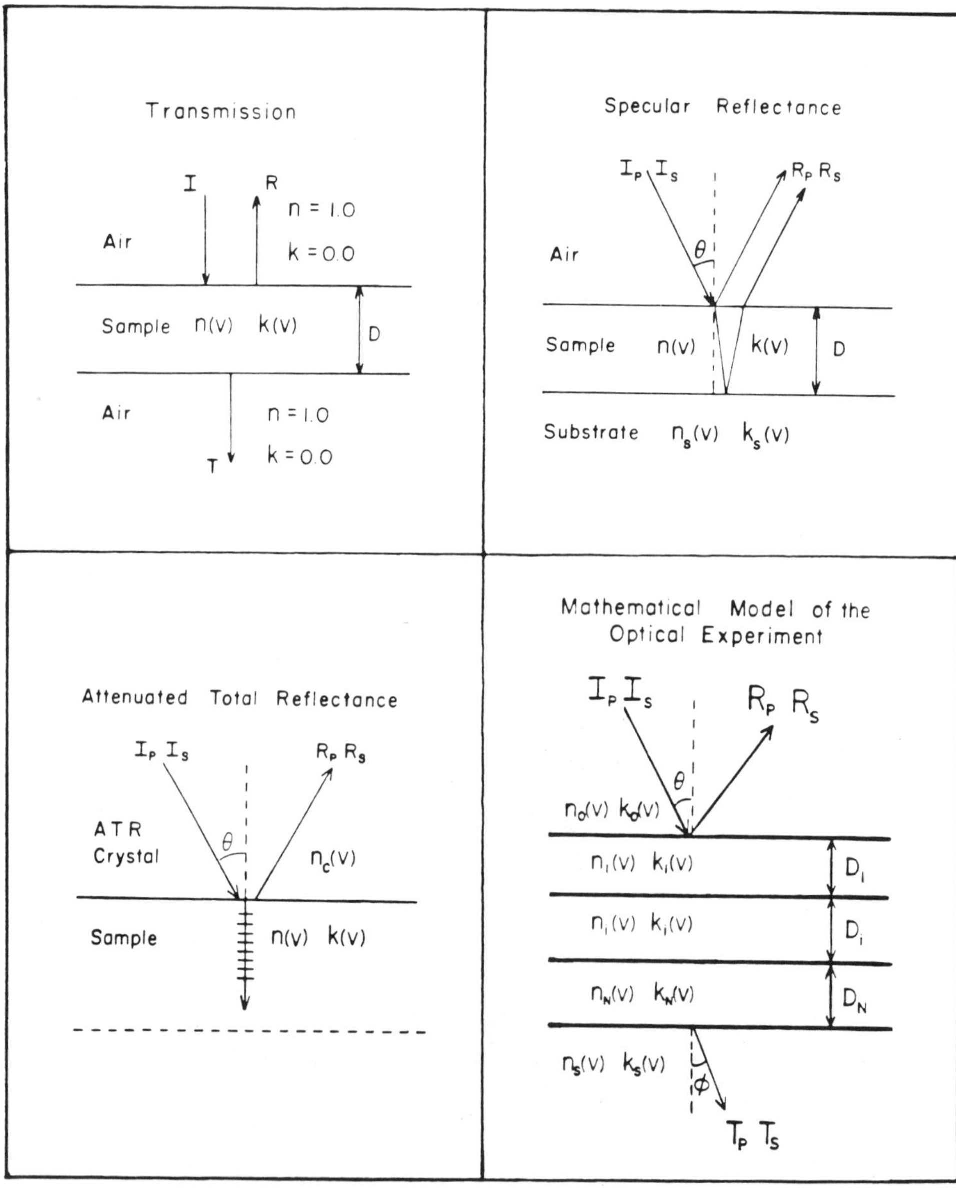

Figure 2 Schematic diagram of the three different spectroscopic experiments and the mathematical algorithm used to simulate them.

that the intensity of the calculated bands closely agrees with the experimental values. For the specular reflectance experiment all three interfaces (air|film, film|Ge, Ge|air) must be included in the calculations to obtain the correct band intensity. For the attenuated total reflectance experiment 13 reflections will occur on the face of the Ge element covered by PMMA. The logarithm of the calculated intensity for a single reflection must therefore be multiplied by 13 to get the final calculated intensity. Thus, three different infrared measurements of a single sample gave three different spectra, but the position, shape, and intensity of those spectra can be accurately predicted using optical constant calculations.

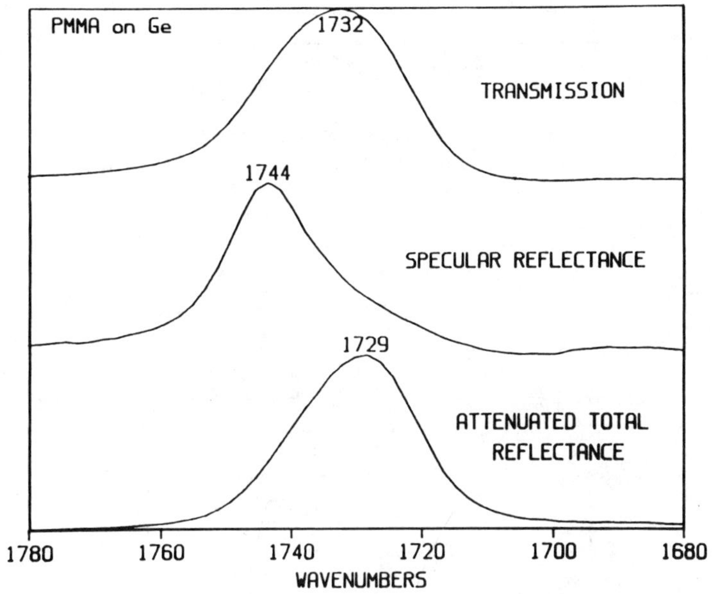

Figure 3. Plot of spectra from three different techniques applied to the same sample. Sample was 500 nm of PMMA on germanium substrate. Transmission: 0° angle, $-\log_{10}(I/I_o)$; Specular reflection: 70° angle, $-\log_{10}(Rp/Rp^{o})$; Attenuated total reflectance: 45° angle, $-\log_{10}(Rs/Rs^{o})$.

Figure 6 shows the severe distortion effects which occur at high angles of incidence for specular reflectance (80° in this case). Note that the relative intensities of the bands around 1200 cm^{-1} have been completely inverted. The upward slope of the reflection spectrum is actually an interference fringe. Both of these effects are correctly predicted by the calculated spectrum.

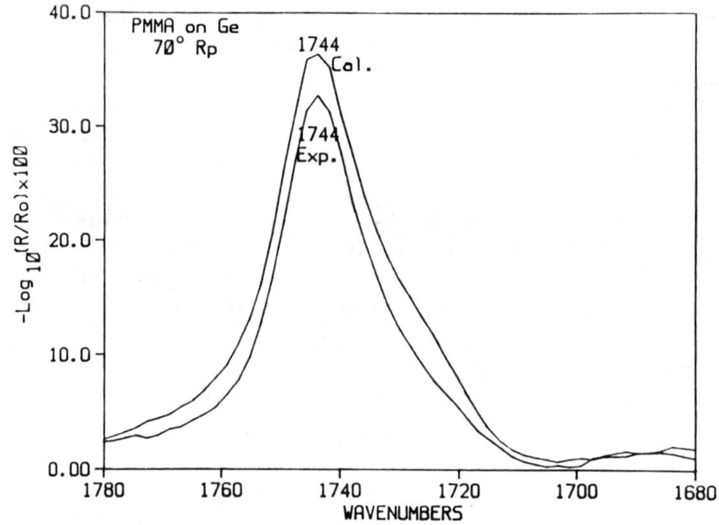

Figure 4 Comparison of experimental and calculated spectra for specular reflectance at 70°. Same sample as in figure 3.

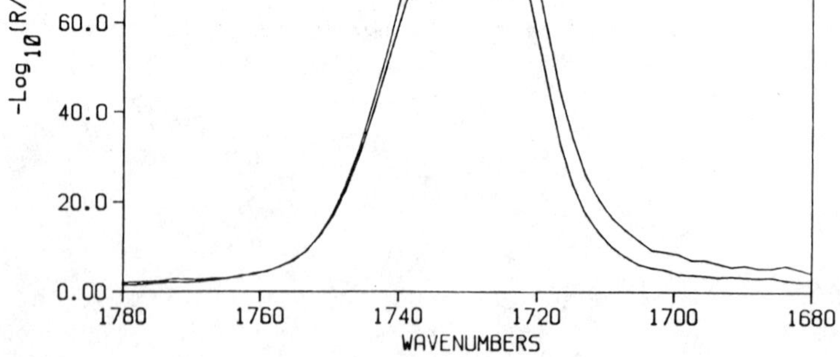

Figure 5 Comparison of experimental and calculated spectra for attenuated total reflectance at 45°. Same sample as in Figure 3.

Interference fringes are also present in the transmission spectrum in figure 6, but the magnitude of the fringes relative to the peaks is greater in reflection than transmission [15]. The relative intensity between the band at 1750 and 1208 cm^{-1} is also correctly predicted by theory. This is important since any orientation present in the sample should lead to a difference between the calculated and observed spectra.

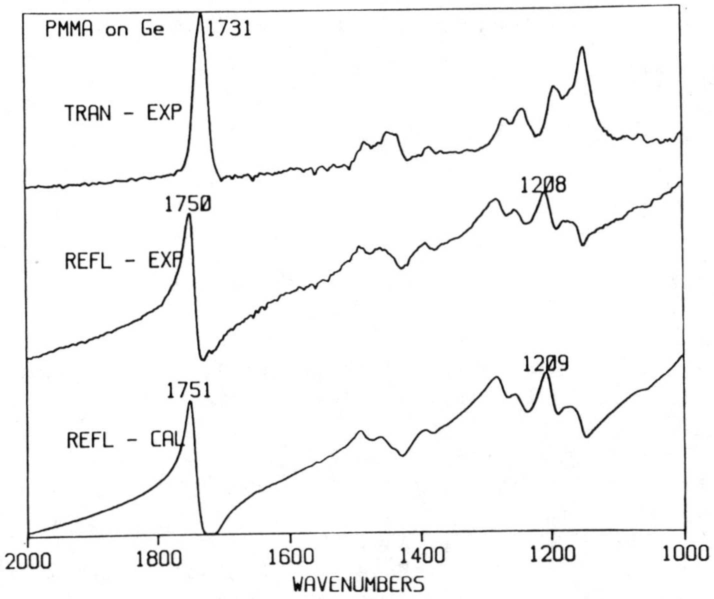

Figure 6. Plot of spectra recorded by transmission at 0° and specular reflectance at 80°, and calculated spectrum for specular reflectance at 80°.

Prest and Luca have shown [24] that solvent cast films will have residual orientation with the chain backbone parallel to the plane of the film. Any preferential orientation of the chain backbone would alter the intensity of the carbonyl vibration. To distinguish between optical artifacts and true orientation, the optical constants generated from transmission at 0° were used to simulate specular reflectance at 70° and ATR at 45°. The optical constants of PMMA were derived from transmission experiments at 0° after annealing the thin film samples to minimize any orientation. Since the agreement is good in both examples, we can conclude that the sample was nearly isotropic. We also did transmission experiments at 45° to verify this conclusion. If the calculated and experimental spectra had differed substantially, then it could be inferred that the sample was oriented.

CONCLUSIONS

Three types infrared spectroscopic experiments were performed on the same sample. The sample consisted of an thin film of PMMA deposited on Ge. The spectra obtained from these experiments differed significantly from each other. These differences were optical artifacts, and not true sample differences. Using the optical constants of the PMMA which had been previously measured from a free standing thin film, the spectra of all three experiments were simulated. These simulations fully accounted for all the observed differences in the three experiments.

ACKNOWLEDGEMENT

The authors gratefully acknowledge the partial financial support of the Office of the Naval Research.

REFERENCES

1. G. Kortum, "Reflectance Spectroscopy", Springer Verlag, New York (1969).

2. W.W. Wendlandt and H.G. Hecht, "Reflectance Spectroscopy", Interscience Publishers, New York (1966), chap. 2.

3. J.D.E. McIntyre, in "Advances in Electrochemistry and Electrochemical Engineering", Wiley, New York (1973) vol. 9, p. 61.

4. H.G. Tompkins, in "Methods of Surface Analysis", A.W. Czanderna, Ed., Elsevier, Amsterdam (1975) vol. 1.

5. N.J. Harrick, "Internal Reflection Spectroscopy", Interscience Publishers, New York (1967).

6. W.N. Hansen, in "Advances in Electrochemistry and Electrochemical Engineering", Wiley, New York (1973) vol. 9, p. 1.

7. R.W. Ditchburn, "Light", Interscience, New York (1953).

8. M. Born and E. Wolf, "Principles of Optics", 3rd. ed., Pergamon Press, Oxford (1965).

9. J.R. Reitz, F.J. Milford, and R.W. Christy, "Foundations of Electromagnetic Theory", 3rd ed., Addison-Wesley, Reading, Mass. (1979).

10. J.P. Hawranek and R.N. Jones, Spectrochim. Acta, 32A, 99 (1976).

11. K. Jezierski, J. Misiewicz, J. Wnuk, and J.M. Pawlikowski, Opt. Appl., 12, 93 (1982).

12. J.F. Blanke, S.E. Vincent, and J. Overend, Spectrochim. Acta, 32A, 163 (1976).

13. R.G. Greenler, R.R. Rahn, and J.P. Schwartz, J. Catalysis, 23, 42 (1971).

14. D.L. Allara, A. Baca, and C.A. Pryde, Macromolecules, 11, 1215 (1978).

15. N.J. Harrick, in "Characterization of Metal and Polymer Surfaces", ed. L.H. Lee, Academic Press, New York (1977) vol. 2, p. 153.

16. E.F. Young and R.W. Hannah, in "Modern Aspects of Reflectance Spectroscopy", W.W. Wendlandt, ed., Plenum Press, New York (1968) p. 218.

17. B. Crawford Jr., T.G. Goplen, and D. Swanson, in "Advances in Infrared and Raman Spectroscopy", R.J.H. Clark and R.E. Hester, eds., Heyden, London (1978) vol. 4, chap. 2.

18. R.N. Jones, "Computer Programs for Infrared Spectroscopy" (National Research Council Canada, Ottawa, 1976), Bulletin No. 14.

19. O.S. Heavens, "Optical Properties of Thin Solid Films", Dover, New York (1965) chap. 4.

20. O.S. Heavens, "Thin Film Physics", Methuen, London (1970), chap. 6.

21. K. Kozima, W. Suetaka, and P.N. Schatz, J. Opt. Soc. Am., 56, 181 (1966).

22. J.P. Hawranek, P. Neelakantan, R.P. Young, and R.N. Jones, Spectrochim. Acta, 32A, 85 (1976).

23. R.T. Graf, J.L. Koenig, and H. Ishida, Appl. Spectrosc., 39, 405 (1985).

24. W.M. Prest, Jr. and D.J. Luca, J. Appl. Phys. 50, 6067 (1979).

QUANTITATIVE ANALYSIS OF NEAT POLYMERIC FIBERS BY DRIFTS USING OPTICAL CONSTANT DATA

R.T.Graf, J.L.Koenig and H.Ishida*

Department of Macromolecular Science
Case Western Reserve University
Cleveland, Ohio 44106

ABSTRACT

Infrared reflectance spectra were obtained of drawn and undrawn poly(ethylene terephthalate) (PET) fibers using a diffuse reflectance attachment. different fiber alignments with respect to the incident beam produced relative intensity changes for the drawn, but not the undrawn fiber spectra. The band positions in the fiber reflection spectra were shifted with respect to their positions in a transmission spectrum. The intensities of the weak bands and overtones was enhanced in the fiber reflectance spectra as compared to transmission spectra. reflection spectra were also obtained of drawn PET film. The film reflection spectra showed the same band shifts as the fiber reflectance spectra, but the overtone bands were not enhanced as in the fiber case. Using the optical constants measured from a solution-crystallized sample of PET, and a well-known equation from the statistical theory of diffuse reflectance, a fiber reflectance spectrum was calculated. This calculated spectrum agreed quite well with the experimental spectrum of undrawn PET fibers in band positions, relative intensities, and absolute intensities.

INTRODUCTION

The information which can be obtained from an infrared spectrum of synthetic and natural fibers depends upon the technique used to obtain that spectrum. Melt pressing or solvent casting [1] is useful only for bulk identification of the fiber material, since orientation and crystallinity may be changed. The KBr pellet technique for fiber samples [2-3] has the advantage of not

*: To whom correspondence should be addressed.

altering the crystalline or amorphous content, but surface and orientation information is still lost. The infrared investigations which have studied neat polymer fibers used transmission [4-5], or attenuated total reflectance [6-9] methods for studying the surface and bulk molecular structure of the fibers. Gillberg and Kemp [8] were able to observe lubricants on the surface of polymer fibers by ATR. However, it is difficult to do quantitative work on neat fiber samples with either ATR or transmission.

Diffuse reflectance Fourier transform infrared spectroscopy (DRIFTS) has recently developed into a versatile technique. This versatility has been possible because of the high energy throughput and signal-to-noise ratio of FTIR, and the quantitative capability of DRIFTS for well characterized samples. A wide variety of samples such as powders [10], adsorbates on silica TLC plates [11], graphite fiber - epoxy matrix composites [12], and surface treated glass fibers [13] have been studied using DRIFTS. The sensitivity of DRIFTS [11], and its quantitative accuracy for powdered samples [10] have been documented by Fuller and Griffiths.

Theories for diffuse reflectance have been broadly classified into either continuum or statistical models [14-15]. Continuum theories involve the use of phenomenological constants, while statistical theories utilize fundamental quantities such as absorptivity, refractive index and particle size [16]. One of the most widely used models is the Kubelka-Munk theory [17]. It is a continuum model that gives a linear relationship between sample concentration and a function of the measured reflectance. However, the concentration range over which this theory gives accurate results is limited [18]. Most theories for diffuse reflectance attempt to model a system of particulate absorbing particles dispersed in a particulate non-absorbing medium [14-18]. For particulate samples, most of the incident radiation is diffusely reflected, while for non-particulate samples a large portion of the radiation may be specularly reflected. For such sample systems the reflected radiation has both diffuse and specular components. Since the statistical models can be tailored to the sample [16], unusual samples such as fibers may be better described by such theories.

In the present study, a diffuse reflectance cell in conjunction with an FTIR spectrometer is used to analyze the external reflectance spectra of polymeric fibers. Both drawn and undrawn fibers are studied to determine the sensitivity of the technique to molecular orientation effects. Polymer film samples are also examined for comparison purposes. External reflectance has the advantage over internal reflectance that there is no optical contact problem and hence it may be easier to quantify the results. The collection efficiency of the diffuse reflectance optics permits spectra of high quality to be obtained.

EXPERIMENTAL

Spectra were collected on a Digilab FTS-20E Fourier transform infrared spectrophotometer equipped with a mercury-cadmium-telluride detector. A Digilab model DRA-100 diffuse reflectance cell was used to collect the fiber reflectance spectra. The incident angle of this diffuse cell is 26° with an angle spread of 18.5^o on each side. All spectra were collected at a resolution of 4 cm^{-1} throughout the range 3800-700 cm^{-1}. All diffuse reflectance spectra are plotted in -log10(R/Ro) format where R is the reflectance spectrum of the fibers wound on a NaCl substrate and Ro is the reflectance spectrum of the bare NaCl substrate.

The poly(ethylene terephthalate) (PET) fibers were supplied by D.D. Wheeler of Allied Corporation. Two sets of fibers were supplied: highly drawn yarn and undrawn yarn. The drawn yarn had a nominal fiber diameter of 18 μm, while the undrawn yarn had a diameter of 30 μm as measured by optical microscopy. The fibers were not altered chemically or physically prior to the infrared analysis other than to wind them around a NaCl substrate. The winding was done by hand such that the fiber axes were parallel to each other. The drawn film was stretched to 400% elongation above Tg [19], and the solution-crystallized PET was recrystallized from dimethylphthalate at 45°C. [20]. The drawn film was approximately 14 μm-thick.

RESULTS AND DISCUSSION

Figure 1 shows a schematic diagram of the fiber reflection experiment. Two quantities must be specified: the orientation of the fiber axes, and the polarization of the radiation. Since the polarization is normally specified as either parallel (p) or perpendicular (s) to the plane of incidence, we will specify the fiber orientation in the same manner. The plane of incidence is defined as the plane which contains both the surface normal vector and the propagation vector. The plane of incidence is thus defined for both figures 1A and 1B as the plane of the paper. In figure 1A the fiber axes are parallel to the plane of incidence, while in figure 1B the fiber axes are perpendicular.

Figure 2 shows a diffuse reflectance spectrum of a highly drawn PET fiber. The orientation of the fibers with respect to the plane of incidence is shown next to each spectrum. The most noticeable aspect of these spectra are the relative intensity changes which occur when the fiber axis orientation is changed. This change in relative intensity appears to result from molecular orientation. The diffuse reflectance spectrum of undrawn PET fibers shown in figure 3 remains constant. The bands at 1745 and

1305 cm^{-1} are about the same intensity for both orientations of the undrawn fibers, while they exhibit intensity inversion for the drawn fibers. It fact, the spectra of the undrawn fibers are remarkably similar despite the difference in fiber geometry. A linear polarizer was not used to collect the spectra in figures 2 and 3; however, (s) polarized radiation has a higher reflectance than (p) for any angle of incidence between 0° and 90°. Therefore, bands with dipole moments oriented perpendicular to the plane of incidence are enhanced relative to bands with dipole moments parallel to the plane of incidence.

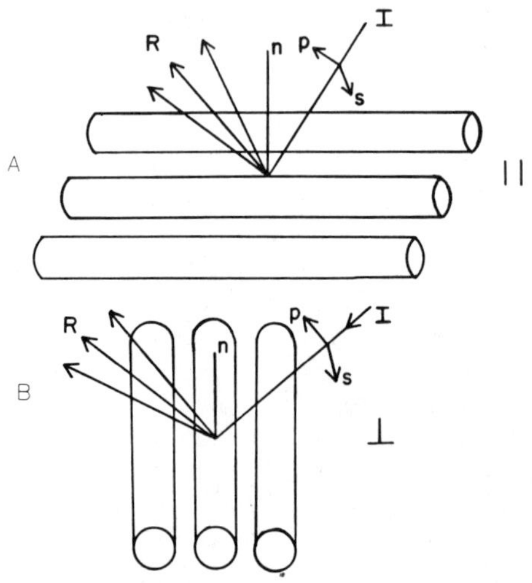

Figure 1. Schematic diagram of the fiber reflection experiment showing two possible orientations of the fibers with respect to the incident radiation. The plane of incidence is the plane of the paper. The fiber axes can be either parallel (//) or perpendicular (⊥) to this plane. The polarization of the incident radiation can also be either parallel (p) or perpendicular (s) to the same plane.

The quality of these spectra should be noted. Gillberg and Kemp have shown [8] the difficulty of obtaining external reflection spectra of fibers. Although the peak positions are shifted to frequencies higher than the true values, all the peaks are readily identifiable. The peaks in a reflection spectrum are shifted relative to their positions in a transmission spectrum because of optical distortion effects [21]. Figure 4 compares a fiber diffuse reflectance spectrum with a transmission spectrum of a PET film. The film sample was solution crystallized from dimethyl phthalate at 45° C. The reflection bands at 1745, 1305, and 738 cm^{-1} correspond to the 1720, 1264, and 727 cm^{-1} transmission bands. The infrared spectrum of PET has received much study, and several researchers have done extensive band assignments

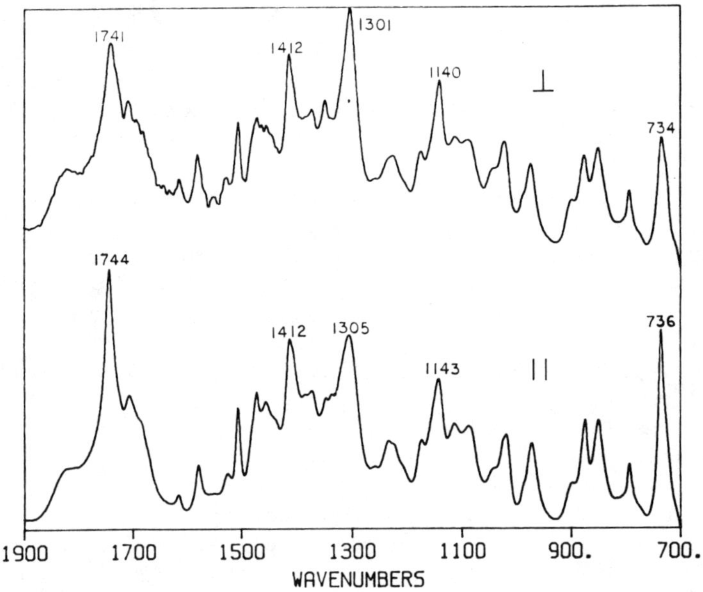

Figure 2 Diffuese reflectance spectra of Drawn PET fibers. Fiber axes were either parallel (//) or perpendicular (⊥) to the plane of incidence.

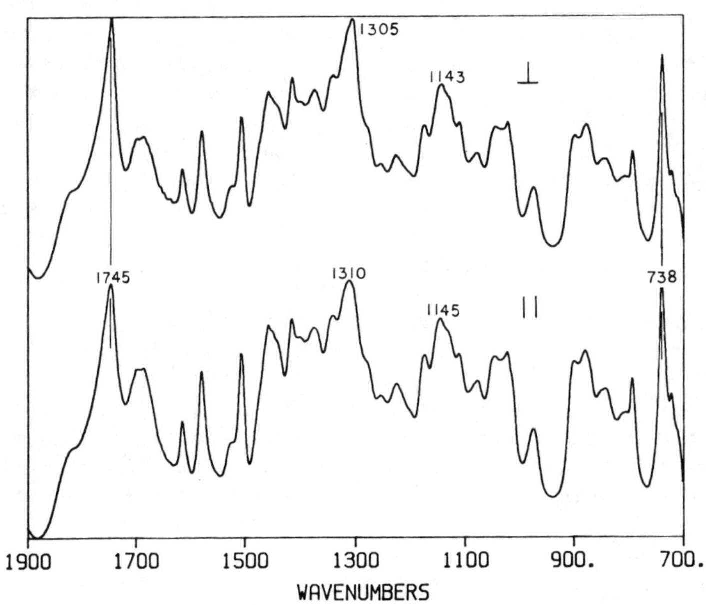

Figure 3 Diffuse reflectance spectra of undrawn PET fibers. Fiber axes were either parallel (//) or perpendicular (⊥) to the plane of incidence.

[22-24]. Boerio and Bahl reported a normal coordinate analysis of PET [22]. They assigned the 1724 cm^{-1} band to the CO stretching vibration. The 1263 cm^{-1} band was assigned to C(O)-O stretching, C-O stretching, and C(O)-O bending vibrations. The 727 cm^{-1} band was assigned to CO and ring CH out-of-plane bending modes.

Another interesting aspect of the spectra in figure 4 is the enhancement of the weak bands relative to the strong ones. Comparison of the transmission and reflection spectra shows the large increase in the intensity of the C-H stretching bands around 3000 cm^{-1} relative to the carbonyl band. The carbonyl overtone at 3431 cm^{-1} and the combination band at 1949 cm^{-1} have also increased in intensity in the reflection spectrum compared to the transmission spectrum. In fact several overtone and combination bands between 1800-2700 cm^{-1} which are clearly visible in reflection are nearly invisible in transmission.

The addition of a linear polarizer to the fiber diffuse reflectance experiment will enhance the measurement of orientation in the drawn fiber spectra. There are four possible combinations of polarizer and fiber axis alignments. Inspection of the diagram in figure 1 will reveal that the incident electric field vector can be resolved into radial and axial components. Thus, it is evident that a perpendicular electric vector with a parallel fiber alignment and a parallel electric vector with perpendicular fiber alignment will both give radially aligned incident electric field vectors. Conversely, it is evident that for near normal incidence, a perpendicular electric vector with perpendicular fiber alignment and a parallel electric vector with parallel fiber alignment will both give axially aligned vectors. Two linear polarizers were used for the actual experiment. One was placed before the entrance to the diffuse reflectance cell and one was placed at the exit of the cell. Both polarizers were always parallel to each other, but their alignment with respect to the plane of incidence was either parallel or perpendicular.

Polarized diffuse reflectance spectra of highly drawn PET fibers are shown in figures 5 and 6. The alignment of the fiber axes and polarizers is listed next to each spectrum. For the spectra in figure 5 the electric vector was aligned axially with respect to the fiber axis, while for the spectra in figure 6 the electric vector was aligned radially with respect to the fiber axis. Furthermore, the axial spectra show an enhancement of the band at 1305 cm^{-1}, and the radial spectra show enhancement of the 1744 cm^{-1} band. This behavior coincides exactly with expectations for a highly drawn fiber, assuming that the 1305 and 1744 cm^{-1} reflection bands correspond to the 1720 (CO) and 1263 [(C(O)-O), (C-O), (C(O)-O)] cm^{-1} absorption peaks of PET. The molecular axis makes an angle of about 5° with the fiber axis for annealed,

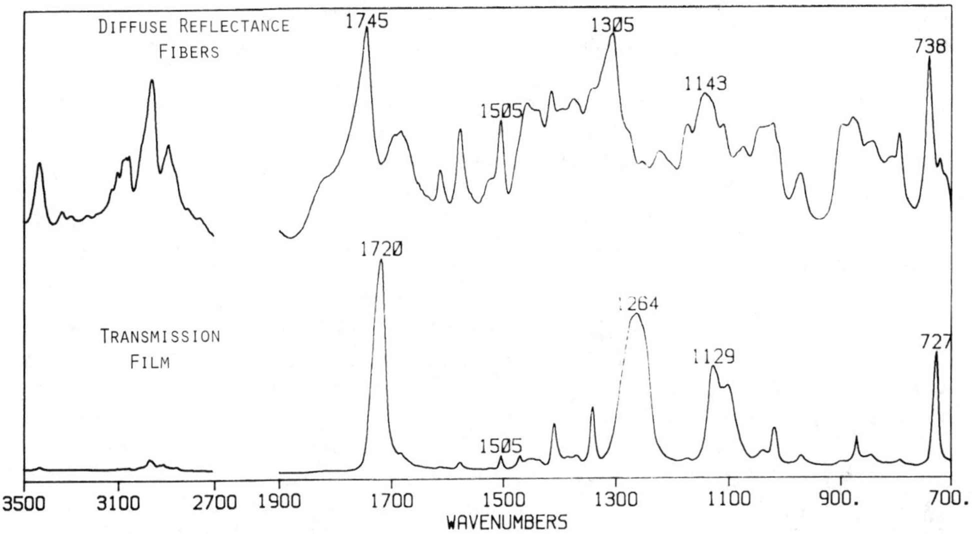

Figure 4 Comparison of a transmission spectrum of a solution crystallized PET film with a diffuse reflectance spectrum of undrawn PET fibers.

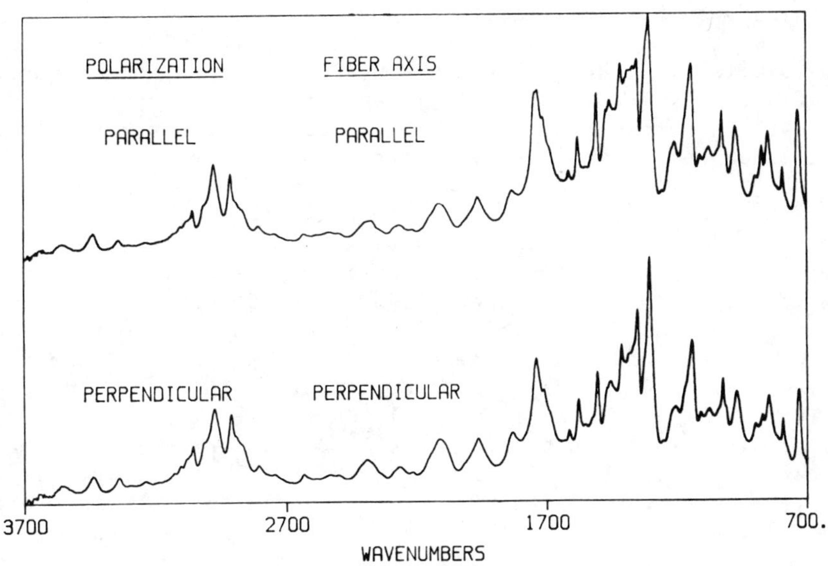

Figure 5 Polarized diffuse reflectance of drawn PET fibers. Axes and polarizers were oriented such that the incident electric field vector was aligned axially with respect to the fiber axes.

drawn PET fibers [25]. Thus, the CO bond makes an angle of 76° with the fiber axis and the C-O bond makes an angle of 47° [24].

The incident polarization will tend to become scrambled by multiple scattering processes in the fiber diffuse reflectance experiment. However, multiple scattering will be attenuated in regions of strong absorption, so for reflection bands which arise from strong absorption peaks simple specular reflectance will dominate the reflection spectrum whereas the diffuse component dominates the weak modes.

The fiber diffuse reflectance experiment will contain contributions from both specular and diffuse scattering. Since the fibers are of the same dimensions as the wavelength of radiation, Mie scattering may also contribute to the measured spectra. In an effort to separate these different contributions, we used a thin PET film as a control. This film sample had been drawn above Tg to 400 percent elongation. The reflectance from such a thin film sample would be almost totally specular. Polarized reflectance spectra of this film sample were collected with the draw axis of the film both parallel and perpendicular to the plane of incidence. These results are shown in figures 7 and 8, along with appropriate fiber reflectance spectra for comparison purposes. The film spectra exhibit the same trend as the fiber spectra. The bands at 1745 and 1306 cm^{-1} undergo intensity inversion when the orientation of the draw axis with respect to the incident radiation is changed. The film and fiber spectra are quite similar in other respects as well. The band shifts noted in the fiber spectra are also present in the film spectra. Minor discrepancies are of course attributed to differences in molecular orientation and crystallinity in these two samples.

The major difference between the film and fiber spectra in figures 7 and 8 is that the overtones and combination bands show moderate intensity in the fiber spectra, but weak intensity in the film spectra. The interference fringes present in the film reflection spectra between 1800-3700 cm^{-1} should not be confused with the overtones and combination bands of the fiber spectra in the same region. The intensity enhancement of these very weak bands in the fiber spectra is probably due to the diffuse scattering component. In regions of little or no absorption the diffuse scattering component would be significant. The regular interference fringes in the film spectra confirm that the reflectance is specular. From this comparison it can be concluded that the fiber reflectance spectrum is dominated by a specular-like reflectance component in regions of strong absorption and in regions of weak absorption the diffuse component is significant.

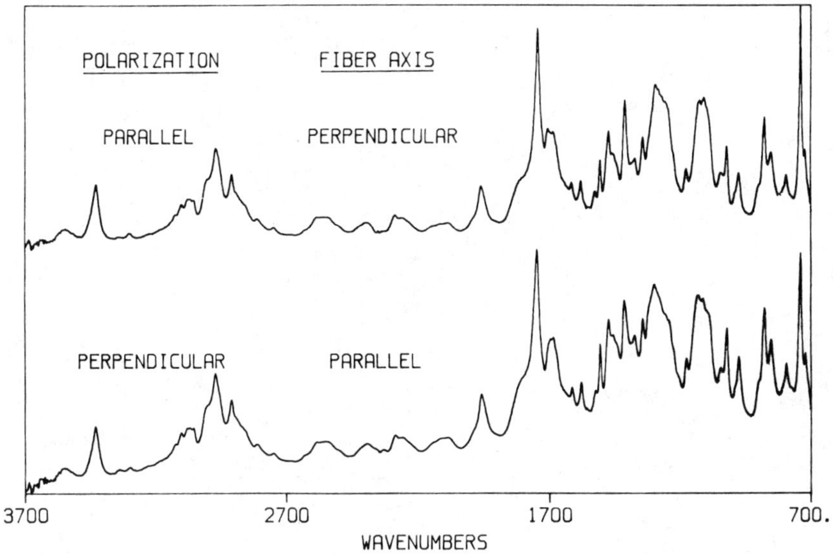

Figure 6 Polarized diffuse reflectance of drawn PET fibers. Fiber axes and polarizers were oriented such that the incident electric field vector was aligned radially with respect to the fiber axes.

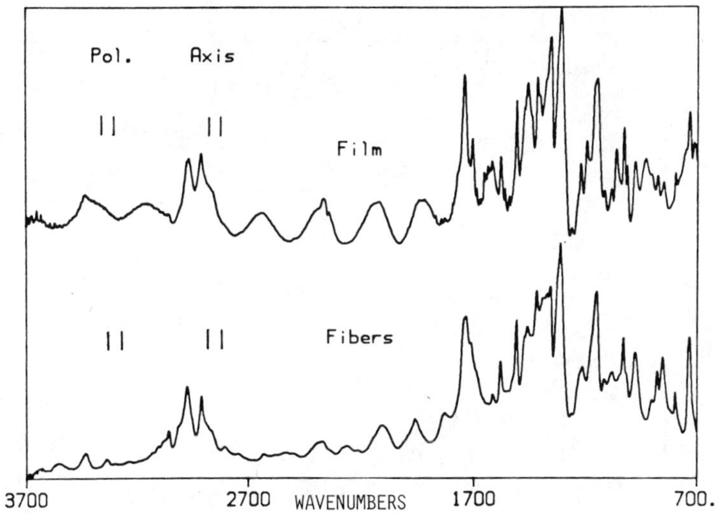

Figure 7 Diffuse reflectance of drawn PET film compared with fiber spectra from figure 4. Fiber axes or the draw axis and the polarizers were oriented such that the incident electric field vector was aligned axially.

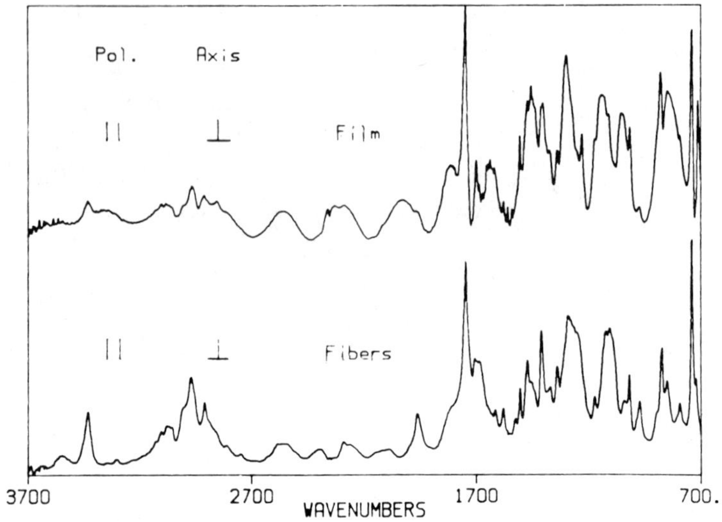

Figure 8 Diffuse reflectance of drawn PET film compared with fiber spectra from figure 4. Fiber axes or the draw axis and the polarizers were oriented such that the incident electric field vector was aligned radially.

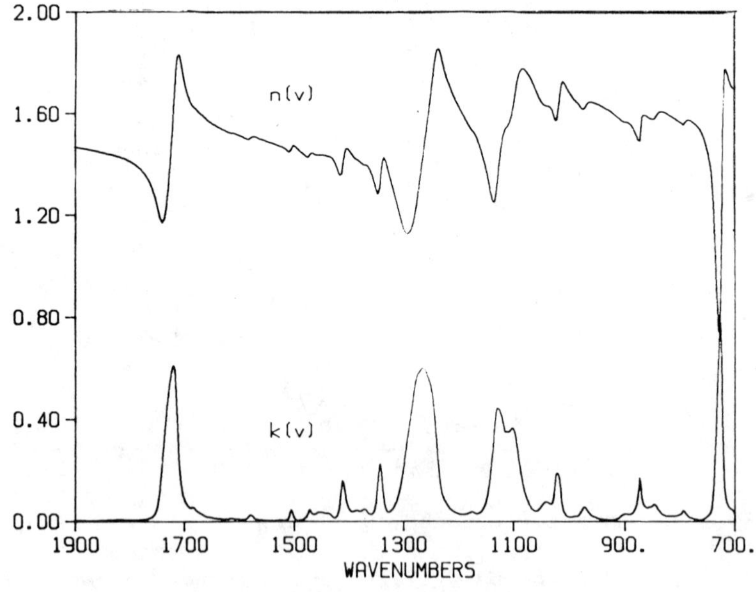

Figure 9 Optical constants of solution crystallized PET.

The optical constants for a solution crystallized sample of PET are given in figure 9. The Harrick method [26] was used to calculate the thickness and refractive index of a PET film from reflection measurements at 15° and 45°, which were the limiting angles of incidence of the diffuse reflectance cell used. The PET peak at 810 cm^{-1} was then used as an internal thickness band [27] to calculate the thickness of a film of solution crystallized PET. The optical constants were then calculated from the solution crystallized PET spectrum by an iterative Kramers-Kronig method [28].

Hecht in a review of diffuse reflectance theories [16] gives the derivation for the following equation:

$$(1)\quad R_{\infty} = \frac{1 + r^2 - t^2}{2r} - \left[\left(\frac{1 + r^2 - t^2}{2r}\right)^2 - 1\right]^{1/2}$$

which he states is fundamental to essentially all statistical theories of diffuse reflectance. The quantity R- is the limiting value of diffuse reflectance such that adding more sample does not change the measured reflectance. The quantities r and t correspond to the reflection and transmission respectively of a single layer. This layer could be modelled in a variety of ways, but is usually thought of as a layer of particles.

This equation was used to model the fiber reflectance spectra. In a procedure similar to the Bodo [29] and Johnson [30] models the quantities r and t were calculated for a thin <u>continuous</u> layer using optical constant data. The r and t values of a single layer were then summed over an infinite number of layers using equation (1). The resulting calculated R- value corresponds to the total reflectance of an infinite stack of thin films separated by air gaps. The thickness of the idealized layer was defined as equal to the nominal diameter of the fibers. This approach ignores the cylindrical geometry of the fibers and is obviously very crude. A schematic diagram of this model procedure is shown in figure 10. Since the fiber diameter is of the same dimension as the wavelength of the infrared radiation, one might expect Mie scattering effects to be important. However for simplicity, effects due to Mie scattering were not included in this model. Furthermore, multiple scattering effects are not allowed other than the multiple reflections which occur in a stack of films. Considering the similarity of the fiber reflectance spectra to the film reflectance spectra, these omissions may not be too serious.

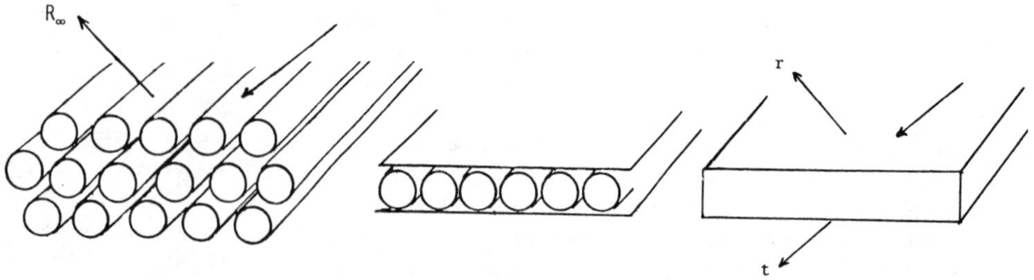

Figure 10. Schematic diagram of: (A) fiber reflection experiment, (B) idealized layer section taken from (A), and (C) thin film model used to approximate (B). The r and t values from (C) are used in equation (1) to calculate R-.

The equations for calculating the reflectance and transmittance of a thin film layer are well known [31-32]. Given the optical constants of the film layer and the ambient and substrate media, r and t can be calculated for any thickness and any angle of incidence. For the present case, the optical constants of the film layer were defined to be those of a solution crystallized sample of PET. The ambient and substrate media optical constants were defined as 1.0 and 0.0 for n and k respectively for all frequencies. The thickness of the layer was defined to be equal to the nominal diameter of the undrawn fibers. Because of the incident angle spread of the diffuse reflectance cell, r and t values were calculated for several angles of incidence and then averaged. The average r and t values were used in equation 1 to obtain the total calculated reflectance. In computing r and t of the film layer interference effects were neglected.

Figure 11 and Table 1 show the results of the model calculations as compared to the actual experimental spectrum. No fitting was done to obtain the calculated spectrum. It was generated strictly from the measured optical constants and equation 1. Despite all the approximations made to obtain a calculated diffuse reflectance spectrum, the absolute intensity as well as the band positions and shapes are in very good agreement with the measured spectrum.

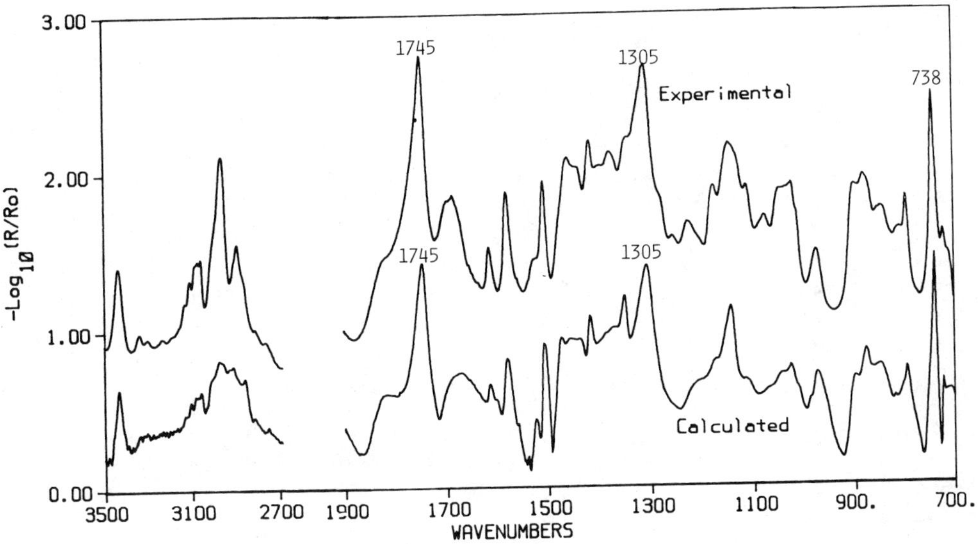

Figure 11. Comparison of experimental diffuse reflectance spectrum of undrawn PET fibers with spectrum calculated from equation 1 and the optical constants in figure 9.

Comparison of figures 4 and 11 shows that the strong reflection bands at 1745 and 1305 cm^{-1} do indeed correspond to the 1720 and 1264 cm^{-1} absorption bands. The weak bands have been enhanced relative to the strong ones in the calculated diffuse spectrum just like the measured diffuse reflectance spectrum. Furthermore, many of the differences that do exist between the calculated and experimental diffuse reflectance spectra can be attributed to material differences. The calculated spectrum used the optical constants of a solution crystallized sample of PET (figure 9), while the experimental spectrum is of undrawn PET fibers.

The good agreement obtained between the experimental and calculated spectra indicates that the diffuse reflectance of closely spaced fibers can be approximated by statistical diffuse reflectance theory. Statistical theory has the advantage that it can be customized to a particular situation by appropriate choice of a physical model. A thin film model was used in this study, and it worked reasonably well. However, a more sophisticated model which takes the cylindrical geometry of fibers into account should give even better results.

PET Band Positions (cm^{-1})

Solution Crystallized (Transmission)			Undrawn Fibers (Diffuse)			Fibers Calculated (Diffuse)	
727	s	***	739	s	***	738	s
793	vw		794	ms		793	ms
846	vw		845	m		848	m
873	w		879	ms		876	ms
899	vw		897	m		898	m
973	w		974	m		974	m
1021	m		1023	ms		1027	m
----			1076	w		----	
1103	s		1110	w		1116	w
1129	s	***	1143	s	***	1143	s
1175	w		1173	m		1171	w
----			1223	m		----	
----			1254	w		----	
1264	vs	***	1305	vs	***	1305	vs
1342	m		1340	w		1348	m
1370	vw		1375	w		----	
1410	m		1414	m		1416	m
1453	vw		1457	ms		----	
1471	w		----			1473	ms
1505	w		1505	ms		1506	ms
1524	vw		1525	w		1524	w
1579	w		1578	ms		1579	ms
1614	vw		1614	m		1614	w
1685	w		1684	ms		1670	ms
1720	vs	***	1745	vs	***	1745	vs
2855	w		----			2855	m
----			2889	ms		----	
2908	w		----			2908	w
2924	w		----			----	
2968	w		2958	s		2971	m
3055	w		3054	ms		3055	w
3102	w		3101	w		3101	w
3431	w		3431	ms		3431	ms

v=very, s=strong, m=medium, w=weak

Table 1: Band positions and relative approximate intensities for PET film and fibers as measured by transmission and diffuse reflectance respectively, and as calculated from optical constant data. Asterisks indicate those bands which experience large frequency shifts when measured in reflectance.

CONCLUSIONS

diffuse reflectance proved a viable technique for obtaining good quality infrared spectra of neat, drawn and undrawn PET fibers. reflection bands due to CO and C-O stretching vibrations underwent intensity inversion indicative of chain axis orientation along the fiber axis for the drawn fibers, but not for the undrawn fibers. The positions of strong bands in the fiber reflection spectra were shifted with respect to their positions in a transmission spectrum. The intensities of the weak bands and overtones was enhanced in the fiber diffuse reflectance spectra as compared to transmission spectra. External reflectance spectra of drawn PET film showed the same band shifts as the fiber reflectance spectra, but the overtone bands were not enhanced as in the fiber case. Using the optical constants measured from a solution-crystallized sample of PET, and a well-known equation from the statistical theory of diffuse reflectance, a fiber reflectance spectrum was calculated. Effects due to Mie scattering were not included in this calculation. Nevertheless, the calculated spectrum agreed quite well with the experimental spectrum of undrawn PET fibers in band position, relative intensity, and absolute intensity.

ACKNOWLEDGEMENTS

This study was supported in part by a grant from the Office of Naval Research. The authors wish to express their appreciation to Dr. D.D. Wheeler of Allied-Signal Corporation for supplying the PET fibers, to L. Fina of our department for supplying oriented and solution cast PET films, and to R.A. Crocombe of Digilab Division of Bio-Rad for supplying information on the diffuse reflectance attachment.

REFERENCES

1. M.C. Grieve and T.M. Kotowski, J. Forens. Sci. 22, 491 (1977).

2. W.O. Statton, J.L. Koenig, and M. Hannon, J. Appl. Phys., 41, 4290 (1970).

3. O. Kirret, P. Koch, L. Lahe, G. Rajalo, and E. Kirjanen, Eesti NSV Tead. Akad. Toim., Keem. 32, 163 (1983).

4. G.A. Tirpak and J.P. Sibilia, J. Appl. Polym. Sci. 17, 643 (1973).

5. D.J. Carlsson, T. Suprunchuk, and D.M. Wiles, Text. Res. J., 47, 456 (1977).

6. J.P. Silibia, in "Surface Characteristics of Fibers and Textiles", Part 1, M.J. Schick, Ed., (Marcel Dekker, New York (1975) chap. 8.

7. A.E. Tshmel, V.I. Vettegren, and V.M. Zolotarev, J. Macromol. Sci.-Phys. B21, 243 (1982).

8. G. Gillberg and D. Kemp, J. Appl. Polym. Sci. 26, 2023 (1981).

9. D.J. Carlsson, T. Suprunchuk, and D.M. Wiles, Can. Text. J., 87, 73 (1970).

10. M.P. Fuller and P.G. Griffiths, Anal. Chem. 50, 1906 (1978).

11. M.P. Fuller and P.G. Griffiths, Appl. Spectrosc. 34, 533 (1980).

12. P.R. Young, B.A. Stein, and A.C. Chang, Proc. 28th Natl. SAMPE Symp. Exhib., 824 (1983).

13. R.T. Graf, J.L. Koenig, H. Ishida, Anal. Chem. 56, 773 (1984).

14. W.W. Wendlandt and H.G. Hecht "Reflectance Spectroscopy", Interscience, New York (1966) chap. 3.

15. G. Kortum "Reflectance Spectroscopy", Springer-Verlag, New York (1969) chap. 4.

16. H. Hecht, J. Res. Natl. Bur. Stand. 80A, 567 (1976).

17. P. Kubelka and F. Munk, Z. Tech. Phys. 13, 593 (1931).

18. H. Hecht, Appl. Spectrosc. 34, 157 (1980).

19. L.J. Fina and J.L. Koenig, to be published.

20. L.J. Fina and J.L. Koenig, Macromolecules, accepted.

21. D.L. Allara, A. Baca, and C.A. Pryde, Macromolecules 11, 1215 (1978).

22. F.J. Boerio and S.K. Bahl, J. Polym. Sci., Polym. Phys. Ed., 14, 1029 (1976).

23. T.R. Manley and D.A. Williams, Polymer 10, 339 (1969).

24. C.Y. Liang and S. Krim, J. Molec. Spectrosc. 3, 554 (1959).

25. R. DE P. Daubeny, C.W. Bunn, and C.J. Brown, Proc. Roy. Soc., A226, 531 (1954).

26. N.J. Harrick, Appl. Opt. 10, 2344 (1971).

27. A. Miyake, J. Polym. Sci. 38, 479 (1959).

28. J.P. Hawranek, P. Neelakantan, R.P. Young, and R.N. Jones, spectrochim. Acta, 32A, 85 (1976).

29. Z. Bodo, Acta Phys. Hung. 1, 135 (1951).

30. P.D. Johnson, J. Opt. Soc. Am. 42, 978 (1952).

31. M. Born and E. Wolf, "Principles of Optics" 6th ed., Pergammon Press, Oxford (1980).

32. J.R. Reitz, F.J. Milford, and R.W. Christy, "Foundations of Electromagnetic Theory" 3rd ed., Addision-Wesley, Reading, Mass. (1979).

FOURIER TRANSFORM POLARIMETRY

Jennifer A. Bardwell and Michael J. Dignam

Department of Chemistry
University of Toronto, Toronto
Canada, M5S 1A1

INTRODUCTION

Electromagnetic radiation propagates through homogeneous isotropic media without change in polarization state. Reflection or transmission through an interface at other than normal incidence, however, does result in a change of polarization state, as does propagation through anisotropic media. Ellipsometry is the art of measurement of the polarization state of fully polarized light, and hence of the measurement of the properties of materials and interfaces to transform fully polarized light. The information sought in the case of transmission through anisotropic materials is the differential dispersion and differential absorption spectra, while from reflection studies, most commonly it is the thickness and optical constants of one or more overlayers. In this paper, we address all of these in connection with Fourier transform UV-VIS and IR spectroscopy of polymers, but in addition, a section on emission spectra, which in general involves making measurements on partially polarized light, has been included, hence the choice of the more general term "polarimetry" over "ellipsometry".

In the following sections are presented in turn: the algebra for handling polarized light and its transformation by optical components and material samples, particularly as it relates to circular and linear anisotropy and specular reflection; ellipsometric and photometric spectroscopy of films; Fourier transform (FT) ellipsometric spectroscopy applied to these same phenomena; and finally, FT emission polarimetry.

ALGEBRA OF POLARIZED LIGHT

a. Jones' Vectors and Matrices

The electric field strength, E_x, of a monochromatic (wavelength λ_o) plane electromagnetic wave travelling in vacuo in the positive z-direction with its electric vector oscillating in the x-direction (i.e. plane polarized in the x-direction) can be represented mathematically by the Eq. (1),

$$E_x = E_x^o \exp\left[i(\omega t - K_o Z + \delta_x^o)\right] \quad (1)$$

where E_x is the field amplitude, $K_o = 2\pi/\lambda_o$ is the wave number in vacuo, ω the angular frequency, δ^o the phase angle for t (time)=0=z, and $i=\sqrt{-1}$. As written, it is understood that only the real part of the right hand side of Eq. (1) is to be taken. If the wave is travelling in an isotropic medium, characterized by a complex refractive index $\hat{n}=n-ik$, the electric field is given by Eq. (1) with K_o replaced by $\hat{n}K_o$. For k=0 (a non-absorbing medium) this leads to a phase velocity of $\omega/nK_o = c/n$, as required, where c is the speed of light; while for $k \neq 0$, the intensity, I, which is proportional to E_x, is given by

$$I = I_o \exp(-2kK_o z) = I_o \text{wzp}(-\beta z)$$

where I_o is the value of I at z=0, and β is the base e chemical extinction coefficient. Thus

$$\beta = 2kK_o = 4\pi k/\lambda_o \quad (2)$$

and one can see that n, the refractive index, characterizes the dispersion properties of the medium, while k, the absorption index, characterizes its absorption properties.

The electric field for any fully polarized monochromatic plane wave travelling in an isotropic, non-absorbing medium is obtained by taking the vector sum of E_x and E_y. Thus, in matrix notation, it is represented by the following column vector,

$$\begin{pmatrix} E_x \\ E_y \end{pmatrix} = \begin{pmatrix} E_x^o \exp(i\delta_x^o) \\ E_y^o \exp(i\delta_y^o) \end{pmatrix} \exp\left[i(\omega t - K_o nz)\right] \quad (3)$$

Since ultimately we are concerned only with the radiation intensity, the factor of unit modulus, $\exp[i(\omega t - K_o nz)]$, is dropped in representing the Jones' vector [2] for polarized light. (All Jones' vectors and matricies differing simply by a factor of unit modulus are considered equivalent.) For fully polarized light, as represented by Eq. (3), E_x^o, E_y^o and $(\delta_x^o-\delta_y^o)$ are constants for any given λ_o. Random or totally unpolarized light cannot be handled by Jones' calculus, but can be thought of as given by Eq. (3) with $E_x^o=E_y^o$ and $(\delta_x^o-\delta_y^o)$ varying randomly on a time scale short compared with that of the measurement.

Transformation of the polarization state via a linear optical element or material is represented by a linear transformation of the Jones' vector, i.e. by a 2x2 matrix. These are generally given relative to the principle axes of the optical element or material, transformation to "laboratory coordinates" being effected by applying a rotational transformation,

$$M(\theta) = \begin{pmatrix} \cos\theta & \sin\theta \\ -\sin\theta & \cos\theta \end{pmatrix} \tag{4}$$

Laboratory coordinates are chosen so that the z-axis is always oriented in the direction of propagation of the radiation at the position in question, while the y-axis may be chosen for convenience, e.g. pointing up, perpendicular to the optical bench, thus settling the x-axis for a right-hand coordinate system. In this convention, θ is defined in the usual sence, counterclockwise looking into the beam.

The principle optical elements used in ellipsometric measurements are linear polarizers and retarders. Choosing the x-axis to coincide with the direction of the electric vector for the transmitted beam, the Jones' matrix for a perfect linear polarizer is given by,

$$M_{px} = \begin{pmatrix} 1 & 0 \\ 0 & 0 \end{pmatrix} \tag{5}$$

while for a perfect retarder it is

$$M_{ret} = \begin{pmatrix} e^{i\Delta/2} & 0 \\ 0 & e^{-i\Delta/2} \end{pmatrix} \tag{6}$$

where the x- and y-axes coincide with the principle axes of the retarder, and Δ is the amount by which the phase of the component of the electric vector in the y-direction is retarted relative to that in the x-direction. A reflecting surface acts as a combined partial or imperfect polarizer and a retarder,

$$M_{ref} = \begin{pmatrix} R_p & 0 \\ 0 & R_s \end{pmatrix} \tag{7}$$

where R_p and R_s are the complex reflection coefficients for the interface for p- and s-polarized light (light polarized in the plane of incidence and perpendicular to it, respectively) and the x-axis is chosen to lie in the plane of incidence. Jones′ matrices relevant to circular and linear anisotropy are presented in the following subsections.

b. Circular-Anisotropic Medium

The Jones′ matrix for a homogeneous, circularly-anisotropic medium of uniform thickness, L, and for the radiation incident normally on the interfaces, is given by [3]

$$M_{CA} = M(\Delta\beta/2i) \cdot M(\Phi)e^{-\beta/2} \tag{8}$$

where Φ is the angle through which the polarized radiation is rotated by the medium, and β and $\Delta\beta$ are the mean and differential chemical extinction coefficients, and no allowance has been made for reflections at the interfaces. In terms of the refractive and absorption indices, Φ, β and $\Delta\beta$ are given by:

$$\Phi = 2\pi[(n_1-n_r)/2]L/\lambda_o \tag{9}$$

$$\beta = \tfrac{1}{2}(\beta_1+\beta_r) = 4\pi[(k_1+k_r)/2]L/\lambda_o \tag{10}$$

$$\Delta\beta = \tfrac{1}{2}(\beta_1-\beta_r) = 4\pi[(k_1-k_r)/2]L/\lambda_o \tag{11}$$

where the subscripts 1 and r refer to left- and right-circularly polarized light respectively, so that n_1 is the refractive index of the medium for left-circularly polarized light, etc.

The equation for $M(\Phi)$ has already been given (Eq. (4) with θ replaced by Φ). Eq. (4) also applies to $M(\Delta\beta/2i)$; however, an alternative, more useful expression for the latter follows,

$$M(\Delta\beta/2i) = \begin{pmatrix} \cos h(\Delta\beta/2) & -i \sin h(\Delta\beta/2) \\ i \sin h(\Delta\beta/2) & \cos h(\Delta\beta/2) \end{pmatrix} \tag{12}$$

c. Linear-Anisotropic Medium

The Jones′ matrix for a homogeneous, linearly-anisotropic medium, with its principle axes oriented in the x-, y- and z-directions respectively, and for the same geometry and approximations as above, is given by

$$M_{LA} = \begin{pmatrix} \exp[\Delta\beta + i\Delta)/2] & 0 \\ 0 & \exp(-\Delta\beta - i\Delta)/2 \end{pmatrix} \exp(-\beta/2) \qquad (13)$$

where

$$\Delta = 4\pi[n_y - n_x)/2]L/\lambda o, \qquad (14)$$

$$\beta = \tfrac{1}{2}(\beta_y + \beta_x) = 4\pi[(k_y + k_y)/2]L/\lambda o \qquad (15)$$

$$\Delta\beta = \tfrac{1}{2}(\beta_y - \beta_x) = 4\pi[(k_y - k_x)/2]L/\lambda o \qquad (16)$$

and n_x is the refractive index for the electric field in the x-direction, etc. The more general (and more complicated) case in which the direction of propagation of the radiation (z-direction) does not correspond to a principle axis of the medium will not be treated here, as the need for it can usually be eliminated through experimental design. Note that if the medium is non-absorbing, it acts as a retarder for $n_x \neq n_y$; while for $2\pi k_y L/\lambda_o >> 1$, and $k_x = 0$, it acts as a linear polarizer. The former is the basis for retardation plates (e.g. quarter wave plates) the latter that for dichroic polarizers such as Polaroid film and gold grid polarizers.

d. Specular Reflection and Transmission for a Film-Covered Surface

The Jones' matrix for reflection from a single planar interface is given by Eq. (7) with R_p and R_s set equal to the Fresnel complex reflection coefficients for the corresponding interface, r_p and r_s. A more interesting situation, however, concerns reflection from a planar surface coated with a thin layer of material of uniform thickness, and in general birefringent or uniaxial, with the optic axis normal to the interface. Polymer films grown on surfaces of isotropic material are likely to satisfy this symmetry condition. That is to say, some preferential molecular orientation with respect to the normal to the surface is to be expected, but on average, the other two orthogonal directions are likely to be equivalent. In such a case, R_p and R_s, the overall or net complex reflection coefficients for the interfacial region, are functions of the complex refractive indices of the three media involved and of the film thickness. For transmission through such an interface, the Jones' matrix is given by Eq. (7) with R_p and R_s replaced by T_p and T_s, respectively. The various reflection and transmission coefficients are given by the following equations

[4]:

$$R_V = \frac{r_{v12} + r_{v23}\exp(-i\chi_v)}{1 + r_{v12}r_{v23}\exp(-i\chi_v)}, \quad v=s \text{ or } p \tag{17}$$

$$T_V = \frac{t_{v12}t_{v2}\exp(-i\chi_v/2)\exp\left[(i2\pi d/\lambda o)n_1 \cos\Phi_1\right]}{1 + r_{v12}r_{23}\exp(-i\chi_v)} \tag{18a}$$

$$r_{s12} = (\hat{n}_1 \cos\Phi_1 - \hat{n}_t \cos\Phi_s)/(\hat{n}_1 \cos\Phi_1 + n_t \cos\Phi_s) \tag{19a}$$

$$r_{p12} = (\hat{n}_t \cos\Phi_1 - \hat{n}_1 \cos\Phi_p)/(\hat{n}_t \cos\Phi_1 + \hat{n}_1 \cos\Phi_p) \tag{19b}$$

$$r_{s23} = (\hat{n}_t \cos\Phi_s - \hat{n}_3 \cos\Phi_3)/(\hat{n}_t \cos\Phi_s + \hat{n}_3 \cos\Phi_3) \tag{19c}$$

$$r_{p23} = (\hat{n}_3 \cos\Phi_p - \hat{n}_t \cos\Phi_3)/(\hat{n}_3 \cos\Phi_p + \hat{n}_t \cos\Phi_3) \tag{19d}$$

$$t_{s12} = 2\hat{n}_1 \cos\Phi_1/(\hat{n}_1 \cos\Phi_1 + \hat{n}_t \cos\Phi_s) \tag{20e}$$

$$t_{p12} = 2\hat{n}_1 \cos\Phi_1/(\hat{n}_t \cos\Phi_1 + \hat{n}_1 \cos\Phi_p)(\cos\Phi_p/\cos\Phi_p') \tag{20f}$$

$$t_{s23} = 2\hat{n}_t \cos\Phi_s/(\hat{n}_t \cos\Phi_s + \hat{n}_3 \cos\Phi_3) \tag{20g}$$

$$t_{p23} = 2\hat{n}_t \cos\Phi_p'/(\hat{n}_3 \cos\Phi_p + \hat{n}_t \cos\Phi_3) \tag{20h}$$

$$\chi_v = (4\pi L/\lambda_o)\hat{n}_t \cos\Phi_v \tag{21}$$

$$\hat{n}_1 \sin\Phi_1 = \hat{n}_t \sin\Phi_s = \hat{n}_n \sin\Phi_p = \hat{n}_3 \sin\Phi_3 \tag{22}$$

The subscripts 1,2 and 3 refer respectively to the medium containing the incident and reflected radiation, the film medium, and the medium containing the transmitted radiation, if any. Thus r_{s12} is the Fresnel reflection coefficient for s-polarized radiation for the boundary between medium 1 and 2, etc. while $\hat{n}_1$ and $\hat{n}_3$ are the complex refractive indices for isotropic media 1 and 3, and $\hat{n}_n$ and $\hat{n}_t$ are the principle components of the complex refractive index tensor for the film for the electric field normal and tangent to the interface respectively. The Fresnel transmission coefficients, t_v, cancel in the final result, but are given here for completeness. The angle of incidence (measured relative to the normal to the surface) is represented by ϕ_1, while ϕ_s, ϕ_p and ϕ_3 are angles (in general complex valued) defined through Eq. (22), and ϕ'_ρ is the angle made by the ray vector with the optic axis in medium 2. Note that Eq. (18a) differs from the more usual expression [5],

$$T_v = \frac{t_{v12}t_{v23}\exp(-iX_v/2)}{1 + r_{v12}r_{v23}\exp(-iX_v)} \tag{18b}$$

by a factor of unit modulus. Eq. (18a) should be used when the absolute phase is required, as for interpreting interferometric measurements.

For reflection from a single interface between ambient (medium 1) and a thick, isotropic polymer sample (medium 2), $R_v=r_{v12}$ with $\phi_s=\phi_p=\phi_2$, and $\hat{n}_s=\hat{n}_p=\hat{n}_2$. As both ellipsometric measurements and photometric measurements (measurements of $|R_p|$ and $|R_s|$) made on such a system readily give values for $\hat{n}_2$ (two independent measurements, two unknowns, n_2 and k_2), further discussion of these equations will be restricted to a film covered substrate.

ELLIPSOMETRIC AND PHOTOMETRIC SPECTROSCOPY OF FILMS

a. Ellipsometric and Photometric Observables

In this section we deal with various strategies for deriving the spectroscopic information from ellipsometric and photometric measurements made on film-covered substrates. We begin by relating the experimental observables to the quantities defined in Eqs. (17) to (22).

Given all of the optical constants and the film thickness, L, Eqs. (17) to (22) completely define R_p, R_s, T_p and T_s. For these equations to be generally valid, however, medium 3 must be wedge--shaped to avoid detecting light that is multiply reflected within it. Experimentally, one can then measure, by photometry, $|R_p|$ and $|R_s|$ and, by ellipsometry, R_p/R_s, i.e. in total three independent quantities at each wavelength and angle of incidence for reflection, and similarly for transmission.

Choosing the film-free surface as the reference state, it is convenient to define the complex optical density function for reflection, D_{Rv}, and for transmission, D_{Tv}, according to

$$D_{Rv} = \ln(\bar{R}_v/R_v) \; ; \qquad D_{rv} = \ln(\bar{T}_v/T_v) \tag{23}$$

where $\bar{R}_v$ and $\bar{T}_v$ are the reflection and transmission coefficients for v-polarized light for the reference state, i.e. for no film present, (so that $\bar{R}_v = r_{v13}$ and $\bar{T}_v = t_{v13}$). It then follows that

$$D_{Qv} = A_{Qv}/2 - i(\delta_{Qv} - \bar{\delta}_{Qv}) \; , \; Q = R \text{ or } T \tag{24a}$$

where

$$A_{Qv} = \ln(\bar{I}_{Qv}/I_{Qv}) \tag{24b}$$

and is the base e absorbance for v-polarized light (reflection absorbance for Q=R, transmission absorbance for Q=T) and $(\delta_{Qv} - \bar{\delta}_{Qv})$ is the phase change introduced by the film. Photometry can be regarded as the art of measuring A_{Qv}, and hence, through Eq. (24), of measuring $\mathrm{Re}(D_{Qv})$, where Re stands for the real-part-of. Ellipsometry, on the other hand, can be regarded as the art of measuring R_p/R_s or T_p/T_s, and hence through Eq. (23), of measuring $(D_{Qp} - D_{Qs})$, both real and imaginary parts.

The observables of ellipsometry are frequently represented by ψ and Δ, where for reflection

$$R_p/R_s = (\tan \psi_R) \exp(i\Delta_R) \tag{25a}$$

$$= |R_p| / |R_s|) \exp(i(\delta_{Rp} - \delta_{Rs})) \tag{25b}$$

so that

$$D_{Qp} - D_{Qs} = -\ln((\tan\psi_Q)/(\tan \bar{\psi}_Q)) - i(\Delta_Q - \bar{\Delta}_Q) \tag{25c}$$

$$= (A_{Qp} - A_{Qs})/2 - i(\Delta_Q - \bar{\Delta}_Q)$$

In summary, then, photometry leads to $\mathrm{Re}(D_{Qp})$ and $\mathrm{Re}(D_{Qs})$, ellipsometry to $(D_{Qp} - D_{Qs})$, a total of three independent data for each λ_o and ϕ_1. Equivalently, they give A_{Qp} and A_{Qs} and $(\Delta_Q - \bar{\Delta}_Q)$ for each λ_o and ϕ_1.

b. Thin Films

Since many systems of interest involve films of thickness small compared to the wavelength of light, we examine this case in some detail to ascertain just what information can and cannot be obtained from photometric and ellipsometric measurements on such systems.

i. Thin Film Equations

For sufficiently thin films, X_v (Eq. (21)) satisfies the condition $|X_v| << 1$, so that D_{Qv} may be expanded as a power series in X_v, with the first few terms only being of importance.

Again using the bare surface as the reference state, the following equations are derived from Eqs. (17) to (22), correct to second order terms in X_v or L for reflection [4], and to first order terms for transmission, following a similar procedure to that employed in reference [4]:

$$D_{R_v} = i(4\pi/\lambda_o)\varepsilon_1^{\frac{1}{2}}\cos\phi_1[L-\Gamma_t-Y_v(\Gamma_t-\Gamma_n)][1 - iX_{2v}] \quad (26)$$

$$D_{T_v} = i(2\pi/\lambda_o)[\Gamma_t[(\varepsilon_3-\varepsilon_1\sin^2\phi_1)^{\frac{1}{2}} - \varepsilon_1^{\frac{1}{2}}\cos\phi_1] + Y_v(\Gamma_t-\Gamma_n)x$$
$$[(\varepsilon_1/\varepsilon_3)(\varepsilon_3-\varepsilon_1\sin^2\phi_1)^{\frac{1}{2}} - \varepsilon_1^{\frac{1}{2}}\cos\phi_1]] \quad (27)$$

where

$$\Gamma_t = L(\varepsilon_t/\varepsilon_1 - 1)/(\varepsilon_3/\varepsilon_1 - 1) = \gamma_{t1}/(\varepsilon_3/\varepsilon_1 - 1) \quad (28a)$$

$$\Gamma_n = L(1 - \varepsilon_1/\varepsilon_n)/(1-\varepsilon_1/\varepsilon_3) = \gamma_{n1}/(1-\varepsilon_1/\varepsilon_3) \quad (28b)$$

$$\gamma_{t1} = L(\varepsilon_t/\varepsilon_1 - 1), \quad \gamma_{n1} = L(1 - \varepsilon_1/\varepsilon_n) \quad (29a)$$

$$Y_p = [\cot^2\phi_1 - \varepsilon_1/\varepsilon_3]^{-1}, \quad Y_s = 0 \quad (30)$$

$$X_{2v} = (2\pi/\lambda o)(\varepsilon_3 - \varepsilon_1\sin^2\Phi_1)^{\frac{1}{2}}[\Gamma_t + Y_v(\varepsilon_1/\varepsilon_3)(\Gamma_t-\Gamma_n)] \quad (31)$$

and $\varepsilon_3=\hat{n}_3^2$, the complex dielectric constant of medium 3, etc.

For sufficiently thin films, the second order term, X_{2v}, may be set to zero. For a thin stratified film, γ_{t1} and γ_{n1} must be replaced by sums of such terms, or in the case of optical constants that vary continuously with distance from the substrate, by integrals [4,6,7], viz

$$\gamma_{t1} = \int_0^L (\varepsilon_t/\varepsilon_1 - 1)dz, \quad \gamma_{n1} = \int_0^L (1 - \varepsilon_1/\varepsilon_n)dz. \tag{32}$$

Plieth and Naegele [8], using integral boundary conditions at the outset to allow for the film optical constants varying with z, obtained these same equations for reflection in the thin film limit, with γ_{t1} and γ_{n1} given by Eq. (32). The physical interpretation of γ_{t1} and γ_{n1} is that they are the principle components of the surface susceptibility tensor, being 4π times the polarization per unit area in the film generated by unit field strength at the surface in medium 1 for the case $\varepsilon_1=1$. For the case $\varepsilon_1 \neq 1$, it is the excess polarization over that in medium 1 that is involved [9].

It is apparent from Eq. (26) that reflection measurements made at a single wavelength and angle of incidence yield $(\Gamma_t - \Gamma_n)$ from $(D_{Rp} - D_{Rs})$ and $Im(\Gamma_t)$ from $Re(D_{Rs}) = A_{Rs}/2$, assuming ε_1 and ε_3 to be known. In a spectral region where the substrate (medium 3) is non-absorbing, transmission measurements also yield $(\Gamma_t - \Gamma_n)$ from $(D_{Tp} - D_{Ts})$ and $Im(\Gamma_t)$ from $Re(D_{Ts})$. Varying the angle of incidence (i.e. varying ϕ_1 and hence γ_p) will give no further information in either case. Thus, in the thin film limit, photometry and ellipsometry alone can only yield information about γ_{t1} and γ_{n1}. Specifically, they cannot yield a value for the film thickness, L. If the second order term, X_{2v}, is significant, this limitation vanishes for both reflection and transmission measurements.

We now examine possible strategies for obtaining values for L, ε_t and ε_n in the thin film limit. Only the reflection case will be considered as that for transmission is similar. The choice between transmission and reflection measurement may be made on the basis of convenience or sensitivity, but in principle all the available information is obtained from either one.

ii. Method Involving Changing ε_1

If medium 1 (usually vacuum or dry N_2, hence $\varepsilon_1=1$) is replaced by a non-absorbing and non-interacting liquid or solid phase of different dielectric constants then ellipsometric measurements for the two cases yield values for $(\Gamma_t-\Gamma_n)$ for each ε_1 value. Now from Eqs. (28a) and (28b),

$$\Gamma_t-\Gamma_n = \gamma_{t1}/(\varepsilon_3/\varepsilon_1 - 1)-\gamma_{n1}/(1-\varepsilon_1/\varepsilon_3), \tag{33a}$$

$$\gamma_{n3}/(1 - \varepsilon_3/\varepsilon_1) - \gamma_{t3}/(\varepsilon_1/\varepsilon_3 - 1), \tag{33b}$$

where γ_{t3} and γ_{n3} are independent of ε_1, and given by

$$\gamma_{t3} = L(\varepsilon_t/\varepsilon_3 - 1), \quad \gamma_{n3} = L(1 - \varepsilon_3/\varepsilon_n). \tag{29b}$$

Thus ellipsometric measurements, made for two or more different values of ε_1, allow one to solve for γ_{t3} and γ_{n3}. Photometry gives no further information, nor does making reflection masurements from the substrate side. Changing the substrate (medium 3) does lead to new information, specifically to γ_{t1} and γ_{n1}, through Eq. (33a) and the two or more independent values for $(\Gamma_t-\Gamma_n)$. From γ_{t3}, γ_{n3}, γ_{t1} and γ_{n1}; ε_t, ε_n and L are then readily calculated.

In general, the option of changing medium 3 is not experimentally accessible. However, if a spectral region exists in which ε_t and ε_n are real-valued and essentially independent of wavelength (i.e. in which the film is non-absorbing) but in which ε_3 varies substantially with wavelength, then γ_{t1} and γ_{n1} can be calculated for this region from the dependence of $\Gamma_t-\Gamma_n$ on ε_3, again using Eq. (33a). Combined with measurements of $\Gamma_t-\Gamma_n$ for different ε_1 values, yielding γ_{t3} and γ_{n3} for the same region; ε_t, ε_n and L can be calculated. Armed, now, with a value for L, ε_t and ε_n can be calculated for the entire spectral region from the spectra of γ_{t3} and γ_{n3}.

Actually, γ_{t1} and γ_{n1} can be calculated from $(\Gamma_t-\Gamma_n)$ measured at any wavelength for which the film is non-absorbing (γ_{t1} and γ_{n1} real-valued) and the substrate absorbing (ε_3 complex-valued) since in that case the real and imaginary parts of $(\Gamma_t-\Gamma_n)$ provide the two independent data required to calculate the two pieces of information, γ_{t1} and γ_{n1}, through Eq. (33a).

In summary, then, if one can find one or more spectral regions in which the film is non-absorbing, while the substrate is absorbing, and furthermore make ellipsometric measurements for two or more ambients, then L and the spectra of ε_t and ε_n can be determined in a computationally simple manner.

The situation is similar for photometric measurements for which the relevant additional equations are

$$\mathrm{Im}(\Gamma_t) = \mathrm{Im}[\gamma_{t1}/(\varepsilon_3/\varepsilon_1 - 1)] = -\mathrm{Im}[\gamma_{t3}/(\varepsilon_1/\varepsilon_3 - 1)] \quad (34a)$$

$$\mathrm{Im}(\Gamma_n) = \mathrm{Im}[\gamma_{n1}/(1 - \varepsilon_1/\varepsilon_3)] = -\mathrm{Im}[\gamma_{n3}/(1 - \varepsilon_3/\varepsilon_1)] \quad (34b)$$

In any spectral region in which the substrate is absorbing, measurement of the reflectance absorbance (and hence $\mathrm{Re}(D_{Rv})$) for three or more angles of incidence leads to values for $\mathrm{Im}(\Gamma_t)$ and $(\Gamma_t - \Gamma_n)$ through Eq. (26) with $X_{2v}=0$. For spectral regions in which the substrate is non-absorbing, but the film absorbing, the imaginary-part-of $(\Gamma_t-\Gamma_n)$ can be obtained from photometry but not its real part. If neither is absorbing, no information is available from photometry to first order terms in X_{2v} or L. The optimum strategy is therefore to determine $(\Gamma_t-\Gamma_n)$ via ellipsometry, $\mathrm{Im}(\Gamma_t)$ via photometry. As only $(\Gamma_t-\Gamma_n)$ is required for the method involving changing ε_1, only ellipsometric measurements need be made. However, measurement of $\mathrm{Im}(\Gamma_t)$ could provide a check on internal consistency. A further such check is to test that ε_t and ε_n satisfy the Kramers-Kronig [10] transformation, or the equivalent.

iii. Application of the Kramers-Kronig Transformation

Any causal function (of which ε is an example) satisfies the Kramers-Kronig integral relationship between its real and imaginary parts [10]. An alternative integral relationship is the conjugate Fourier transform relationship [11,12]. These relationships, which are mathematically equivalent, require data over a wide spectral range. The Fourier transform method, which is computationally faster, and can be performed by the software provided with FT-IR spectrometers, is given by:

$$\varepsilon_i(\omega) = \frac{2}{\pi}\int_0^\infty dt\ \sin \omega t \int_0^\infty [\varepsilon_r(\omega') - \varepsilon_r(\infty)]\cos\omega' t\ d\omega' \quad (35a)$$

$$\varepsilon_r(\omega) - \varepsilon_r(\infty) = \frac{2}{\pi}\int_0^\infty dt\ \cos \omega t \int_0^\infty \varepsilon_i(\omega')\ \sin\omega' t\ d\omega' \quad (35b)$$

where $\varepsilon_r(\omega)$ and $\varepsilon_i(\omega)$ are the real and imaginary parts of ε.

To evaluate the right hand side of Eq. (35b) using the software present in a commercial FT spectrometer, the substitutions $t \rightarrow \delta$, and $\omega' \rightarrow 2\pi\nu'$ are made so that

$$\varepsilon_r(\bar{\nu}) - \varepsilon_r(\infty) = 4\int_0^\infty d\delta \cos 2\pi\bar{\nu}\delta \int_0^\infty \varepsilon_i(\bar{\nu}')\sin 2\pi\bar{\nu}'\delta \, d\bar{\nu}' \qquad (35c)$$

where δ is the optical path diference. The inner integral is proportional to the imaginary part of the single-sided Fourier transform with $\varepsilon_i(\bar{\nu}')$ as the "interferogram", as performed by an FT spectrometer [13]. The second integral is proportional to the real part of the single-sided Fourier transform of the result of the first Fourier transform. Because the spectrometer always integrates with respect to δ, a multiplicative constant will arise from the variable changes.

In practice, one must insure that the data is compatible with the spectrometer's conditions for an interferogram, including appropriate settings of the software flags. In addition, apodization routines must be circumvented.

Since Γ_t is a single valued function of L, ε_t, ε_1 and ε_3 and furthermore, ε_t, ε_1, and ε_3 are all causal functions (i.e. functions that satisfy the Kramers-Kronig or conjugate Fourier transform relationships), then Γ_t is also a causal function [11]. Thus, $Re(\Gamma_t)-Re(\Gamma_{t\infty})$ can be calculated from $Im(\Gamma_t)$, using Eq. (35b) or (35c), where $\Gamma_{t\infty}$ is the value of Γ_t at $\omega = \infty$, and is unknown as it depends on L. However, for various trial values of $\Gamma_{t\infty}$, Γ_t and Γ_n can now be calculated, and hence also γ_{tl} and γ_{nl} using Eqs. (28a) and (28b). The spectra of $Im(\gamma_{tl})$ and $Im(\gamma_{nl})$ should display only the absorption properties of the film (assuming ε_1 real valued). If, however, $\Gamma_{t\infty}$ is chosen incorrectly, then they will also display the absorption features of ε_3. Thus, provided the substrate has absorption features, $\Gamma_{t\infty}$ can be chosen to minimize these in the spectra of $Im(\gamma_{tl})$ and $Im(\gamma_{nl})$.

We see, then, that use of the Fourier transform reciprocal relationships can often allow one to determine the spectra of γ_{tl} and γ_{nl}, but not a value for L. An additional measurement of some kind is required to obtain L, and hence the spectra of ε_t and ε_n from the spectra of γ_{tl} and γ_{nl}.

Naegele and Plieth [14] have determined the anisotropy of the optical constants of electrochemically formed platinum oxide. Their approach was to measure the change in reflectivity due to formation of the oxide layer for s- and p-polarized light. By using the Kramers-Kronig relationships to calculate the phase shifts and estimating the thickness, they were able to obtain $\hat{n}_n$ and $\hat{n}_t$, for various estimated thicknesses.

The Kramers-Kronig relationships have been used extensively to calculate the phase shift from the measured reflectivity to yield the optical constants. We reiterate the equivalence of the conjugate Fourier transform relationships and the Kramers-Kronig relationships, and the relative convenience of the FT relationships. Experiments where the reflectance is measured in an FT spectrometer (see for example ref. 15) are particularly amenable to this approach.

iv. Isotropic Thin Films

If the film is isotropic, so that $\varepsilon_t = \varepsilon_n = \varepsilon_2$, then from Eqs. (29a) and (29b),

$$L = \gamma_{tj}\gamma_{nj} / \gamma_{tj} - \gamma_{nj}), \qquad j = 1 \text{ or } 3$$

Thus L and the spectrum of ε_2 can be determined given the spectrum of Γ_t and Γ_n. If the assumption of film isotropy is correct, then L calculated from Eq. (36) must of course be real valued and independent of wavelength to within experimental error.

Roth et al. [10] made use of an approach combining those outlined in subsections iii and iv in which ε_2 was calculated from ellipsometric data for trial values of L. The "correct" L was taken to be that which lead to the closest adherence of ε_2 to the Kramers-Kronig relation. Figure 1 shows this optimization for trial values of L, applied to an oxide film on silver, grown electrochemically. Excellent agreement was obtained with the film thickness determined coulometrically. The optical constants for the film of Figure 1 are shown in Figure 2, calculated using L=96 nm [16].

Finally we note that for $|\varepsilon_3| >> 1$, as for metals in the IR, $\Gamma_n >> \Gamma_t$ so that from ellipsometry and/or photometry one can determine only Γ_n and hence only γ_{n1}. In such a case, there is no way of determining whether or not the film is anisotropic.

c. Thick Films (X_{2v} not negligible)

It is clear from the form of Eq. (26), including X_{2v}, that when this term is not negligible, measurements covering a range of angles of incidence provide additional information, sufficient in principle to calculate ε_n, ε_t and L. The conditions, that L is independent of λ_o, and that ε_n and ε_t each satisfy the Kramers-Kronig or conjugate Fourier transform relationship, can be used in a computer search for the best set of optical constants, but in

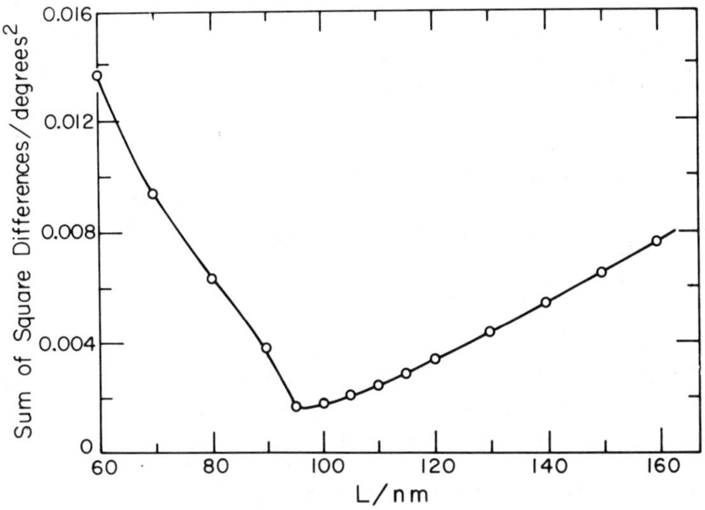

Figure 1 A plot of square differences between the measured and calculated values of the optical constants against trial film thickness, L, for an $A_{G2}O$ film on silver (10).

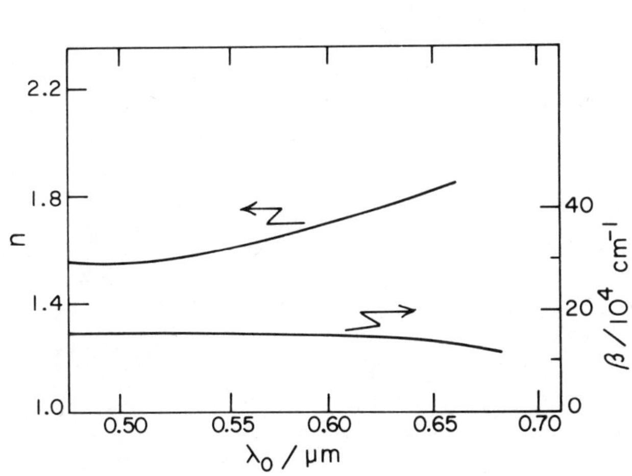

Figure 2 Spectra of the refractive index, n, and the extinction coefficient, β, calculated for the film of Fig. 1, using a thickness of 960 Å (16).

general the computational procedure is not trivial. For such calculations, the "exact" equations, Eqs. (18) to (22), should be used, as inclusion of the term X_{2v} only extends the range of validity of Eq. (22) to somewhat thicker films, but still ignores terms of order 3 and higher in L.

FOURIER TRANSFORM ELLIPSOMETRIC SPECTROSCOPY

a. Based on a Conventional FT Spectrometer

i. Michelson Interferometer

In a conventional Fourier transform IR or UV-VIS spectrometer, the radiation from the source is divided by the interferometer into two beams by division of amplitude, then recombined coherently, with the optical path length traversed by one of the beams being scanned. In the ideal spectrometer, the mean radiation density (averaged over the optical path length difference) at any particular wavelength, λ_o, in the recombined beam would be half that in the source beam and 100% modulated as the optical path length is scanned through a range $>\lambda_o$. In practice, the Michelson interferometer (MI) generally used in commercial instruments leads to mean radiation densities considerably less than this due to losses at the beam splitter. A phase compensated beam splitter presents three parallel, partially reflecting interfaces to the source beam, only the central of which acts to divide and recombine the radiation. The remaining two surfaces lead to a reduction in radiation density by a factor $K_v=(1-r_v^2)^4$ due to reflection losses, where r_v is the Fresnel reflection coefficient for its surfaces, and is, of course, different for s- and p-polarized light. The mean radiation density after recombination of the beam is then given by $2K_vR'^2_v(1-R'^2_v)I_o$ where R'_v is the reflection coefficient for the beam-splitting-and-recombining interface, and is dependent on both λ_o and the polarization state of the light, and I_o is the radiation density in the source beam. Ideal behaviour is recovered only for $K_v=1$ and $R'^2_v=1/2$; otherwise less than half is recovered, and furthermore the recombined beam is partially polarized. The condition $R'^2_v\simeq1/2$ is achieved by coating the support with a film of thickness selected to best satisfy this condition for the wavelength range of interest (e.g. KBr coated with a thin Ge film). However, R'_v is then strongly wavelength dependent so that the optimum condition on R'_v can be maintained over a relatively small wavelength range only ($\lambda_{maximum}<2\lambda_{minimum}$). This is a particular problem in the far IR (~400 cm^{-1} to ~10 cm^{-1}) where several Mylar beam splitters are required for a region in which a single detector (liquid He cooled bolometer) is optimum.

We now consider two methods by which full ellipsometric measurements can be made using a conventional MI.

ii. Modulated Retarder

In the first method, a linear polarizer and modulated retarder are placed after the MI in the manner of Nafie [17,18]. The high frequency polarization modulation leads to side bands on the interferogram, which contain circular or linear dichroism information depending on the amplitude of the phase modulation. Placing a second linear polarizer following the sample would lead to additional side bands and full ellipsometric information (differential absorption and dispersion). An advantage of this arrangement is that the polarization state of the radiation can be varied rapidly using a stress modulated retarder [17], the disadvantage being that the retarder is highly non-achromatic and cannot operate at wavelengths longer than about 7 μm, since materials for stress modulation do not exist beyond this range.

iii. Modulated Polarizer

A second arrangement is achieved by introducing a rotating linear polarizer following the sample, in addition to one in front of the sample [19-22] and possibly also a fixed retarder immediately next to the sample. This method, while applicable to the entire spectral region, causes problems due to the fact that linear polarizers connot be rotated much faster than ~100 rps, at which rate, the signal modulation cannot be separated from that resulting from the scanning interferometer. Thus either the polarizer must be stepped and the interferometer scanned, or the interferometer stepped and the polarizer rotated continuously. Neither solution is entirely satisfactory when high sensitivity is required as they both lead to long scan times and hence possible drift problems. There is a modulation approach, however, which overcomes all of these problems, i.e. allows rapid scanning, covers the full frequency range, and involves optical components, all of which are achromatic. This approach, which involves replacing the Michelson interferometer (MI) by a polarizing Michelson interferometer (PMI), will be discussed shortly, but first we give several examples of the determination of optical constants using a conventional Fourier transform spectrometer with no polarization modulation.

iv. Ellipsometry Using the Intensity Method

Using the experimental design shown in Figure 3, with a retarder and linear polarizers in a common configuration for ellipsometry, the optical constants of fused quartz were determined by Röseler and Molgedey[22] (see Fig.4). The intensity

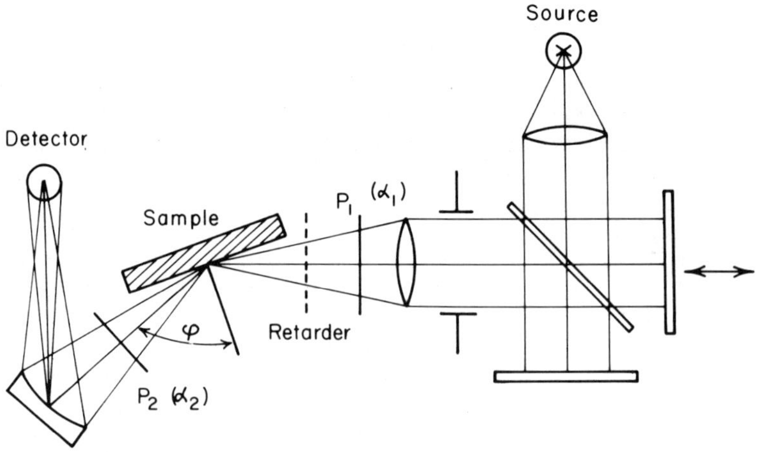

Figure 3 Fourier spectrometer-ellipsometer as constructed by Rössler, (22). P_1 and P_2 are polarized set at angles α_1 and α_2 respectively.

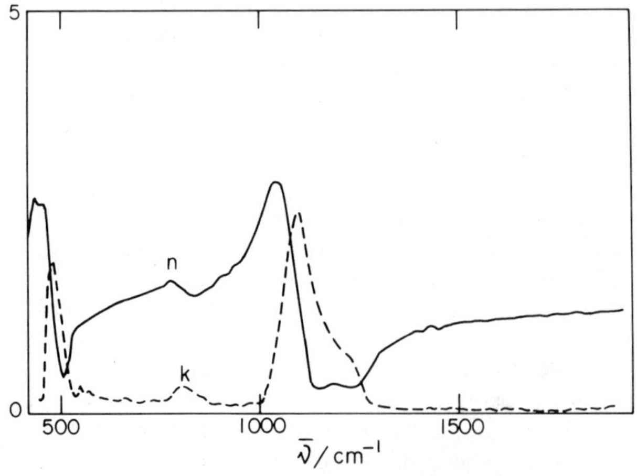

Figure 4 Optical constants of fused silica, ϕ = 70°, determined using the experimental arrangement shown in Figure 3 (22). β is in units of cm^{-1}.

at the detector as a function of the polarizer angles, α_1 and α_2 is given by:

$$I(\alpha_1) = \frac{m(\alpha_1)}{2} I_o[|r_p|^2\cos^2\alpha_2 + |r_s|^2\sin^2\alpha_2]\{1 + \cos2\Psi'\cos2\alpha_1 + \sin2\Psi'\cos(\Delta + \delta)\sin2\alpha_1\} \quad (37a)$$

where

$$\tan\Psi' = \tan\Psi.\tan\chi.\tan\alpha_2 \quad (37b)$$

and $\tan\psi$ is defined by [25], $\tan\chi$ and δ are the amplitude and relative phase changes introduced by the retarder, I_o is the intensity of the incident light, and $m(\alpha_1)$ is a calibration function. Röseler and Molgedey measured $I(\alpha_1)$ (or rather the corresponding detector signal) versus wavelength for several values of α_2. Taking intensity ratios , they were able to calculate $\tan\psi$ and Δ and hence the optical constants.

Particularly simple equations result if the retarder is ommited, the final polarizer set at 45° ($\alpha_1=\pi/4$) and the second polarizer set in turn to -45°, 0°, 45°, and 90° ($\alpha_2=-\pi/4$, 0, $\pi/4$ and $\pi/2$), yielding measured spectra S_{-45}, S_0, S_{45}, S_{90}, where $S_{45} \propto I(\pi/4)$. Eq. (37) then gives

$$\tan^2\Psi = S_{90}/S_0 \quad (38a)$$

$$2\tan\Psi\cos\Delta/(1 + \tan^2\Psi) = (I_{45} - I_{-45})/(I_{45} + I_{-45}) \quad (38b)$$

Thus from these four spectra, simple algebra gives the spectra of $\tan\psi$ and Δ, and hence n and k for the reflecting medium.

A limitation of this method is that the sign of Δ is not immediately established, a problem which, however, is usually easily overcome, e.g. by choosing the sign so that the Kramers-Kronig relation is satisfied.

The same method can, of course, be applied to film-covered surfaces. In that case one may wish to determine the spectrum of A_{Rs} in addition to those of $\tan\bar{\psi}/\tan\psi$ and $(\Delta-\bar{\Delta})$ in order to provide maximum data. This is obtained from spectral measurements made with and without the film, and with $\alpha_1=\pi/2=\alpha_2$.

v. Dispersive FT-IR

Dispersive Fourier transform spectroscopy is a technique for the determination of the optical constants of a solid, liquid or gas in which the sample is introduced into one arm of the interferometer [23]. For solid, opaque samples, the sample must be optically flat, and replaces the stationary mirror. For transparent solids or liquids or gases, the sample is placed into one of the active arms of the interferometer (in a cell if necessary). In this case, the sample must be optically uniform in thickness. In either case, the two arms of the interferometer are no longer identical, and so both the attenuation and phase shift are present in the interferogram, and recourse to a Kramers-Kronig transformation or Fourier tranformation, is not necessary for isotropic samples. Experimentally, this technique is quite difficult, because one must avoid systematic errors in the measured phase spectra. This method has been used to determine the optical constants of low loss polymers between 4 and 40 cm^{-1} [24]. The optical constants of polymethylmethacrylate and polystyrene measured in this by J.R.Birch et al. are shown in Figure 5 as an example.

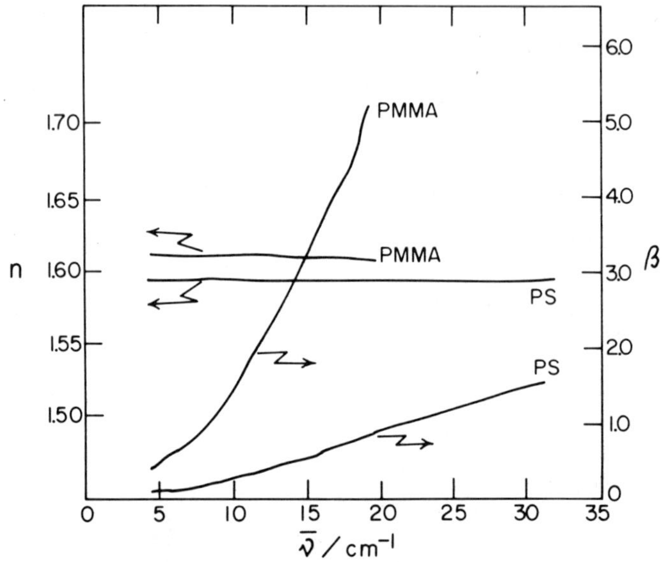

Figure 5. Optical constants of polystyrene (PS) and polymethylmethacrylate (PMMA) determined by Birch et al. [24] using dispersive Fourier transform apectroscopy.

b. Based on a Polarizing Interferometer

i. Description of A PMI

A polarizing Michelson interferometer (PMI) divides the radiation from the source into orthogonally, linearly polarized components, then recombines them coherently, with the optical path length traversed by one of them being scanned. PMI's have recently been reviewed by Martin [25]. One type of polarizing interferometer is obtained by replacing the beam splitter in a conventional MI by a grid polarizer, and the planar retromirrors by roof-top mirrors oriented to rotate the beams 90°. This type of PMI is referred to as a Martin-Puplett interferometer (MPI) [26] and is shown schematically in Figure 6. The incident radiation is devided into components linearly polarized orthogonal to each other. The component that is first reflected at the polarizer (polarized with its electric vector parallel to the grid lines of the polarizer), upon reversal at the rooftop retromirror, is transmitted by the polarizer, since its polarization direction is rotated 90° by the roof-top mirror. Similarly, the component first transmitted, on second encounter is reflected by the polarizer. Thus in the ideal limit, all of the incident radiation is recovered in the recombined beam. Allowing for reflection losses, etc. for a phase compensated MPI (but neglecting mirror losses as before) the radiation density after recombination of the beams is given approximately by $K_{ps}=(1-r^2_p)^2(1-r^2_s)^2$ [3], where r_v has been defined already. Except for reflection losses, therefore, essentially all of the incident radiation is recovered, but clearly the emerging radiation at any given λ_o is not modulated in intensity as the interferometer is scanned, since coherent addition of orthogonally polarized light can only lead to modulation in the polarization state, not the intensity, as the relative phase is changed. Furthermore, if the source beam is totally unpolarized, so is the recombined beam, so that scanning the interferometer in that case results in no modulation in any observable property of the beam. To obtain modulation in the signal reaching the detector, the source beam must be at least partially polarized, and furthermore, a dichroic sample or optical component (e.g. a linear polarizer) must lie between the MPI and the detector [3]. The choice as to how to achieve these conditions depends upon the intended application.

Partially polarized light can be treated as an incoherent superposition of totally unpolarized light, and fully polarized light. The influence of the MPI on unpolarized light is given above, while that on fully polarized light is given by its Jones' matrix [3].

$$M_{MPI} = K_{MPI} \begin{pmatrix} 0 & \exp(i\pi d/\lambda_o) \\ \exp(-i\pi d/\lambda_o) & \end{pmatrix} \qquad (39)$$

where $K_{MPI} \simeq (1-r^2_p)(1-r^2_s)$, d is the optical path length difference between the two beams, and the x-axis has been chosen to be parallel to the plane of incidence of the polarizing beam splitter.

FT spectrometers based on MPI's have been constructed by Burton and Akimoto [27] as well as by other groups [28-30]. Most applications have been to transmission spectroscopy or as spectrometers to analyze atmospheric emission [25]. These measurements can be made on conventional MI's, the advantage of the MPI being that it is achromatic and operates with high efficiency over the entire far IR frequency range, whereas, as already noted, several Mylar film beamsplitters are required to adequately cover the same range using an MI.

Placing grid polarizers in the two arms of a MI, oriented to polarize the light parallel and perpendicular respectively to the plane of incidence of the beam splitter, converts it to a PMI, but one of less than half the efficiency of a MPI. Furthermore, the polarizers must be of the same, uniform optical thickness to maintain coherence in the combined beam, a condition fairly easily met in the far IR, where unsupported grids or grids on Mylar films can be used, but not so easily met in the near and mid IR where e.g. KBr supported grid polarizers must be used.

Another polarizing interferometer, that has been used for conventional Fourier transform transmission spectroscopy and dispersive Fourier transform spectroscopy, is shown in Figure 7 [25]. The input polarizer, P, is oriented at 45° to the beam splitter. It is roughly equivalent in performance to a MPI placed between linear polarizers set at 45°, but is a less flexible arrangement for ellipsometric and polarimetric measurements and suffers from the further disadvantage that the two beams have different histories at the beam splitter, i.e. the beam at M_1 is reflected twice, while the beam at M_2 is transmited twice by the beam splitter.

As much of the existing work on PMI's has been recently reviewed by Martin [25], we will not expand further on the subject here. Instead, we concentrate on the use of MPI's for polarimetric measurements. This application, which follows clearly from the theory of MPI's, has nevertheless not been exploited experi-

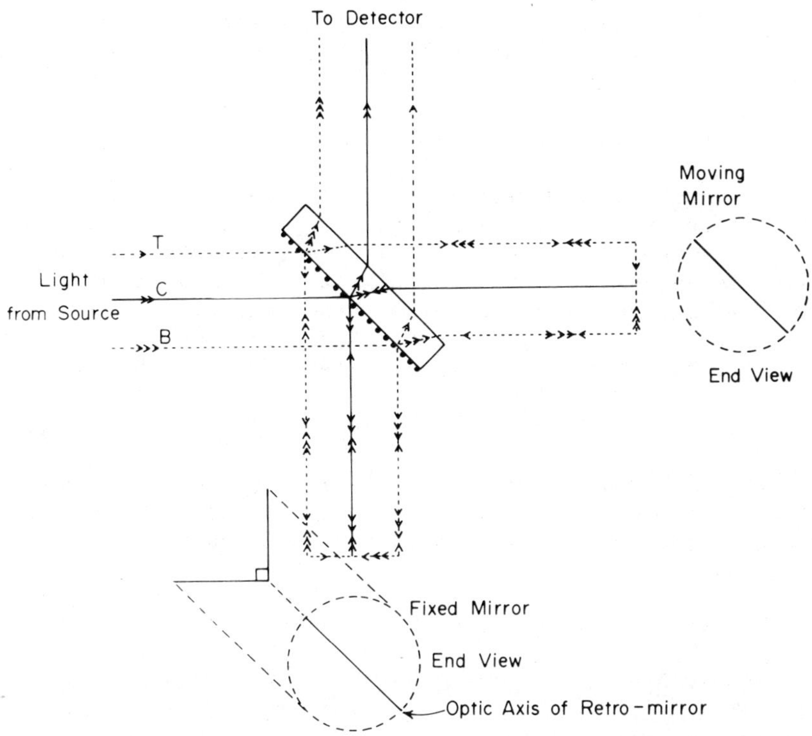

Figure 6 Layout of a polarizing Michelson Interferometer (2) reproduced by permission of Appl. Spectros. The wire grid lines on the front of the beam splitter are exaggerated for clarity. The rays designated T and B lie in the top right and bottom left of the incident beam and are interchanged by the retro-mirrors. The ray designated C is contrally positioned and sent back along itself by the retromirrors.

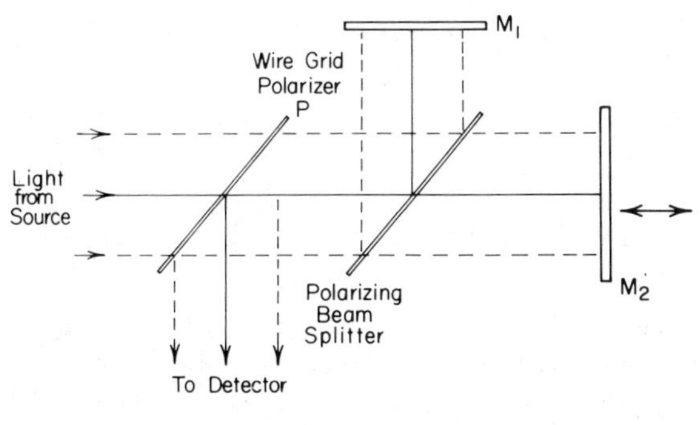

Figure 7 Schematic diagram of a polarizing interferometer used by Ledsham, Chambers

mentally. The theory behind these polarimetric measurements was first developed by Dignam and Baker [3], and has been briefly discussed by Martin [25].

ii. Application to Conventional Transmission Spectroscopy

Choosing the x-axis parallel to the plane of incidence of the beam splitter, then on placing linear polarizers both in front of and following the MPI, and set at $\pm 45^{\circ}$, results in a fully intensity modulated, linearly polarized beam of mean intensity roughly 1/4 that of the source beam, hence suitable for conventional spectroscopy. Using the second polarizer as a beam splitter results in a dual-beam FT-IR spectrometer [3].

iii. Application to Circular Dichroism and Dispersion Spectra

Placing a linear polarizer ($\pm 45^{\circ}$) in front of the MPI, and circularly dichroic material after it, results in a modulated signal at the detector, contributed by radiation of wavelength λ_o, given by [3]

$$S(\omega) = -K(\omega)\ [\exp(-\beta)][\sinh\ \Delta\beta]\sin\omega t \qquad (40)$$

where $\omega t=2\pi d/\lambda_o$ and $K(\omega)$ is an instrument constant. Placing an additional linear polarizer immediately following the absorption cell, at 0°, results in the signal [3]

$$S_f(\omega) = \frac{1}{2}TK(\omega)[\exp(-\beta)][\sin 2\Phi)\cos\omega t-(\sin\Delta\beta)\sin\omega t] \qquad (41)$$

where T is the transmittance of the polarizer for light polarized in the direction transmitted by the polarizer. Placing the additional linear polarizer immediately in front of the absorption cell, at 45°, gives [3]

$$S_b(\omega) = \frac{1}{2}TK(\omega)[\exp(-\beta)](\cosh\Delta\beta)\cos\omega t \qquad (42)$$

Thus choosing S_b as the reference signal, or effective I_o, we have

$$S/S_b = -2iT^{-1}\tanh\Delta\beta \qquad (43)$$

and

$$S_f/S_b = [(\sin 2\Phi)/(\cosh\Delta\beta)]-i[\tanh\Delta\beta] \qquad (44)$$

Thus Eq. (43) gives $-2T^{-1}\tanh\ \Delta\beta$ as the imaginary part of the Fourier transform, while Eq. (44) gives $(\sin\ 2\Phi)/(\cosh\ \Delta\beta)$ and $-\tanh\Delta\beta$ as the real and imaginary parts respectively of the Fourier transforms. Hence even when phasing errors are a problem, $\Delta\beta$ can be measured accurately via the procedure leading to

Eq. (43), while $\Delta\beta$ can be obtained from that leading to Eq. (44), since in general $\Delta\beta <<1$, giving $S_f/S_b \simeq \sin 2\Phi$.

These equations for S are obtained from the Jones´ vector $|a_i>$, representing the fully polarized light passing through the initial polarizer in front of the MPI, set at 45°,

$$|a_i> = \frac{1}{\sqrt{2}}\begin{pmatrix}1\\1\end{pmatrix}A_o$$

where $|A_o|^2=I_o$, the intensity transmitted by this polarizer. This vector is transformed by multiplication by the Jones´ matrices corresponding to the optical components through which the light passes. The signal is calculated by taking the dot product of the resulting vector with its complex conjugate. Only the A.C. component will be detected, and when two signals are ratioed, the real part of the Fourier transform corresponds to the in-phase term, while the imaginary part corresponds to the out-of-phase term.

iv. Application to Linear Dichroism and Specular Reflection

As the Jones´ matrix for a reflecting surface and a linearly dichroic material are equivalent, only specular reflection will be treated. To obtain the results for linear dichroism, the following substitutions are made,

$$|R_s| \rightarrow \exp[-(\beta + \Delta\beta)/2] \tag{45a}$$

$$|R_p| \rightarrow \exp[-(\beta - \Delta\beta)/2] \tag{45b}$$

Orienting the reflecting surface with its plane of incidence at 45°, and placing a linear polarizer in front of the MPI at 45° leads to the result [3],

$$S(\omega) = \frac{1}{2}K(\omega)[|R_s|^2 - |R_p|^2]\cos\omega t \tag{46}$$

Placing an additional polarizer immediately following the reflection cell, set at 0°, gives [3]

$$S_f(\omega) = TK(\omega)|R_pR_s|(\sin\Delta)\sin\omega t \tag{47}$$

Placing the additional polarizer immediately before the reflection cell, set at -45°, gives [3]

$$S_b(\omega) = \tfrac{1}{2}TK(\omega)|R_s|\cos\omega t. \quad (48)$$

Thus choosing S_b as the reference signal before, one obtains

$$S/S_b = T^{-1}[1 - |R_p/R_s|^2] \quad (49)$$

$$S_f/S_b = \frac{1}{2}[1 - |R_p/R_s|^2] + i\frac{1}{2}|R_p/R_s|\sin\Delta \quad (50)$$

Other polarizer arrangements are useful in particular situations [3], however it is clear from Eq. (49) and (50) that the data will yield the spectra of $|R_p|/|R_s|$ and Δ.

EMISSION POLARIMETRY

The emission from e.g. a polymer sample will be partially polarized in certain wavelength regions if the emitting dipoles are not randomly oriented. At any given wavelength, the emitted radiation can be represented as a superposition of random and fully polarized components. Using the specimen as the radiation source for the MPI, only the polarized component will be detected (Care must be taken, however, to avoid changing the polarization state of the light with any columnating optics used.) The Jones' vector for this fully polarized component, choosing one of the axes of the polarization ellipse as the x-axis, is given by

$$J = a\begin{pmatrix} \cos\alpha \\ \pm i\sin\alpha \end{pmatrix} \quad (51)$$

where the upper sign applies for right hand polarized light, the lower for left. On rotating the sample about the z-axis through an angle θ, and placing a linear polarizer following the MPI at an angle of 45°, the resulting signal is given by

$$S(\omega) = K'(\omega)\, a^2[(\sin 2\theta)(\cos 2\alpha)\cos\omega t \pm (\sin 2\alpha)\sin\omega t] \quad (52)$$

where $K'(\omega)$ is an instrumental function. Measurements made for several values of the orientation angle, θ, provide data to calculate $K'(\omega)a^2$, α, and the ellipse orientation and sense (right or left polarized). Calibration with a black body radiator (placing a linear polarizer in front of the MPI) then gives $K'(\omega)$ and hence full characterization of the polarized part of the emission. The unpolarized part of the emission can also be fully characterized by repeating the measurements with a linear polarizer, set at $\pm 45^{\circ}$, placed between the polymer specimen and the MPI. However, the main advantage of using the MPI in such an application is that in the initial configuration, the instrument is sensitive to the polarized emission only.

Fluorescence polarization measurements could be carried out in the same manner.

SUMMARY AND CONCLUSIONS

The potential of FT spectroscopy for ellipsometric and polarimetric measurement, to produce the optical constants in the case of thick specimens, the optical constants and film thickness in the case of thin film systems, has not been realized. In particular, the authors believe that a great opportunity is being missed by not developing spectrometers based on the Martin-Puplett interferometer for use in conventional, dual beam, circular dichroism, linear dichroism, reflection ellipsometry, and emission and fluorescence polarimetry covering the entire visible and infrared range. Such a system has major advantages over those based on a conventional Michelson interferometer.

Until such systems are available, the best approach for circular and linear dichroism is that based on a stress modulated retarder, an approach that cannot be extended to the far IR, however. Such a system can also be applied to reflection ellipsometric measurements, but seems to offer little, if any, advantages over the much simpler approach of running conventional reflection spectra with grid polarizers placed before and after the reflecting surface. For such spectra run for the bare, and film-covered surface provide essentially all the data that can be obtained for a given angle of incidence.

Finally, we draw attention to the fact that the Kramers--Kronig transformation can be effected through a double Fourier transform, a fact which has been known for a long time, but neglected. The use of this procedure in connection with FT spectroscopy, to provide additional information for evaluating optical constants from such spectroscopic data, is clearly natural, yet has to date never been applied.

The opportunity for major instrumental and software developments in this field appear great.

ACKNOWLEDGEMENTS

The authors thank the Natural Sciences and Engineering Research Council of Canada for supporting this work.

REFERENCES

1. For a general reference to optical equations see M.Born and E.Wolf, "Principles of Optics", 2nd Ed., Pergamon, London (1964). The sign conventions used here, however, are those suggested by R.H.Muller, Surf. Sci., 16, 14 (1969).

2. R.C.Jones, J. Opt. Soc. Am., 31, 23 (1941).

3. M.J.Dignam and M.D.Baker, Appl. Spectrosc., 33, 186 (1981).

4. M.J.Dignam, M.Moskovits and R.W.Stobie, Trans. Faraday Soc., 67, 3306 (1971).

5. O.S.Heavens, "Optical Properties of Thin Solid Films", Butterworths, London (1955) p.55.

6. M.J.Dignam and M.Moskovits, J. Chem. Soc., Faraday Trans. 2, 69, 56 (1973).

7. R.W.Stobie, B.Rao and M.J.Dignam, Surf. Sci., 56, 334 (1976).

8. W.J.Plieth and K.Naegele, Surf. Sci., 64, 484 (1977).

9. C.J.F.Bottcher and P.Bordewijk, "Theory of Electric Polarization", Vol.II, 2nd Ed., Elsevier, New York (1978) p.427.

10. J.Roth, B.Rao and M.J.Dignam, Trans. Faraday Soc., 71 (1975).

11. C.W.Peterson and B.W.Knight, J. Opt. Soc. Am., 63, 1238 (1978).

12. F.W.King, J. Opt. Soc. Am., 68, 994 (1978).

13. R.J.Bell, "Introductory Fourier Transform Spectroscopy", Academic, New York (1972) p.157.

14. K.Naegele and W.J.Plieth, Surf. Sci., 61, 504 (1976).

15. J.Zwinkels, canadian Research, Oct. 1983, 16.

16. D.B.Gibbs, B.Rao, R.A.Griffin and M.J.Dignam, J. Electrochem. Soc., 122, 1187 (1975).

17. L.A.Nafie and M.Diem, Appl. Opt., 33, 130 (1979).

18. L.A.Nafie, E.D.Lipp and C.G.Zimba, Proc. SPIE, Fourier Transform Spectroscopy, 289, 457 (1981).

19. R.W.Stobie, B.Rao and M.J.Dignam, J. Opt. Soc. Am., 65, 25 (1975).

20. R.W.Stobie, B.Rao and M.J.Dignam, Appl. Opt., 14, 199 (1975).

21. A.Roseler, Infrared Phys., 21, 349 (1981).

22. A.Roseler, Infrared Phys., 24, 1 (1984).

23. J.R.Birch, Proc. SPIE, Fourier Transform Spectroscopy, 289, 362 (1981).

24. J.R.Birch, J.D.Dromey and J.Lesurf, Infrared Phys., 21, 225 (1981).

25. D.H.Martin, in "Infrared and Millimeter Waves", Vol.6, K.J.Button, Ed., Academic, New York (1982) pp.65-148.

26. D.H.Martin and E.Puplett, Infrared Phys., 10, 105 (1969).

27. T.J.Parker, D.A.Ledsham and W.G.Chambers, Infrared Phys., 18, 179 (1978); and D.A.Ledsham, W.G.Chambers and T.J.Parker, Infrared Phys., 16, 515 (1976).

28. C.H.Burton and Y.Akimoto, Infrared Phys., 20, 115 (1980).

29. K.Yoshihara, A.Kitade and T.Matsushita, J. Appl. Phys. Japan, 21, L206 (1982).

30. F.Sugawara and H.Nagai, J. Appl. Phys. Japan, 18, 1659 (1979).

31. D.K.Lambert and P.L.Richards, Appl. Opt., 17, 1595 (1978).

AUTHOR INDEX

SUBJECT INDEX